AF411441

Innovative Techniques for Safety, Reliability, and Security in Control Systems

Innovative Techniques for Safety, Reliability, and Security in Control Systems

Editors

Francisco-Ronay López-Estrada
Guillermo Valencia-Palomo

MDPI • Basel • Beijing • Wuhan • Barcelona • Belgrade • Manchester • Tokyo • Cluj • Tianjin

Editors
Francisco-Ronay
López-Estrada
Tecnológico Nacional de
México, IT Tuxtla Gutiérrez
Tuxtla Gutierrez
Mexico

Guillermo Valencia-Palomo
Tecnológico Nacional de
México, IT Hermosillo
Hermosillo
Mexico

Editorial Office
MDPI
St. Alban-Anlage 66
4052 Basel, Switzerland

This is a reprint of articles from the Special Issue published online in the open access journal *Processes* (ISSN 2227-9717) (available at: https://www.mdpi.com/journal/processes/special_issues/safety_ Reliability_Security).

For citation purposes, cite each article independently as indicated on the article page online and as indicated below:

LastName, A.A.; LastName, B.B.; LastName, C.C. Article Title. *Journal Name* **Year**, *Volume Number*, Page Range.

ISBN 978-3-0365-8044-9 (Hbk)
ISBN 978-3-0365-8045-6 (PDF)

Contents

About the Editors

Francisco-Ronay

Francisco-Ronay received his Ph.D. degree in automatic control from the University of Lorraine, France, in 2014. He has been a lecturer at the National Technological Institute of Mexico, IT Tuxtla Gutiérrez since 2008. He is part of the Editorial Board of the *Mathematical and Computational Applications* and *Int. J. of Applied Mathematics and Computer Science* journals. He is the author/co-author of more than 50 research papers published in ISI Journals and several international conferences. He has led several funded research projects, and the funding for these projects comprises a mixture of sole investigator funding and collaborative grants. Furthermore, he collaborates with European universities, such as the University of Lorraine, University of Paris 2, University of Stavanger, and Polytechnique University of Catalonia. His research interests are fault diagnosis and fault-tolerant control based on convex LPV and Takagi-Sugeno models and their applications, and the extension of these techniques for the safety and management of UAV, mobile robots, and water distribution systems.

Guillermo Valencia-Palomo

Guillermo Valencia-Palomo was born in Merida, Mexico, in 1980. He received his Engineering degree in Electronics from the Tecnológico Nacional de México-IT Mérida, Mexico, in 2003, and his M.Sc. and PhD degrees in Automatic Control from the National Center of Research and Technological Development (CENIDET) in 2006 and the University of Sheffield, U.K., in 2010, respectively. Since 2010, Dr. Valencia-Palomo has been a Full-time (Tenured) Professor at Tecnológico Nacional de México-IT Hermosillo. He is the author/co-author of over 60 research papers published in JCR-indexed journals and several international conferences. As a product of his research, he has one patent in commercial exploitation. He has led a number of funded research projects and the funding for these projects is comprised of a mixture of sole investigator funding, collaborative grants, and funding from industry. His research interests include predictive control, linear parameter varying systems, fault diagnosis, fault-tolerant control systems, and their applications in different physical systems. He is an Associate Editor of *IEEE ACCESS*, the *International Journal of Aerospace Engineering*, and *IEEE LATIN AMERICA TRANSACTIONS*. He is a member of the Editorial Board of *Mathematical and Computational Applications*.

 *processes*

Editorial

Innovative Techniques for Safety, Reliability, and Security in Control Systems

Francisco-Ronay López-Estrada [1] and Guillermo Valencia-Palomo [2],*

[1]   TURIX Diagnosis and Control Group, Tecnológico Nacional de Mexico, IT Tuxtla Gutiérrez,
Carretera Panamericana km 1080, SN, Tuxtla Gutierrez 29050, Mexico; frlopez@tuxtla.tecnm.mx

[2]   Tecnológico Nacional de México, IT Hermosillo, Av. Tecnológico y Periférico Poniente S/N,
Hermosillo 83170, Mexico

*   Correspondence: gvalencia@hermosillo.tecnm.mx

Citation: López-Estrada, F.-R.;
Valencia-Palomo, G. Innovative
Techniques for Safety, Reliability, and
Security in Control Systems. *Processes*
2023, *11*, 1795. https://doi.org/
10.3390/pr11061795

Received: 5 June 2023

Accepted: 8 June 2023

Published: 13 June 2023

Control systems have become a critical component in the advancement of many engineering and science fields. With the increasing demand for safety and reliability, fault diagnosis (FD) and fault-tolerant control (FTC) systems have been developed and play a paramount role in safety-critical systems. These systems are utilized in various fields, such as water distribution networks, unmanned aerial vehicles (UAVs), aircraft, spacecraft, chemical and biochemical plants, and nuclear power plants, where even minor faults can lead to catastrophic consequences.

FD is of primary importance, as it enables online monitoring processes and facilitates the implementation of active FTC systems. In active FTC systems, the FD module determines which component exhibits abnormal behavior and feeds this information to the controller, which then redistributes or adapts the control law to maintain stability while controlling the system's performance degradation. For instance, if a UAV with faulty actuators is considered, the FTC system would redistribute the motors' thrust and torque to avoid collisions that could result in economic losses, harm to people, or loss of human life.

Recently, the scientific community has become more interested in the increasing complexity of modern control systems and has started researching systems with specific structures, such as multi-agent systems (MASs), networked control systems (NCSs), and cyber–physical systems (CPSs). These systems consider the interactions between different agents and the presence of discrete information processing and communication channels. Modern MASs, NCSs, and CPSs have transformed numerous areas and sectors, increasing overall efficiency and performance. However, they have also introduced the risk of faults due to non-reliable communication channels, non-neglectable time delays, and cyber attacks. These malicious actions, motivated by terrorism, criminality, or cyber warfare, exploit the system's vulnerabilities and result in some form of damage.

Therefore, there is a growing interest in developing new techniques or adapting existing fault diagnosis and fault-tolerant methods to secure the aforementioned systems. Control theory is continuously evolving, producing new theoretical results that can be exploited in innovative fault diagnosis and fault-tolerant control techniques.

We are excited to introduce the Special Issue that presents recent advances in developing and applying innovative safety, reliability, and security techniques in control systems. It comprises 13 high-quality papers divided into 3 literature reviews and 10 research articles. They are summarized in the following.

The first paper by Solís et al. [1] offer an extensive review of fault diagnosis in the stators of brushless direct current motors. Given their extensive range of applications, these motors are often required to operate at their limit, provoking faults, such as turn-to-turn short circuits, coil-to-coil short circuits, phase-to-phase short circuits, and phase open circuits, among others. The review covers signal processing, model-based, and data-based methods.

Li et al. [2] deeply analyze, in their review, the research status and developments in hydraulic synchronous control systems that are widely used in various industrial fields. It gives an introduction to the research significance, control theory, and methods used in these systems and classifies the synchronization control systems. The equivalent, master–slave and cross-coupling control modes and their different related control algorithms are studied thoroughly. Finally, the paper presents some trends in this research field.

Hua et al. [3] present a review about automobile break-by-wire (BBW) control technology, an important safeguard mechanism in active automobile safety. The article summarizes the BBW development history, its structure, classification, and operating principles. The most important control strategies are analyzed and it also presents this technology's research status and trends taking the electromechanical brake in the braking system as an example to discuss the current challenges and the way forward.

Son and Du [4] consider a general system model where independent random variables replace the parameters with arbitrary probability distributions to quantify uncertainties. Each random variable is approximated by a polynomial depending on a Gaussian variable. The authors apply the generalized dimension reduction method, where the integrand is decomposed into a sum of functions depending on lower-dimensional spaces. The method is applied to the reliability analysis of mechanical systems.

Borja et al. [5] present a fault detection and isolation method based on non-linear sliding mode observers for a wind turbine mathematical model. Faults on the pitch and drive train systems are considered since these subsystems are prone of faults in these systems due to the operating conditions. The method is based on designing independent observers sensitive to one fault. Subsequently, the observers are arranged in a bank to isolate the failure by binary logic. The tests involve practical considerations, such as disturbance, uncertainty, and measurement noise.

Bermúdez et al. [6] propose a predictive control strategy for pressure management and reduction in leakages in water distribution networks. The control algorithm considers a pressure reduction valve (PRV) for regulating the pressure in the water distribution system. The control scheme proposes a strategy to manage the PRV's high non-linearity and considers the demand profile throughout the day and the leaks that occur in the pipeline. An Extended Kalman Filter estimates the leak's magnitude and location and, with the aid of a rule-based set point manager, reduces the fluid loss in the event of a leak.

Milošević et al. [7] demonstrate that continuous monitoring and recording of the data of the pumping stations operation processes (electrical parameters, such as electrical power, pressure or flow in the pipelines, water levels in the tanks, changes in various discrete states, etc.) represent a significant resource that can be used to develop various hybrid models using the appropriate data-driven techniques. The obtained models are used for diagnosis and tolerant control of the irregular operation modes, as well as the fault-tolerant control in water supply systems.

Shi et al. [8] study the application of the Cobb–Douglas production function on optimizing safety inputs to reduce accident losses of a coal mine enterprise. The entropy weight method and the analytical hierarchy process determined each safety input indicator's weight order. The Cobb–Douglas production function was used to calculate the accident loss function of the safety input structure, and the accident loss function was obtained by multiple regression analysis. The optimal configuration of safety inputs was obtained by fitting the accident loss function. Finally, the optimal loss and mean squared error of the corresponding functions of the two safety input structures were compared.

Luo et al. [9] conduct an in-depth analysis of the failure modes of intelligent ships to optimize their design and ensure regular and safe navigation. In particular, the fixed-weight Failure Mode Effects and Criticality Analysis (FMECA) is combined with the decision-making trial and evaluation laboratory method to analyze intelligent ship positioning systems' failure modes and effects. This combined method overcomes the traditional FMECA methods to differentiate between severity, incidence, and detection rates. It allows the correlation of failure causes to be analyzed, bringing the results closer to reality.

González-Rodríguez et al. [10] compare two control techniques, computational torque control and sliding mode control, when applied to a PUMA robotic arm with six degrees of freedom. Both control algorithms cope with the non-linear behavior of the robotic arm, parameter uncertainties, and perturbations. In order to validate the controllers experimentally, the authors developed a hardware interphase based on a pic microcontroller that interacts with the control algorithms programmed in Matlab.

Gómez-Coronel et al. [11] propose a two-step algorithm to calibrate the parameters of a chlorine decay model in water distribution systems (WDS) based on genetic algorithms (GA). In the first step, the GA employs historical chlorine concentration measurements at some nodes to estimate the unknown values of the bulk and wall reaction coefficients. Once the parameters have been estimated, in the second step, the decay model is used to predict the chlorine decay concentration in for each node of the WDS for any concentration input at the pumping station. Finally, a second GA-based algorithm is implemented to obtain the minimal chlorine concentration needed at the input to ensure that every node in the system meets the official normativity requirements for free chlorine in a WDS.

Borja-Jaimes et al. [12] tackle the issue of fault detection and isolation in actuators for the pitch and drive train systems of a wind turbine benchmark. The method uses a bank of sliding-mode observers that analyzes the non-linear output error injection signal required for keeping the observer in a sliding motion.

Ríos-Ruiz et al. [13] focus on developing a versatile functional observer for Takagi–Sugeno descriptor systems. A novel structure is proposed for estimating linear functions of the states in non-linear descriptor systems represented in Takagi–Sugeno descriptor form. Its distinctive feature lies in the extra flexibility in the observer design, thereby enabling enhanced estimation performance in the presence of parametric uncertainties. The designed approach is applied to a non-linear model of a single-link robotic arm with a flexible link to demonstrate its effectiveness.

The articles featured in this Special Issue are a testament to the rapid growth and exciting potential of a particular field in engineering. With its applications expanding across various domains, pursuing work in this area demands proficiency in control engineering, systems design, numerical analysis, mathematics, and process engineering, among other areas. We believe that this Special Issue will help bridge these communities and highlight the benefits of collaboration across interdisciplinary domains.

We extend our heartfelt appreciation to the enthusiastic authors, reviewers, and editorial staff who contributed to making this Special Issue possible. Their passion and expertise have enabled us to showcase the latest advancements in this field and inspire further innovation in the future.

Funding: This research has been supported by Tecnológico Nacional de México under the program *Proyectos de Investigación Científica y Desarrollo Tecnológico e Innovación*.

Conflicts of Interest: The authors declare no conflict of interest.

References

1. Solís, R.; Torres, L.; Pérez, P. Review of Methods for Diagnosing Faults in the Stators of BLDC Motors. *Processes* **2023**, *11*, 82. [CrossRef]
2. Li, R.; Yuan, W.; Ding, X.; Xu, J.; Sun, Q.; Zhang, Y. Review of Research and Development of Hydraulic Synchronous Control System. *Processes* **2023**, *11*, 981. [CrossRef]
3. Hua, X.; Zeng, J.; Li, H.; Huang, J.; Luo, M.; Feng, X.; Xiong, H.; Wu, W. A Review of Automobile Brake-by-Wire Control Technology. *Processes* **2023**, *11*, 994. [CrossRef]
4. Son, J.; Du, Y. Modified Dimension Reduction-Based Polynomial Chaos Expansion for Nonstandard Uncertainty Propagation and Its Application in Reliability Analysis. *Processes* **2021**, *9*, 1856. [CrossRef]
5. Borja-Jaimes, V.; Adam-Medina, M.; López-Zapata, B.Y.; Vela Valdés, L.G.; Claudio Pachecano, L.; Sánchez Coronado, E.M. Sliding Mode Observer-Based Fault Detection and Isolation Approach for a Wind Turbine Benchmark. *Processes* **2022**, *10*, 54. [CrossRef]
6. Bermúdez, J.R.; López-Estrada, F.R.; Besançon, G.; Valencia-Palomo, G.; Santos-Ruiz, I. Predictive Control in Water Distribution Systems for Leak Reduction and Pressure Management via a Pressure Reducing Valve. *Processes* **2022**, *10*, 1355. [CrossRef]
7. Milošević, M.; Radić, M.; Rašić-Amon, M.; Litričin, D.; Stajić, Z. Diagnostics and Control of Pumping Stations in Water Supply Systems: Hybrid Model for Fault Operating Modes. *Processes* **2022**, *10*, 1475. [CrossRef]

8. Shi, X.Z.; Zhu, J.Y.; Zhang, S. Research on the Optimization Method of Safety Input Structure in Coal Mine Enterprise. *Processes* **2022**, *10*, 2497. [CrossRef]
9. Luo, X.; He, H.; Zhang, X.; Ma, Y.; Bai, X. Failure Mode Analysis of Intelligent Ship Positioning System Considering Correlations Based on Fixed-Weight FMECA. *Processes* **2022**, *10*, 2677. [CrossRef]
10. González-Rodríguez, A.; Baray-Arana, R.E.; Rodríguez-Mata, A.E.; Robledo-Vega, I.; Acosta Cano de los Ríos, P.R. Validation of a Classical Sliding Mode Control Applied to a Physical Robotic Arm with Six Degrees of Freedom. *Processes* **2022**, *10*, 2699. [CrossRef]
11. Gómez-Coronel, L.; Delgado-Aguiñaga, J.A.; Santos-Ruiz, I.; Navarro-Díaz, A. Estimation of Chlorine Concentration in Water Distribution Systems Based on a Genetic Algorithm. *Processes* **2023**, *11*, 676. [CrossRef]
12. Borja-Jaimes, V.; Adam-Medina, M.; García-Morales, J.; Guerrero-Ramírez, G.V.; López-Zapata, B.Y.; Sánchez-Coronado, E.M. Actuator FDI Scheme for a Wind Turbine Benchmark Using Sliding Mode Observers. *Processes* **2023**, *11*, 1690. [CrossRef]
13. Ríos-Ruiz, C.; Osorio-Gordillo, G.L.; Astorga-Zaragoza, C.M.; Darouach, M.; Souley-Ali, H.; Reyes-Reyes, J. Generalized Functional Observer for Descriptor Nonlinear Systems—A Takagi-Sugeno Approach. *Processes* **2023**, *11*, 1707. [CrossRef]

processes

Review

Review of Methods for Diagnosing Faults in the Stators of BLDC Motors

Ricardo Solís [1], Lizeth Torres [1,*] and Pablo Pérez [2]

[1] Instituto de Ingeniería, Universidad Nacional Autónoma de México, Mexico City 04510, Mexico
[2] Facultad de Ingeniería, Universidad Nacional Autónoma de México, Mexico City 04510, Mexico
* Correspondence: ftorreso@iingen.unam.mx

Abstract: A brushless direct current (BLDC) motor is a type of permanent magnet machine that is highly efficient and powerful and requires occasional maintenance. Thanks to these fortunate characteristics, this type of motor has various applications in high-tech industries. However, since BLDC motors are often required to operate at high-speed rotations and under extreme conditions, temperature overshoots can appear during operation, provoking damage to the windings. The purpose of this review is to present the results of a recent investigation and recollection of different methods used for the diagnosis of electrical faults in the stator, such as turn-to-turn short circuits, coil-to-coil short circuits, phase-to-phase short circuits and phase open circuits. In particular, this review presents an analysis of the available diagnosis methods according to the type of fault, the method or technique used for the diagnosis, the evaluated physical variables and the context in which the methods were evaluated (in simulations or in experimental tests). Based on this analysis, the following classifications of diagnostic methods are proposed: signal-based, model-based and data-based methods. Then, the pros and cons of each method class are described and discussed.

Keywords: fault detection and isolation; brushless DC motor; permanent motors

1. Introduction

Some desirable characteristics of permanent magnet motors make them very attractive candidates for various classes of industrial and commercial applications, such as robotic applications, motion control systems, aerospace systems and means of transport [1,2]. Permanent magnet motors can be classified according to where the magnets are placed: (1) on the stator, (2) on the rotor or (3) on both parts (see Figure 1). According to [3], motors with magnets in the stator can be classified into two groups: (1) the flux uncontrollable (FU) group and (2) the flux controllable (FC) group. The FU group can be classified into three types of *stator-PM machines*: doubly salient PM (DSPM) motors, flux-reversal PM (FRPM) motors and flux-switching PM (FSPM) motors. The FC group can be classified into flux-mnemonic permanent magnet motors and hybrid-excited permanent magnet motors.

Regarding the motors that have the magnets on the rotor, the most popular are the brushless direct current (BLDC) motors, permanent magnet synchronous (PMS) motors and permanent magnet stepper (PMST) motors, and they are known as *rotor-PM machines*. Since the operating principles of *rotor-PM machines* are essentially identical, the material presented in this review is applicable to this type of machine but with special emphasis on BLDC motors, due to the great interest of high-tech industries in their use in position and speed control applications, which for many years were commonly carried out by DC motors.

BLDC motors were developed to avoid the constant brush and commutator maintenance of DC motors, and it is thanks to this fact that BLDC motors have been gaining popularity in applications that require low-maintenance operations. This popularity can be verified by googling BLDC in `Google Trends`. The results of this search are shown in

Citation: Solís, R.; Torres, L.; Pérez, P. Review of Methods for Diagnosing Faults in the Stators of BLDC Motors. *Processes* **2023**, *11*, 82. https://doi.org/10.3390/pr11010082

Academic Editor: Jie Zhang

Received: 1 November 2022
Revised: 16 December 2022
Accepted: 16 December 2022
Published: 28 December 2022

Figure 2. In this figure, it can be noted that the number of searches has been increasing exponentially over the last few years.

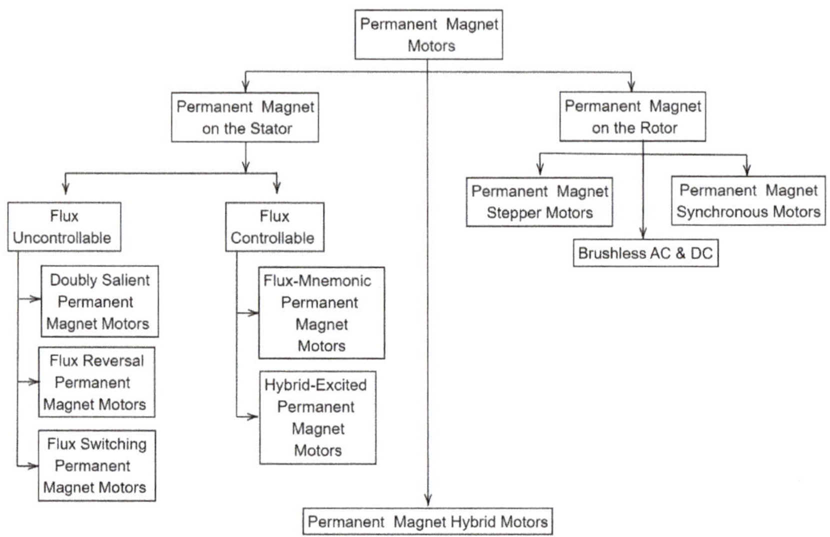

Figure 1. Classification of permanent magnet motors.

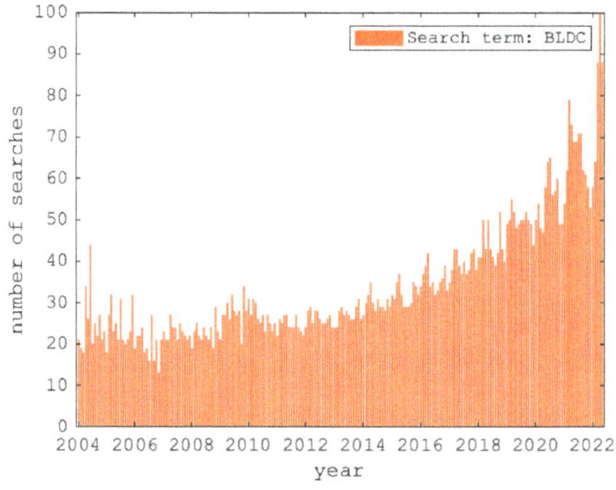

Figure 2. Trend of Google searches for the term BLDC.

The BLDC motor has electrical characteristics similar to those of a conventional DC motor but with the particularity that its reliability has been enhanced by the replacement of a mechanical commutation with an electronic one. The investigations that have been carried out with respect to these motors have allowed identifying the advantages that they present, among them being the following:

- A greater speed range due to non-dependence on the mechanical characteristics of the brushes;
- A better torque–speed characteristic due to the absence of friction produced by the brushes, which reduces the useful torque of the motor; in other words, a better capacity of heat dissipation that allows the generation of a higher torque;

- Better dynamic response due to the fact that the rotor has less inertia;
- Less noise due to the absence of discharges during the switching process;
- A smaller size, which widens its range of application.

Like most machines, BLDC motors are frequently used to work in harsh conditions, withstanding overloads and overheating. Thus, it is normal that faults occur over time. Winding-related faults are the most common faults in BLDC motors. Their main causes are high temperatures, loss of insulation, aging and contamination. If winding-related faults are detected at an early stage, it is possible to prevent catastrophic situations that can affect the environment and life. For this purpose, however, it is necessary to embed fault diagnosis systems in processes or applications that use BLDC motors.

According to [4,5], faults in BLDC motors can be classified as mechanical faults, electrical faults and magnetic faults, as shown Figure 3. According to [6–9], 30–40% of faults in BLDC motors occur in the stator. The main faults in the stator are shown in Figure 4, and they are insulation faults that provoke short circuits. A short circuit between turns is usually named an inter-turn fault. A short circuit between two coils bellowing to the same phase is called a coil-to-coil fault. A short circuit between two turns or two coils of different phases is called a phase-to-phase fault or inter-phase fault, respectively. A break in the wiring (or an abnormal operation that changes the resistance value to an extremely high value, ceasing the current flow) is an open-circuit fault [6].

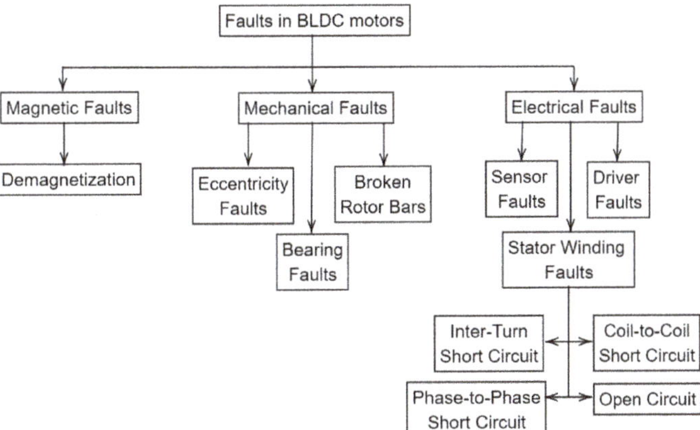

Figure 3. Classification of faults in a BLDC motor.

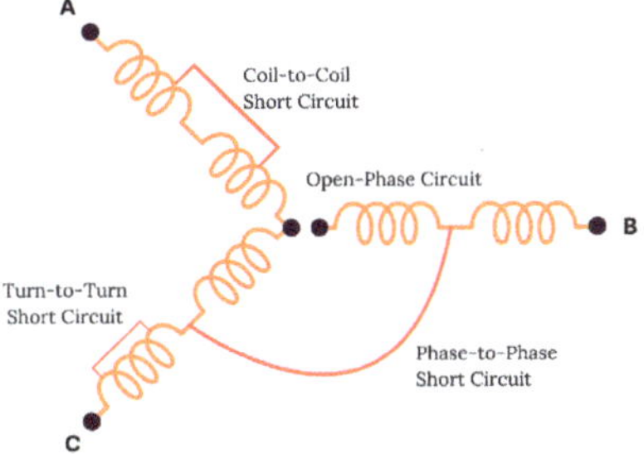

Figure 4. Common faults in the stator of a BLDC motor.

A review of health monitoring, fault diagnosis and failure prognosis techniques for brushless permanent magnet machines was presented in [4]. This review discusses in a general and brief way the different methods to diagnose and predict faults of various kinds in permanent magnet machines: electrical, mechanical and magnetic. To delve deeper into a topic that arouses more interest every day, the purpose of this work is to present a deep investigation of different methods and techniques for the diagnosis and detection of faults, specifically in the stator of BLDC motors. In particular, this review presents an analysis of the available diagnosis methods according to the type of stator fault, the method or technique used for the diagnosis (based on the signal, models or data), the diagnostic physical variables (current, position and frequency) and the context in which the methods are evaluated (in a simulation or in real time).

Section 2 presents a classification of the more common faults that happen in the stators of BLDC motors. Section 3 proposes a classification of the available methods for the diagnosis of stator faults, as well as a description of some of the most relevant methods. Finally, in Section 4, some conclusions are given.

2. Faults in the Stator

The identification of each stator fault can be accomplished through changes in the behavior of the machine quantities. These changes are known as *symptoms*. For example, if there is a significant reduction in phase current, a coil-to-coil fault is likely occurring. If there is a distinctly large cycle of Back-EMF, and the speed does not drop below the actual speed, the cause is likely to be a stator inter-turn failure. If the current is twice the healthy current, and the torque increases to three times the actual torque, a phase-to-phase fault may be occurring. Lastly, if zero current passes to any phase, then it is a winding fault or open circuit of such a phase [6,10].

2.1. Turn-to-Turn Short Circuit (Inter-Turn Fault)

The stator winding is the weakest part of BLDC motors. These windings are covered with insulating materials to prevent short circuits between adjacent windings. Organic materials used for insulation in electrical machines are subject to deterioration from a combination of thermal overload and cycling, voltage transients in the insulating materials, mechanical stress and contamination. If a fault occurs in the inter-turn stator windings, then shorted windings are produced, followed by extreme heat due to PM-induced current in the shorted windings. This type of fault is called an inter-turn fault (ITF) [8,11]. ITFs significantly affect the electromagnetic properties of a motor, such as variations in harmonic characteristics, inrush current, magnetic saturation, cross magnetization and back electromotive force (EMF). Furthermore, the propagation of ITFs can rapidly lead to total motor failure within a few seconds by causing excessive heat that is proportional to the square of the circulating current in the shorted turns [12,13]. The main causes of winding failures are excessive heat, loose insulation, aging due to operation and contamination, among others [5,14].

2.2. Phase Coil-to-Coil Fault (Coil-to-Coil Fault)

Compared with open-circuit faults, short circuits are more common faults that result from the failure of the insulation system, especially the insulation between turns. According to statistics, about 80% of electrical failures in the stator are due to weak insulation between turns. The build-up of ohmic heat can further deteriorate the surrounding insulation and subsequently develop other serious faults, such as coil-to-coil, phase-to-phase or phase-to-ground faults [15]. In addition, the great increase in currents, and especially the appearance of negative sequence currents during a coil-to-coil fault, are the causes of the appearance of another type of fault in the machine known as a demagnetization fault. Taking into account the above, research has been carried out involving implementing coil-to-coil fault diagnosis techniques and methods [10].

2.3. Fault between Phases (Phase-to-Phase Fault)

A short-circuit fault occurs when any two phases of a stator winding are shorted. For example, in a three-phase motor with Phase A, Phase B and Phase C, two of its phases are short-circuited, leading to a change in the performance of the machine. This results in a change in current at twice its nominal value for one phase, while in the other, it is less significant. This distinguishes it from coil-to-coil faults, where the overall change in phase current balances out and has no significant increase in current in either phase, although the sinusoidal nature is distorted. On the other hand, the back electromotive force also undergoes a significant change, and therefore the fluctuating magnitude goes beyond the nominal value [6]. Taking into account the above, research has been carried out for implementing phase-to-phase fault diagnosis techniques and methods [10].

2.4. Open-Circuit Fault of a Phase (Open-Circuit Fault)

An open-circuit winding fault occurs due to high inrush currents and sometimes due to high mechanical vibration opening the stator windings. Such a type of fault in any phase directly makes the current in that phase zero, since that phase is totally disconnected. Due to the open winding in one phase, two other phases are also affected, and the current increases. This type of fault is distinguished from the other faults mentioned above by a zero-current flow in any of the phases which are open stator windings. The back EMF of the machine decreases and fluctuates within the rated limits of the motor. There is not much of an increase in speed due to the depletion of the counter-electromotive force, but there is a non-stability, and the speed is no longer constant [6]. Taking into account the above, research has been carried out which implements open-circuit fault diagnosis techniques and methods.

A fault very similar to the open-circuit fault is the high-resistance connection (HRC) fault [16], which is the result of insulation aging, poor artistry and damaged surfaces due to corrosion. This type of fault makes the resistance increase and current decrease. When a motor runs with an HRC fault, the stator winding is asymmetric, which may lead to increased torque ripple, local overheating and additional loss, in addition to even more serious faults [17].

3. Classification of Methods for Diagnosing Faults in BLDC Motors

A fault is an impermissible deviation of at least one characteristic (feature) of a process from the usual and acceptable standard condition. A fault can be attributed to many causes, and sooner or later, it can lead to breakdowns if corrective measures are not taken [18]. The purpose of a fault diagnosis method is to determine the type, size and location of the most likely fault, as well as its detection time. The methods that have been used for diagnosing faults in BLDC motors can be classified into three categories: data-based, signal-based and process model-based methods [9,19].

3.1. Signal-Based Methods

A signal is a manifestation (visual, auditory, electrical, magnetic, etc.) of a physical phenomenon that can evolve in time or space. For practical purposes, signals provide information about the phenomenon or about the system (source) that is provoking it. To capture this information in the form of physical quantities (vibration, current and temperature), sensors are required. If a fault occurs in a system, it is very likely that some changes in its behavior will occur. These changes are symptoms that indicate that the system is not in a healthy condition. The usual symptoms are functions in the time domain, such as magnitudes, arithmetic or quadratic mean values, limit values, trends and statistical moments, or functions in frequency domain such as spectral power densities, frequency spectral lines and the cepstrum, among others [20].

The purpose of a fault diagnosis method based on a signal(s) is to extract information from one or more signals about a possible fault or set of faults occurring in a system. For extracting this information, the process illustrated in Figure 5 must be executed. The basic process only includes four tasks: acquisition, transformation, extraction and recognition. However, more than one signal is sometimes used for diagnosis. Therefore, an additional task is required: signal separation.

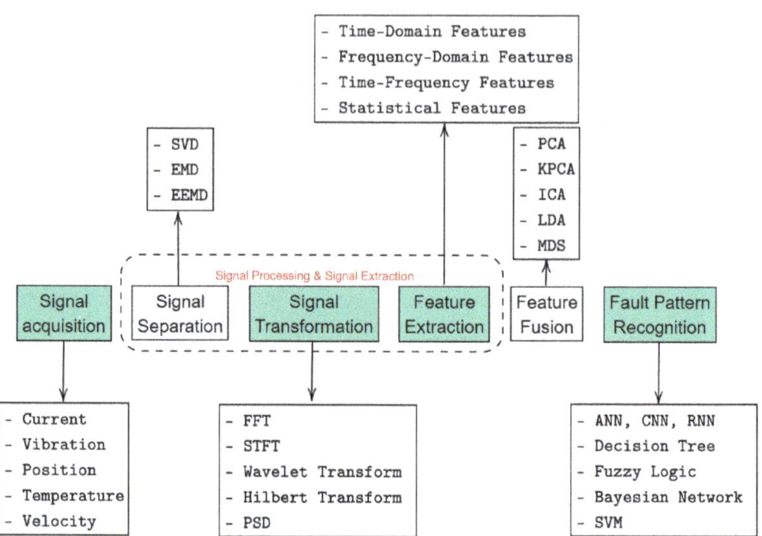

Figure 5. Process of fault diagnosis based on signals. The basic tasks of this process are signal acquisition, signal transformation, feature extraction and fault pattern recognition.

Many measured process signals show oscillations that are either harmonic or stochastic in nature, or both. If changes in these signals are related to faults in the actuators, process and sensors, fault detection methods based on signal models can be applied. Especially for machine vibration, the signals of position, velocity or acceleration allow one to detect, for example, imbalances or bearing failures (turbo machines), detonations (gasoline engines) and vibrations (metal grinding machines). However, the signals from many other sensors, such as the electric current, position, velocity, force, flow and pressure, also often contain oscillations with a variety of frequencies higher than the process dynamics [18].

Signal models can be divided into non-parametric models, such as frequency spectra or correlation functions, or parametric models, such as amplitudes for different frequencies or autoregressive–moving-average (ARMA) process-type models. Another way to classify signal-based methods is according to the nature of the signal: periodical, stochastic or non-stationary. Table 1 lists a classification of methods for the analysis of stationary periodic signals such as bandpass filtering or Fourier analysis for non-stationary periodic signals, such as wavelet transforms, and for stochastic signals, such as correlation functions, CUSUM and Kalman filters.

Table 1. Classification of methods based on signals according to the nature of the signal to be used for the diagnostic.

	Periodic Signals	Stochastic Signals	Non-Stationary Signals
Methods	Bandpass Filtering, Fourier Analysis, Parametric Spectral Estimation, Correlation Analysis	Correlation Analysis, CUSUM, Kalman Filter, ARMA	ARMA, STFT, Wavelet Analysis, Detrend Fluctuation Analysis

In [12], the authors used the harmonic analysis of the line currents to find the existence of a third harmonic that indicates the existence of an inter-turn fault. To evaluate the method, they used an FEM model to generate the current signals as well as an experimental set-up.

A recent method for diagnosing inter-turn faults was presented in [21]. The method employs the measured three-phase currents for the diagnostic. The method starts with the normalization of the measured currents, and then a modal current is calculated from these currents. Having the modal current, three different moving indices, namely the mean-based index (MBI), variance-based index (VBI) and energy-based index (EBI), are calculated in parallel to recognize the inter-turn fault condition. Application of these three indices enables the method to investigate the signal from three different aspects and increase the reliability and quickness of the fault diagnosis process. In addition, these indices are easy to calculate with simple mathematical operators. Hence, the method can be easily implemented online. To differentiate the healthy cases, such as load change, from faults, an auxiliary index is also computed. To justify the method with more details, the following four steps are taken into consideration.

In Table 2, some works that proposed signal based-methods for diagnosing inter-turn faults are listed together with the following particularities: the signal techniques used for the signal treatment, the measured variable(s) used for the diagnosis and the test environment in which the method was evaluated (simulated or experimental).

Table 2. Signal based-methods for diagnosing inter-turn faults. E = Experimental environment; S = simulation environment.

Reference	Main Methods	Measured Variables	S or E
[22]	PSD	Phase currents	S and E
[12]	HA	Line currents	S and E
[23–25]	HA, FEM	Phase currents, Back-EMF, electromagnetic torque, velocity	S
[26]	FFT	Phase currents	E
[27]	FFT, DWT	Phase currents	E
[21]	MBI, VBI, EBI	Phase currents	S and E
[10]	Bispectrum analysis	Phase currents	E
[28]	FFT	Phase currents	S and E

3.2. Model-Based Methods

Different methods for fault diagnosis using mathematical models have been developed for BLDC motors. These methods can be roughly classified into (1) methods without estimation error feedback and (2) methods with estimation error feedback. In the first method, the model is fed with the input information of the BLDC motor (voltages and torque). The response of the model (currents, angular displacement and angular velocity) is compared with the response of the BLDC motor. If there is a discrepancy between the responses, this is probably because there exists a fault or a set of faults. For obtaining good results with this type of method, it is necessary that the BLDC motor model is well-calibrated in healthy conditions. On the contrary, false alarms will appear. A drawback of this class is that a disturbance can be mistaken for a fault. Another drawback of methods without estimation error feedback is that they only serve to detect faults and not to locate or isolate them.

The methods with estimation feedback errors are more advantageous. The error between the model response and the BLDC motor response is injected into the model as an additional input and then multiplied by a gain that causes this error to move toward zero over time. These methods are (1) usefulness in detecting, locating and isolating faults, (2) usefulness for applications in real time and (3) robustness against disturbances. These methods are also known as observer based-methods, since a model with estimation error feedback is called a state observer. Usually, state observers are also called *virtual sensors* or *soft sensors*.

A state observer is an algorithmic tool that estimates variables such as the state variables, unknown inputs, disturbances, parameters and faults of a process (e.g., a BLDC motor). The parts of a state observer are (1) a mathematical model and (2) an error term (correction term) for ensuring the convergence of the algorithm. A state observer is fed with the available measurements of the process (inputs and outputs).

To derive a general structure of a state observer, let us consider the general structure of the continuous model of a system in a state-space representation, which is given as follows:

$$\dot{x}(t) = f(x(t), u(t)),$$
$$y(t) = h(x(t)), \tag{1}$$

where $x(t) \in \mathbb{R}^n$ is the state vector, $\dot{x}(t) \in \mathbb{R}^n$ is the state derivative vector, $u(t) \in \mathbb{R}^m$ is the external (exogenous) input vector or control signal, $y(t) \in \mathbb{R}^p$ represents the output vector (i.e., the measured states (variables) acquired by the sensors), $f \in \mathbb{R}^n$ represents the vector field and $h \in \mathbb{R}^p$ is the continuous output function. Since a state observer is the model of the system plus a correction (adaptation term), this can be expressed as follows:

$$\dot{\hat{x}}(t) = \underbrace{f(\hat{x}(t), u(t))}_{\text{Model Copy}} + \underbrace{K(\hat{x}(t))(y(t) - \hat{y}(t))}_{\text{Correction Term}},$$
$$\hat{y}(t) = h(\hat{x}(t)), \tag{2}$$

where $\hat{x}(t)$ and $\hat{y}(t)$ are the online estimations of $x(t)$ and $y(t)$, respectively, and $K(\hat{x}(t))$ is the gain of the observer. Thus, the design of the state observer consists of choosing an appropriate gain $K(\hat{x}(t))$ so that the estimation error tends toward zero when $t \to \infty$ with the desired properties of time convergence and robustness. If the observation error $e(t)$ is defined as follows:

$$e(t) = x(t) - \hat{x}(t),$$

then the dynamics of the error observation can be derived from Equations (1) and (2) and expressed as

$$\dot{e}(t) = f(\hat{x}(t) + e(t), u(t)) - f(\hat{x}(t), u(t)) - K(\hat{x}(t))(h(\hat{x}(t) + e(t)) - h(\hat{x}(t))).$$

An observer connected to a BLDC motor has the structure of the block diagram shown in Figure 6. The inputs can be the the voltages or the torques. These inputs, or at least a subset of them, must be registered to be injected into the state observer. The state, which is the smallest possible subset of system variables that can represent the complete state of a system at any time, can be either the currents or the angular velocity. The measured outputs are the measurements provided by in situ sensors (position sensors or current sensors).

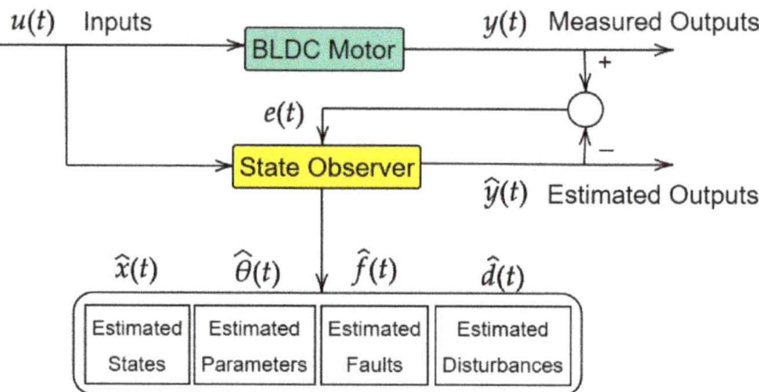

Figure 6. Architecture of a state observer.

The observer-based methods can be classified into three categories: (1) methods based on residual generation, (2) methods based on a bank of state observers and (3) methods based on parameter estimation. This classification is illustrated in Figure 7.

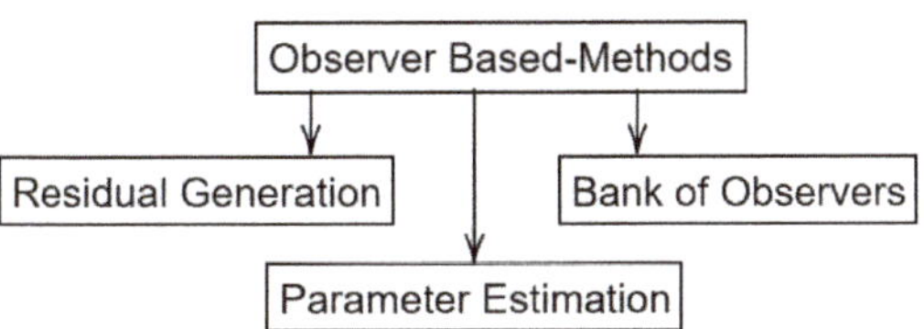

Figure 7. Methods based on state observers for diagnosing faults in the stators of BLDC motors.

Methods based on residual generation: A residual is an estimable quantity that can be used to warn if a fault occurs in a system. Usually, a residual is designed to be zero (or small enough in a realistic case where the process is subjected to noise and the model is uncertain) in the fault-free case and deviate significantly from zero when a fault occurs [29]. These methods comprise two stages: residual generation and residual evaluation. The easiest way to generate a residual is by estimating a variable (or set of variables) with the help of a state observer at the same time that this variable (or set of variables) is measured with hard sensors. Residuals can then be generated by subtracting the estimated variables from the measured variables such that

$$
\begin{aligned}
r_1(t) &= y_1(t) - \hat{y}_1(t) \\
r_2(t) &= y_2(t) - \hat{y}_2(t) \\
\vdots\;\; &= \qquad \vdots \\
r_n(t) &= y_n(t) - \hat{y}_n(t)
\end{aligned}
$$

where r_i represents the residuals, y_i represents the measured outputs and $\hat{y}_i$ represents the estimated outputs $\forall i = 1, 2, \ldots, n$.

To evaluate a residual, it is necessary to establish some metric. The easiest way to evaluate a residual is by setting a threshold. If the residual exceeds this threshold, then there is a fault in the system; otherwise, there is not. A schema of residual based-methods is shown in Figure 8.

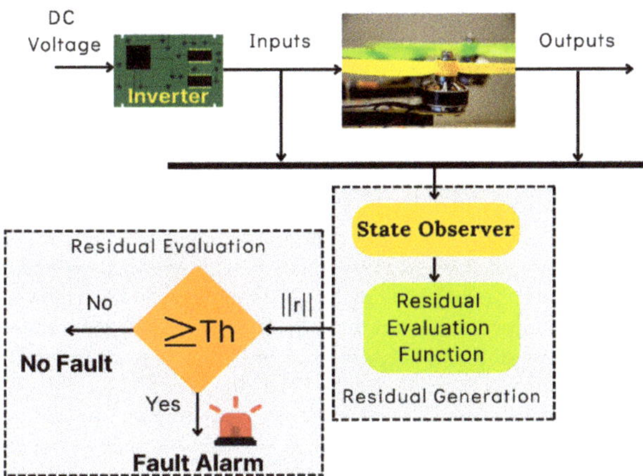

Figure 8. Usual architecture of methods based on residuals for diagnosing faults in the stators of BLDC motors.

In [30,31], the authors proposed fault diagnosis approaches based on a sliding mode observer that estimates the phase currents. The estimation error is used to generate residuals for the detection and location of a short-circuit fault in the stator winding turns.

Methods based on a bank of state observers: The architecture of these methods is illustrated in Figure 9. A bank of observers is made up of a set of state observers that work (estimate) in parallel. Each state observer is different from the other because each observer is constructed from a model involving a particular and different fault. For illustration purposes, let us consider a bank composed of three state observers. The first observer can be designed from a model with an open-circuit fault, and thus this observer will detect this kind of fault. The second observer can be constructed from a model with an inter-turn fault, and the third observer can be designed from a model with a phase-to-phase fault. The three observers receive the same information from the system (inputs and outputs), and the three observers compute an error estimation. The errors are evaluated by using suitable metrics to determine the smallest error. The fault will be the involved in the observer that produced the smallest error.

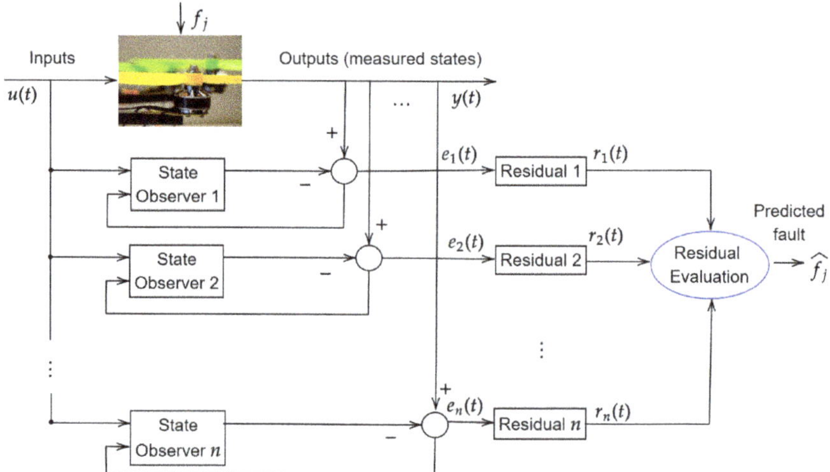

Figure 9. Bank of n observers, where $u(t)$ and $y(t)$ are the inputs and outputs (measured states) of the BLDC motor, respectively, $e_i(t)$ ($\forall i = 1, 2, \ldots, n$) is the error computed by the ith observer, $r_i(t)$ is the residual calculated from the error $e_i(t)$, f_j is the type of fault affecting the system and $\hat{f}_j$ is the estimated fault.

Methods based on the parameter estimation: When a fault happens in a BLDC motor, some parameters change. For example, damaged or broken bearings may augment the friction, or an increasing temperature in the stator may increase the phase resistance of all coils [32,33]. These parameter changes can be estimated by different algorithms, among them being the means of the state observers designed from mathematical models that involve the parameters of the BLDC motor. The architecture of these methods is illustrated in Figure 10. The fundamental idea of these methods is to directly estimate the parameters of the system by means of a state observer. Then, the estimated parameters are subtracted from the nominal parameters (i.e., from the values of the parameters under normal conditions). This is performed to calculate the error, which is evaluated using a predefined threshold. If the error exceeds this threshold, then there is a fault in the system. Depending on which parameter is evaluated, the type of fault can be determined.

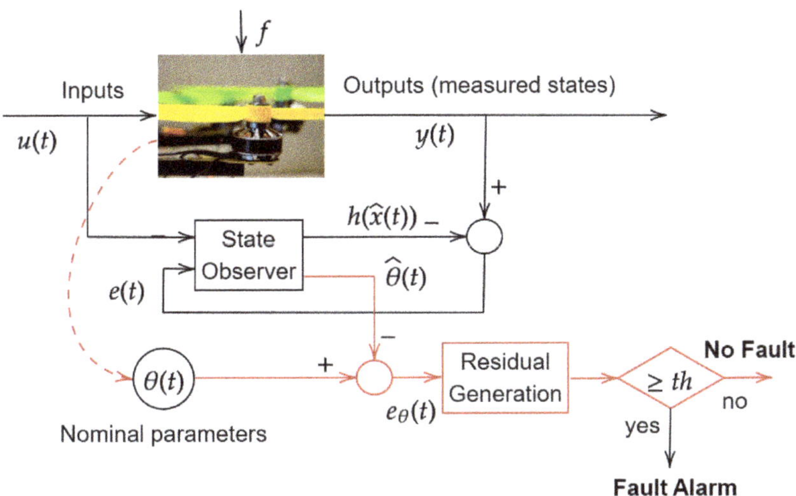

Figure 10. Methods based on parameters estimation, where $u(t)$ and $y(t)$ are the inputs and outputs (measured states) of the BLDC motor, respectively. $e(t)$ is the observation (estimation) error, $\theta(t)$ represents the parameters in nominal (normal) operation conditions, $h(\hat{x}(t))$ is the function output of the observer which is dependent on the estimated states, $\hat{\theta}$ represents the estimated states, $e_\theta(t)$ is the error between the nominal and estimated parameters and th is the threshold to surpass when there is a fault.

3.3. Data-Based Methods

Methods based on models are appropriate to be used when the dimension of the process (under diagnosis) is low and it can be modeled with low-order models. However, to diagnose faults in more complex systems, these methods are no longer recommended. An alternative to diagnose faults without using models is the use of data-based methods, which use information acquired from the process (under diagnosis). It can be said that these methods are the recent alternative for active supervision of systems too complex to have an explicit analytic model or signal symptoms of faulty behavior for.

The application of these diagnosis methods essentially consists of two stages: (1) training, in which the historical datasets are presented as a priori knowledge of the process under monitoring and transformed to a diagnostic system or algorithm, and (2) online running, in which the online measurement data are processed in the diagnostic system or by using the diagnostic algorithm for reliable fault detection and identification [20].

Data-based fault diagnosis methods can be categorized into two categories: supervised and unsupervised learning methods [34]. Unsupervised methods require the model to be initially developed using normal operation data. The faults are then detected as deviations from the normal behavior. Supervised methods require the training of a classifier on historical data which comprise both normal and faulty states. The trained model is then used for the detection of future faults.

In [35], Park et al. proposed a method that uses a database containing data on the phase currents and voltages of a BLDC motor under fault conditions. These data were obtained through numerical simulations using a model based on the FEM or using a WFT. From these data, the values of some input impedances under fault conditions were calculated. For diagnosis, the authors proposed comparing these fault impedances with the impedances obtained from the voltages and currents that were measured for diagnosis.

In [11], Hosseini et al. introduced a supervised data-based method to detect and classify faults in BLDC motors, namely stator inter-turn faults, rotor dynamic imbalance, rotor static imbalance and different combinations of these. The current signal of the BLDCM is used together with the motor torque and motor speed to achieve the classification of a broad range of faults. The fault features of the measured signals are extracted using a packet wavelet transform (PWT). These features, which include the energy, in the two modes of BLDCM operation, without load and with load, are used as input data for an ANN. The ANN weights are updated by particle swarm optimization (PSO) and a genetic algorithm (GA).

In [36], Borja et al. proposed a supervised learning method to classify different types of faults: bearing inner ring damage, inter-turn faults and holes in the rotor. The method uses k-nearest neighbors and a DWT for feature extraction.

Table 3 lists some of the most important contributions proposed for diagnosing inter-turn faults. Table 4 details some of the main characteristics of data-based methods.

Table 3. Data-based methods for diagnosing inter-turn faults.

Reference	Main Methods	Measured Variables	S or E
[11]	Wavelets, ANN, PSO, Gf	Motor torque, motor speed	S
[13]	Logic comparison, FEM	Back-EMF	S and E
[35]	WFT, FEA	Phase currents and voltages	S and E

Table 4. Comparison between supervised and unsupervised learning methods.

	Supervised Learning	Unsupervised Learning
Methods	Bayesian Network, Random Forest, Decision Tree, k-Nearest Neighbors, Fisher's Discriminant Analysis, Artificial Neural Networks, Support Vector Machine	Principal Component Analysis, Partial Least Square, Independent Component Analysis, Autoencoders
Features	(1) Require both input and output system data. (2) Require labeled data. (3) Predict the output. (4) Requiere supervision for training.	(1) Require only input system data. (2) Employ unlabeled data. (3) Find hidden patterns in data. (4) Do not requiere supervision for training.
Computational Complexity	(1) Very complex. (2) Require feedback to improve the accuracy of the prediction.	(1) Less complex. (2) No feedback is needed.
Learning	Offline learning (usually)	Online learning
End Goal	Predict an output. Develop a model to (1) predict new values or (2) understand existing relationship between input and output data.	Gain insight from data. Develop a model to (1) place observations from a dataset into a specific cluster or (2) create rules to identify associations between variables.
Subtypes	(1) Regression and (2) classification.	(1) Clstering and (2) association.
Performance	More accurate	Less accurate
Fault Classes	Known in advance	Not known in advance

3.4. Summary

Table 5 summarizes the advantages and disadvantages that various authors have found during application of the methods reviewed in this article. Table 6 summarizes the investigation of different techniques and methods for fault detection and diagnosis. The research task was carried out by collecting information from different documents, books, magazines, states of the art and reviews that allowed detailing characteristics about the type of fault to diagnose, the physical variables used to perform the diagnosis and if the method was executed in a simulation or online.

Table 5. Advantages and disadvantages of the main methods used for diagnosing stator faults in BLDC motors.

Method	Advantages	Disadvantages
Model-based	(1) Can be naturally integrated into a fault-tolerant control scheme. (2) Can be highly accurate. (3) Requires less data than data-based methods. (4) All phases of diagnosis (detection, isolation and identification) can be conceived using the same model. (5) Few hard sensors are required with respect to the other methods.	(1) They requiere well-calibrated models. (2) Real-life system physics is often too stochastic and complex to model. (3) Sample time (or sample frequency) is important.
Data-based	(1) Do not require some model based on the physics of the BLDC motor. (2) Therefore, they are suitable for applications where a model is not available. (3) Suitable for processes with many sensed variables (i.e., when a large quantity of data is available). (4) Sample time (or sample frequency) is not important.	(1) Diagnosis accuracy relies on data quantity and quality. (2) Historical data on the behavior of the BLDC throughout its active life is required. (3) A sizable quantity of sensors is required.
Signal-based	(1) Do not require some model based on the physics of the BLDC motor. (2) Historical information about the BLDC motor is not required.	(1) Risk of false alarms due to disturbances and changes in the BLDC operating conditions. (2) High-speed computing power for transforming signals in real time. (3) The accuracy of the diagnosis depends on the quality of the sensors that provide the signals. (4) Sample time (or sample frequency) is important.

Table 6. Description of diagnostic methods for stator faults in BLDC motors. For the implementation aspects, S = tested in simulation, E = experimentally tested, ON-L = it can work online (i.e., in real time) and OFF-L = it can work offline (i.e., it works with stored data).

Type of Fault	Method	Physical Variables	Implementation	Comp. Burden
Inter-Turn	FEM, KF, HA [12]	Current	S, E, ON-L	High
	Spectral methods, SVM methods [4]	Current	S, OFF-L	Medium
	WFT, FEM [8]	Inductance, torque, voltage	S, E, OFF-L	Medium
	WFT, FEM, [37]	Back-EMF, current	S, E, OFF-L	Medium
	WFT, FFT [35]	Impedance, current, voltage, coil resistance	S, E, OFF-L	Low

Table 6. *Cont.*

Type of Fault	Method	Physical Variables	Implementation	Comp. Burden
	Search coils, FEA [38]	Magnetic flux, voltage	S, E, OFF-L	High
	Hybrid analytical-numerical approach, ECC, FEMM [23]	Current	Co-S, OFF-L	High
	FFT, Park's phasor analysis [26]	Current	S, OFF-L	Medium
	ECC, MEC, numerical methods, hybrid models, IWFT [25]	Current	S, Co-S, E, OFF-L	High, medium (depending on the method)
	RUL, RNN, LSTM [39]	Voltage, torque, temperature, velocity, current	S, E, OFF-L	High
	Technique based on undulations in velocity [40]	Velocity	E, OFF-L	Low
	Self-encoding convolutional network model [41]	Phase currents	S, OFF-L (applicable online)	High
	LS, AE [42]	Resistivity, inductance, voltage	S, OFF-L (applicable online)	High
	Co-simulation multidomain technique, FEM [43]	Flux density, current, torque vibrations	Co-S, E, OFF-L	Medium
	SVW, FFT [44]	Currents	S, E, OFF-L	Medium
	FEI, FFT, WDT [27]	Current	S, E, OFF-L	High
	FEM [45]	Current, voltage, electromagnetic torque, temperature	S, E, OFF-L	Medium
	CNN [46]	Current	S, E, OFF-L	High
	Current observer [47]	Voltage, current, position, velocity	S, E, OFF-L	Low
	Sliding mode observer (SMO) [48]	Current, voltage	S, ON-L	Low

Table 6. *Cont.*

Type of Fault	Method	Physical Variables	Implementation	Comp. Burden
	SSAE, Siamese neural networks [49]	Current, impedance, torque	S, E, ON-L	High
	DTCRV [50]	Current	S, E, OFF-L	Low
	Wave packet transform [51]	Current, vibration signal of the stator	E, OFF-L	High
	TDF, WVD, CWD [52]	Current	S, E, OFFL	Medium
	KF [53]	Current, BEMF	S, E, OFF-L	Medium
	FFT [54]	Phase voltages	S, OFF-L	Medium
	KF [55]	Voltage (voltage residuals)	S, ON-L	High
	Spectral density estimator (PSD), Welch and Burg method [56]	Current	S, OFF-L	High
	PSD, MCSA [57]	Current	S, OFF-L	Medium
	Maxwel II. 2D, numerical methods, FEM [24]	eletromagnetic magnitude, phase currents, BFEM, Electromagnetic torque, velocity, magnetic flux density, flux linkage	Co-S, OFF-L	High
	Math models [58]	Current, torque	S, ON-L	Medium
	Arithmetic mean [59]	Current, position	S, OFF-L	Low
	FEM [60]	Electromagnetic torque	S, E, OFF-L	High
	KF [9]	Phase currents	S, ON-L	Low High
	FEM, EMD, WVD [61]	Current	S, OFF-L	High

Table 6. *Cont.*

Type of Fault	Method	Physical Variables	Implementation	Comp. Burden
Coil-to-coil	Fast Kurtogram autogram, MCSA [5]	Vibration, current	E, OFF-L	High
	MCSA [62]	Current	S, E, OFF-L	Medium
Phase-to-phase	FEM numerical methods [15]	BFEM, magnetic flux density, phase current	S, OFF-L	High
	FEM [63]	None	S, OFF-L	High
	Math model [64]	Self-inductions, mutual inductance, current	S, OFF-L	Medium
	DWT, ANN [65]	Voltage, current, sequence current negative	S, OFF-L	High
Open circuit	Wavelet transform, SVM, DWT, NCA [66]	Current	S, OFF-L	High
	FOC, current comparison method (voltage error) [67]	Current	S, E, OFF-L	Medium
	MPCC [68]	Current	S, E, OFF-L	Medium
	RMS [69]	Normalized current	S, OFF-L	Low
	DWT, NN [70]	Current, velocity	S, OFF-L	High
	Current residual [71]	Current, torque, voltage	S, ON-L	Low
	FFT [16]	Impedances	S, E, OFF-L	Low
	FEM and function wavelets [72]	Current	S, OFF-L	High

From the classification table, it can be noticed that most of the diagnostic methods designed for BLDC motors detect and locate inter-turn faults, and most of these methods are signal-based methods, which can be seen as a drawback if the application is subject to disturbances and noise, since signal-based methods are less robust than the other classes.

4. Conclusions

To carry out this work, information on diagnostic methods to detect faults in the stators of BLDC motors was collected. The purpose of this collection was to classify and analyze the information in order to convey to the reader a detailed summary of the current state of the subject. It is worth saying that all the methods presented, classified and discussed in this review are also useful for permanent magnet synchronous motors, since the only difference is that synchronous motors develop a sinusoidal back EMF instead of a trapezoidal back EMF. The collected information was structured in the following way: (1) a classification of the main methods to diagnose stator faults in BLDC motors, (2) a subclassification of each method class, (3) a table with the advantages and disadvantages of each method class, (4) a comparison between supervised and unsupervised learning methods, which are subclasses of data-based methods, and (5) a table that organized the methods according to the type of stator fault to be diagnosed, the physical variables used by each method, the environment in which the method was tested (simulated or experimental) and the computational burden of each method. It was observed that there are some data-based methods that use a lot of computational consumption for data processing, more computation time and more memory to store data, among other factors, which makes these methods not so appropriate for applications in which only a BLDC motor is involved, but they are recommended for industries that have a large number of motors. It was also noticed that most of the fault diagnosis methods use neural networks, wavelets and signal-based methods in general. However, these methods are not very robust to disturbances and unknown inputs, and thus speed and load variations can influence the fault diagnosis, leading to poor fault location or fault identification. Therefore, it is advisable to continue designing methods based on models, which are more robust, or methods that combine both approaches to take advantage of the benefits that both offer. Finally, most of the techniques use current and voltage as diagnostic variables and can only diagnose one or two faults. Therefore, it is advisable to design methods that incorporate more variables in order to detect more types of faults with the same algorithm.

Author Contributions: Conceptualization, R.S., L.T. and P.P.; methodology, L.T.; formal analysis, R.S. and L.T.; investigation, R.S.; resources, L.T.; data curation, R.S.; writing—original draft preparation, R.S.; writing—review and editing, L.T.; visualization, L.T.; supervision, L.T.; project administration, P.P.; funding acquisition, P.P. All authors have read and agreed to the published version of the manuscript.

Funding: R. Solís thanks CONACYT for his PhD scholarship (CVU: 856691). The authors thank DGAPA-UNAM for the financial support provided through the IT101322 project *Diágnostico en tiempo real de motores BLDC.*

Institutional Review Board Statement: Not applicable.

Informed Consent Statement: Not applicable.

Data Availability Statement: Not applicable.

Conflicts of Interest: The authors declare no conflict of interest.

Abbreviations

The following abbreviations are used in this manuscript:

AE	Autoencoder
ANN	Artificial Neural Network
ARMA	Autoregressive–Moving-Average
CNN	Convolutional Neural Network
CWD	Choi–Williams Distribution
DC	Direct Current
DTCRV	Drive-Tolerant Current Residual Variance
DWT	Discrete Wavelet Transform
EBI	Energy-Based Index
ECC	Electrical Equivalent Circuit
EEMD	Ensemble Empirical Mode Decomposition
EMD	Empirical Mode Decomposition
EMF	Electromotive Force

FEA	Finite Element Analysis
FEM	Finite Element Method
FEMM	Finite Element Method Magnetic
FEI	Fasor Espacial Instantaneo
FFT	Fast Fourier Transform
FOC	Field-Oriented Control
GA	Genetic Algorithm
HA	Harmonic Analysis
ICA	Independent Component Analysis
IWFT	Improved Winding Function Theory
KF	Kalman Filter
KPCA	Kernel Principal Component Analysis
LDA	Linear Discriminant Analysis
LS	Least Squares
LSTM	Long Short-Term Memory
MBI	Mean-Based Index
MDS	Multi-Dimensional Scaling
MEC	Magnetic Equivalent Circuit
MCSA	Motor Current Signature Analysis
MPCC	Model Predictive Current Control
NCS	Neighborhood Component Analysis
NN	Neural Network
PWT	Packet Wavelet Transform
PSD	Power Spectral Density
PCA	Principal Component Analysis
PSO	Particle Swarm Optimization
RMS	Root Mean Square
RNN	Recurrent Neural Network
RUL	Remaining Useful Life
STFT	Short-Time Fourier Transform
SVD	Singular Value Decomposition
SVM	Support Vector Machine
SMO	Sliding Mode Observer
SSAE	Stacked Sparse Autoencoders
TDF	Time–Frecuency Distribution
VBI	Variance-Based Index
WFT	Winding Function Theory
WVD	Wigner–Ville Distribution

References

1. Moosavi, S.S.; Djerdir, A.; Amirat, Y.A.; Khaburi, D.A. Demagnetization fault diagnosis in permanent magnet synchronous motors: A review of the state-of-the-art. *J. Magn. Magn. Mater.* **2015**, *391*, 203–212. [CrossRef]
2. Mohanraj, D.; Aruldavid, R.; Verma, R.; Sathyasekar, K.; Barnawi, A.; Chokkalingam, B.; Mihet-Popa, L. A Review of BLDC Motor: State of Art, Advanced Control Techniques, and Applications. *IEEE Access* **2022**, *10*, 54833–54869. [CrossRef]
3. Chau, K.T. Stator-Permanent Magnet Motor Drives. In *Electric Vehicle Machines and Drives: Design, Analysis and Application*; Wiley-IEEE Press: Hoboken, NJ, USA, 2015; pp. 147–194. [CrossRef]
4. Da, Y.; Shi, X.; Krishnamurthy, M. Health monitoring, fault diagnosis and failure prognosis techniques for Brushless Permanent Magnet Machines. In Proceedings of the 2011 IEEE Vehicle Power and Propulsion Conference, Chicago, IL, USA, 6–9 September 2011; pp. 1–7.
5. Shifat, T.A.; Hur, J.W. An effective stator fault diagnosis framework of BLDC motor based on vibration and current signals. *IEEE Access* **2020**, *8*, 106968–106981. [CrossRef]
6. Usman, A.; Doiphode, N.T.; Rajpurohit, B.S. Stator Winding Faults investigation in Permanent Magnet Synchronous Motor using Motor Signatures: Part I. In Proceedings of the 2019 International Conference on Electrical Drives & Power Electronics (EDPE), The High Tatras, Slovakia, 24–26 September 2019; pp. 160–168.
7. Chen, Y.; Liang, S.; Li, W.; Liang, H.; Wang, C. Faults and diagnosis methods of permanent magnet synchronous motors: A review. *Appl. Sci.* **2019**, *9*, 2116. [CrossRef]
8. Kim, K.T.; Park, J.K.; Hur, J.; Kim, B.W. Comparison of the fault characteristics of IPM-type and SPM-type BLDC motors under inter-turn fault conditions using winding function theory. *IEEE Trans. Ind. Appl.* **2013**, *50*, 986–994. [CrossRef]

9.	Vanchinathan, K.; Valluvan, K.R.; Gnanavel, C.; Gokul, C.; Albert, J.R. An improved incipient whale optimization algorithm based robust fault detection and diagnosis for sensorless brushless DC motor drive under external disturbances. *Int. Trans. Electr. Energy Syst.* **2021**, *31*, e13251. [CrossRef]
10.	Pietrzak, P.; Wolkiewicz, M. Stator Winding Fault Detection of Permanent Magnet Synchronous Motors Based on the Short-Time Fourier Transform. *Power Electron. Drives* **2022**, *7*, 112–133. [CrossRef]
11.	Hosseini, F.; Abedi, M.; Mohammad Hosseini, S. Classification of Multiple Electromechanical Faults in BLDC Motors Using Neural Networks and Optimization Algorithms. *Res. Technol. Electr. Ind.* **2022**, *1*, 114–124.
12.	Lee, S.T.; Hur, J. Detection technique for stator inter-turn faults in BLDC motors based on third-harmonic components of line currents. *IEEE Trans. Ind. Appl.* **2016**, *53*, 143–150. [CrossRef]
13.	Lee, S.T.; Kim, K.T.; Hur, J. Diagnosis technique for stator winding inter-turn fault in BLDC motor using detection coil. In Proceedings of the 2015 9th International Conference on Power Electronics and ECCE Asia (ICPE-ECCE Asia), Seoul, Republic of Korea, 1–5 June 2015; pp. 2925–2931.
14.	Antunes, H.; Fonseca, D.; Cardoso, A. Symmetrical Six-Phase Induction Motor Stator Faults Diagnostics Approach, Immune to Unbalanced Supply Voltage, Based on the Analysis of the Midpoint Electrical Potential of the Stator Star. *Eng. Proc.* **2022**, *24*, 22.
15.	Usman, A.; Doiphode, N.T.; Rajpurohit, B.S. Finite Element Modeling of Stator Winding Faults in Permanent Magnet Synchronous Motor: Part II. In Proceedings of the 2019 International Conference on Electrical Drives & Power Electronics (EDPE), The High Tatras, Slovakia, 24–26 September 2019; pp. 169–176.
16.	Wang, H.; Qian, G.; Cao, W.; Lu, S. A Two-Step Online Fault Detection Method for High-Resistance Connection Faults in a Vehicular BLDC Motor. In *Advanced Sensors and Sensing Technologies for Electric Vehicles*; AIP Publishing LLC Melville: New York, NY, USA, 2022; p. 1.
17.	Gupta, A.; Jayaraman, K.; Reddy, R. Performance Analysis and Fault Modelling of High Resistance Contact in Brushless DC Motor Drive. In Proceedings of the IECON 2021—47th Annual Conference Of The IEEE Industrial Electronics Society, Toronto, ON, Canada, 13–16 October 2021; pp. 1–6.
18.	Isermann, R. *Fault-Diagnosis Applications: Model-Based Condition Monitoring: Actuators, Drives, Machinery, Plants, Sensors, and Fault-Tolerant Systems*; Springer Science & Business Media: Berlin/Heidelberg, Germany, 2011.
19.	Suawa, P.; Meisel, T.; Jongmanns, M.; Huebner, M.; Reichenbach, M. Modeling and Fault Detection of Brushless Direct Current Motor by Deep Learning Sensor Data Fusion. *Sensors* **2022**, *22*, 3516. [CrossRef]
20.	Ding, S.X. *Data-Driven Design of Fault Diagnosis and Fault-Tolerant Control Systems*; Springer: Berlin/Heidelberg, Germany, 2014.
21.	Jafari, A.; Faiz, J.; Jarrahi, M.A. A simple and efficient current-based method for interturn fault detection in BLDC motors. *IEEE Trans. Ind. Inform.* **2020**, *17*, 2707–2715. [CrossRef]
22.	Park, J.K.; Seo, I.M.; Hur, J. Fault type detection using frequency pattern of stator current in IPM-type BLDC motor under stator inter-turn, dynamic eccentricity, and coupled faults. In Proceedings of the 2013 IEEE Energy Conversion Congress and Exposition, Denver, CO, USA, 15–19 September 2013; pp. 2516–2521.
23.	Usman, A.; Rajpurohit, B.S. Detection and Identification of Stator Inter-turn Faults and Demagnetization effects in Hybrid Analytical-Numerical model of a BLDC Motor using Electromagnetic Signatures. In Proceedings of the 2020 IEEE Texas Power and Energy Conference (TPEC), College Station, TX, USA, 6–7 February 2020; pp. 1–7.
24.	Usman, A.; Rajpurohit, B.S. Numerical Analysis of Stator Inter-turn Fault and Demagnetization effect on a BLDC Motor using Electromagnetic Signatures. In Proceedings of the 2020 IEEE International Conference on Power Electronics, Smart Grid and Renewable Energy (PESGRE2020), Cochin, India, 2–4 January 2020; pp. 1–6.
25.	Usman, A.; Rajpurohit, B.S. Modeling and Classification of Stator Inter-Turn Fault and Demagnetization Effects in BLDC Motor Using Rotor Back-EMF and Radial Magnetic Flux Analysis. *IEEE Access* **2020**, *8*, 118030–118049. [CrossRef]
26.	Acevedo, R.J.; Piña, F.J.V.; Salas, R.Á.; Zárate, C.H.S. Diagnóstico de falla eléctrica de estator en motores bldc de vehículo ligero en diferentes regímenes de velocidad mediante la transforma discreta de fourier (stator's electric fault diagnosis in bldc motors of light vehicle in different velocity regimes by means of the discret fourier's transform). *Pist. Educ.* **2020**, *42*, 337–347.
27.	Maldonado Ruelas, V.A.; Villalobos Piña, F.J.; Alvarez Salas, R.; Morones Alba, J.A.; Ortiz Medina, R.A. Detección de fallos multicriterio de un motor BLDC (Brushless Direct Current). *DYNA-Ing. Ind.* **2018**, *93*, 556–562.
28.	Wang, H.; Wang, J.; Wang, X.; Lu, S.; Hu, C.; Cao, W. Detection and Evaluation of the Inter-turn Short Circuit Fault in a BLDC-Based Hub Motor. *IEEE Trans. Ind. Electron.* **2022**, *70*, 3055–3068. [CrossRef]
29.	Frisk, E. Residual Generation for Fault Diagnosis. Ph.D. Thesis, Linköpings Universitet, Linköping, Sweden, 2001.
30.	El Mekki, A.; Saad, K.B. A BLDC fault diagnosis approach based on a super-twisting sliding mode observer. In Proceedings of the 2016 4th International Conference on Control Engineering & Information Technology (CEIT), Hammamet, Tunisia, 16–18 December 2016; pp. 1–5.
31.	El Mekki, A.; SAAD, K.B. Diagnosis based on a sliding mode observer for an inter-turn short circuit fault in brushless dc motors. *Rev. Roum. Sci. Tech. Electrotech. Energétique* **2018**, *63*, 391–396.
32.	Moseler, O.; Isermann, R. Model-based fault detection for a brushless DC motor using parameter estimation. In Proceedings of the IECON'98. Proceedings of the 24th Annual Conference of the IEEE Industrial Electronics Society (Cat. No. 98CH36200), Aachen, Germany, 31 August–4 September 1998; Volume 4, pp. 1956–1960.
33.	Moseler, O.; Isermann, R. Application of model-based fault detection to a brushless DC motor. *IEEE Trans. Ind. Electron.* **2000**, *47*, 1015–1020. [CrossRef]

34. Taqvi, S.A.A.; Zabiri, H.; Tufa, L.D.; Uddin, F.; Fatima, S.A.; Maulud, A.S. A review on data-driven learning approaches for fault detection and diagnosis in chemical processes. *ChemBioEng Rev.* **2021**, *8*, 239–259. [CrossRef]

35. Park, J.K.; Jeong, C.L.; Lee, S.T.; Hur, J. Early detection technique for stator winding inter-turn fault in BLDC motor using input impedance. *IEEE Trans. Ind. Appl.* **2014**, *51*, 240–247. [CrossRef]

36. Borja, C.A.; Tisado, K.J.; Ostia, C. Fault Diagnosis of a Brushless DC Motor Using K-Nearest Neighbor Classification Technique with Discrete Wavelet Transform Feature Extraction. In Proceedings of the 2022 IEEE 14th International Conference on Computer and Automation Engineering (ICCAE), Brisbane, Australia, 25–27 March 2022; pp. 122–126.

37. Kim, T.; Lee, H.W.; Kwak, S. The internal fault analysis of brushless DC motors based on the winding function theory. *IEEE Trans. Magn.* **2009**, *45*, 2090–2096.

38. Da, Y.; Shi, X.; Krishnamurthy, M. A new approach to fault diagnostics for permanent magnet synchronous machines using electromagnetic signature analysis. *IEEE Trans. Power Electron.* **2012**, *28*, 4104–4112. [CrossRef]

39. Shifat, T.A.; Jang-Wook, H. Remaining Useful Life Estimation of BLDC Motor Considering Voltage Degradation and Attention-Based Neural Network. *IEEE Access* **2020**, *8*, 168414–168428. [CrossRef]

40. Hadef, M.; Mekideche, M.R.; Djerdir, A.; N'diaye, A.O. Turn-to-Turn Short Circuit Faults between Two Phases in Permanent Magnet Synchronous Motor Drives. *Jordan J. Electr. Eng.* **2017**, *3*, 208–222.

41. Qi, J.; Wan, H. A Detection Method of Phase Failure of PMSM Based on Deep Learning. In Proceedings of the 2020 IEEE Chinese Automation Congress (CAC), Shanghai, China, 6–8 November 2020; pp. 5591–5595.

42. Bao, L.; Yao, G.; Chen, S.; Wang, Z.; Hu, X.; Huang, Y. An On-line Detection Method for Single-Phase Inter-Turn Fault Occurring in High-Speed PMSM. In Proceedings of the 2020 IEEE 23rd International Conference on Electrical Machines and Systems (ICEMS), Hamamatsu, Japan, 24–27 November 2020; pp. 1095–1100.

43. Ullah, Z.; Hur, J. Analysis of Inter-Turn-Short Fault in an FSCW IPM Type Brushless Motor Considering Effect of Control Drive. *IEEE Trans. Ind. Appl.* **2020**, *56*, 1356–1367. [CrossRef]

44. Pietrzak, P.; Wolkiewicz, M. Application of Support Vector Machine to stator winding fault detection and classification of permanent magnet synchronous motor. In Proceedings of the 2021 IEEE 19th International Power Electronics and Motion Control Conference (PEMC), Gliwice, Poland, 25–29 April 2021; pp. 880–887.

45. Zhao, J.; Guan, X.; Li, C.; Mou, Q.; Chen, Z. Comprehensive Evaluation of Inter-Turn Short Circuit Faults in PMSM Used for Electric Vehicles. *IEEE Trans. Intell. Transp. Syst.* **2021**, *22*, 611–621. [CrossRef]

46. Maraaba, L.S.; Milhem, A.S.; Nemer, I.A.; Al-Duwaish, H.; Abido, M.A. Convolutional neural network-based inter-turn fault diagnosis in LSPMSMs. *IEEE Access* **2020**, *8*, 81960–81970. [CrossRef]

47. Palavicino, P.C.; Lee, W.; Sarlioglu, B. Detection and Compensation of Inter-turn Short Circuit in Interior Permanent Magnet Synchronous Machines with 2-level Voltage Source Inverter. In Proceedings of the 2020 IEEE Energy Conversion Congress and Exposition (ECCE), Detroit, MI, USA, 11–15 October 2020; pp. 4460–4465.

48. Vasilios, I.C. Detection of PMSM Inter-Turn Short-Circuit Based on a Fault-Related Disturbance Observer. *Int. J. Simul. Syst. Sci. Technol.* **2020**, *21*. [CrossRef]

49. Zhang, J.; Wang, Y.; Zhu, K.; Zhang, Y.; Li, Y. Diagnosis of Interturn Short-Circuit Faults in Permanent Magnet Synchronous Motors Based on Few-Shot Learning Under a Federated Learning Framework. *IEEE Trans. Ind. Inform.* **2021**, *17*, 8495–8504. [CrossRef]

50. Park, C.H.; Lee, J.; Kim, H.; Suh, C.; Youn, M.; Shin, Y.; Ahn, S.H.; Youn, B.D. Drive-Tolerant Current Residual Variance (DTCRV) for Fault Detection of a Permanent Magnet Synchronous Motor Under Operational Speed and Load Torque Conditions. *IEEE Access* **2021**, *9*, 49055–49068. [CrossRef]

51. Liang, H.; Chen, Y.; Liang, S.; Wang, C. Fault detection of stator inter-turn short-circuit in PMSM on stator current and vibration signal. *Appl. Sci.* **2018**, *8*, 1677. [CrossRef]

52. Attestog, S.; Van Khang, H.; Robbersmyr, K.G. Improved Quadratic Time-frequency Distributions for Detecting Inter-turn Short Circuits of PMSMs in Transient States. In Proceedings of the 2020 IEEE International Conference on Electrical Machines (ICEM), Gothenburg, Sweden, 23–26 August 2020; Volume 1, pp. 1461–1467.

53. Otava, L.; Buchta, L. Integrated diagnostic system for winding fault detection of the three-phase PMSM. In Proceedings of the 2020 IEEE 12th International Congress on Ultra Modern Telecommunications and Control Systems and Workshops (ICUMT), Brno, Czech Republic, 5–7 October 2020; pp. 33–40.

54. Gao, F.; Zhang, G.; Li, M.; Gao, Y.; Zhuang, S. Inter-turn fault identification of surface-mounted permanent magnet synchronous motor based on inverter harmonics. *Energies* **2020**, *13*, 899. [CrossRef]

55. Mansouri, B.; Idrissi, H.J.; Venon, A. Inter-Turn Short-Circuit Failure of PMSM Indicator based on Kalman Filtering in Operational Behavior. In *Annual Conference of the PHM Society*; 2019; Volume 11. Available online: https://pdfs.semanticscholar.org/bdb7/5c0d0e604ebc7353d71078c6bf51b7f78c0b.pdf (accessed on 1 December 2022).

56. Zerdani, S.; El Hafyani, M.L.; Zouggar, S. Inter-Turn Stator Winding fault Diagnosis for Permanent Magnet Synchronous Motor based Power Spectral Density Estimators. In Proceedings of the 2020 IEEE International Conference on Smart Grid and Clean Energy Technologies (ICSGCE), Kuching, Malaysia, 4–7 October 2020; pp. 137–142.

57. Yassa, N.; Rachek, M. Modeling and detecting the stator winding inter turn fault of permanent magnet synchronous motors using stator current signature analysis. *Math. Comput. Simul.* **2020**, *167*, 325–339. [CrossRef]

58. Cui, R.; Fan, Y.; Li, C. On-line inter-turn short-circuit fault diagnosis and torque ripple minimization control strategy based on OW five-phase BFTHE-IPM. *IEEE Trans. Energy Convers.* **2018**, *33*, 2200–2209. [CrossRef]
59. Yang, Y.; Chen, Y.; Hao, W. Online detection of inter-turn short-circuit fault in dual-redundancy permanent magnet synchronous motor. *IET Electr. Power Appl.* **2021**, *15*, 104–113. [CrossRef]
60. Gao, C.; Lv, K.; Si, J.; Feng, H.; Hu, Y. Research on Inter-turn Short Circuit Fault Indicators for Direct-drive Permanent Magnet Synchronous Motor. *IEEE J. Emerg. Sel. Top. Power Electron.* **2021**, *10*, 1902–1914. [CrossRef]
61. Rosero, J.; Romeral, L.; Ortega, J.; Rosero, E. Short-circuit detection by means of empirical mode decomposition and Wigner–Ville distribution for PMSM running under dynamic condition. *IEEE Trans. Ind. Electron.* **2009**, *56*, 4534–4547. [CrossRef]
62. Barendse, P.; Pillay, P. A new algorithm for the detection of faults in permanent magnet machines. In Proceedings of the IECON 2006-32nd Annual Conference on IEEE Industrial Electronics, Paris, France, 6–10 November 2006; pp. 823–828.
63. Urresty, J.; Riba, J.; Romeral, L.; Saavedra, H. Analysis of demagnetization faults in surface-mounted permanent magnet synchronous with inter-turns and phase-to-ground short-circuits. In Proceedings of the 2012 IEEE XXth International Conference on Electrical Machines, Marseille, France, 2–5 September 2012; pp. 2384–2389.
64. Xiangli, K.; Ma, R.; Zhang, Q.; Wang, W. Modeling and Simulation of Short Circuit Faults in Stator Coils of Brushless DC Motor. In *Mechatronics and Automatic Control Systems*; Springer: Berlin/Heidelberg, Germany, 2014; pp. 35–45.
65. Wang, Z.; Yang, J.; Cao, C.; Gu, Z. Phase-phase short fault analysis of permanent magnet synchronous motor in electric vehicles. *Energy Procedia* **2016**, *88*, 915–920. [CrossRef]
66. Shao, M.; Yang, G.; Sun, G.; Su, J. A method of open circuit fault diagnosis for five-phase permanent magnet synchronous motor based on wavelet analysis. In Proceedings of the 2019 IEEE 22nd International Conference on Electrical Machines and Systems (ICEMS), Harbin, China, 11–14 August 2019; pp. 1–6.
67. Kontarček, A.; Bajec, P.; Nemec, M.; Ambrožič, V.; Nedeljković, D. Cost-effective three-phase PMSM drive tolerant to open-phase fault. *IEEE Trans. Ind. Electron.* **2015**, *62*, 6708–6718. [CrossRef]
68. Huang, W.; Du, J.; Hua, W.; Lu, W.; Bi, K.; Zhu, Y.; Fan, Q. Current-based open-circuit fault diagnosis for PMSM drives with model predictive control. *IEEE Trans. Power Electron.* **2021**, *36*, 10695–10704. [CrossRef]
69. Slimen, S.B.; Bourogaoui, M.; Sethom, H.B.A. Easy and effective multiple faults detection and localization method for PMSM drives. In Proceedings of the 2019 IEEE International Conference on Advanced Systems and Emergent Technologies (IC_ASET), Hammamet, Tunisia, 19–22 March 2019; pp. 311–316.
70. Abed, W.; Sharma, S.; Sutton, R. Fault diagnosis of brushless DC motor for an aircraft actuator using a neural wavelet network. In Proceedings of the IET Conference on Control and Automation 2013: Uniting Problems and Solutions, Birmingham, UK, 4–5 June 2013.
71. Neethu, S.; Sreelekha, V. Hysteresis controller based open phase fault tolerant control of BLDC motor drives. In Proceedings of the 2014 IEEE International Conference on Power Signals Control and Computations (EPSCICON), Thrissur, India, 6–11 January 2014; pp. 1–6.
72. Fu, Z.; Liu, X.; Liu, J. Research on the fault diagnosis of dual-redundancy BLDC motor. In Proceedings of the ICPE 2020—The International Conference on Power Engineering, Guangzhou, China, 4–6 December 2020; pp. 17–22.

Review

Review of Research and Development of Hydraulic Synchronous Control System

Ruichuan Li [1], Wentao Yuan [1,*], Xinkai Ding [1,*], Jikang Xu [2], Qiyou Sun [1] and Yisheng Zhang [1]

[1] School of Mechanical Engineering, Qilu University of Technology (Shandong Academy of Sciences), Jinan 250353, China
[2] College of Mechanical and Electronic Engineering, Shandong University of Science and Technology, Qingdao 266590, China
*   Correspondence: 10431210002@stu.qlu.edu.cn (W.Y.); 10431200080@stu.qlu.edu.cn (X.D.)

Abstract: Hydraulic synchronous control systems are widely used in various industrial fields. This paper deeply analyzes the research status and development trend of the hydraulic synchronous control system. Firstly, it gives a brief introduction of the research significance control theory and control methods of the hydraulic synchronous control system. Secondly, the hydraulic synchronization control system is classified, the synchronization error is analyzed, and some solutions to synchronization error are given. Then, according to the classification of the hydraulic synchronous control system, relevant research is carried out. In this paper, three control modes (equivalent, master–slave and cross-coupling) and related control algorithms (fuzzy PID control, sliding mode control, robust control, machine learning control, neural network control, etc.) of closed-loop hydraulic synchronous control systems are studied in detail. Finally, the development trend of the hydraulic synchronization control system is predicted and prospected, which can provide some reference for promoting the research and application of hydraulic synchronization technology in the future industrial field.

Keywords: hydraulic synchronization; closed loop control; synchronization error; control mode; control algorithm

Citation: Li, R.; Yuan, W.; Ding, X.; Xu, J.; Sun, Q.; Zhang, Y. Review of Research and Development of Hydraulic Synchronous Control System. *Processes* **2023**, *11*, 981. https://doi.org/10.3390/pr11040981

Academic Editors: Francisco Ronay López-Estrada and Guillermo Valencia-Palomo

Received: 4 March 2023
Revised: 17 March 2023
Accepted: 21 March 2023
Published: 23 March 2023

1. Introduction

Hydraulic transmission has the advantages of large output power, stability and reliability, fast response speed, compact and flexible installation, and can achieve a wide range of stepless speed regulation. It is often widely used in large-scale mechanical equipment, metal processing equipment, large-scale metallurgical equipment, agricultural machinery equipment, aerospace, and other related industrial fields [1–4] and has become an important technical force to promote the development of mechanical equipment. Additionally, the hydraulic synchronous control system is an important component in the field of hydraulic transmission, which has a great influence on the research and development of this field.

The development of the hydraulic synchronous control system has gone through a long process. With the unremitting efforts of many scholars and experts, the hydraulic synchronous control system is also developing and progressing [3,5–8]. Joseph Bramah [9], an Englishman, invented the world's first hydraulic press in 1795, and the hydraulic field began to develop gradually. Around the 1870s, the steam engine-driven water pump was gradually applied to hydraulic equipment such as extruders and shears. In the 1930s, the integrated hydraulic system composed of a pump, valve, and actuator was gradually applied to the hydraulic synchronization equipment, and the hydraulic synchronization control system developed rapidly. In the late 1970s, the feedback element began to be applied in the hydraulic closed-loop synchronous control system. Compared with the previous open-loop control, the synchronization accuracy and response speed have been significantly improved. At the same time, the first fuzzy controller developed by the University of London in the United Kingdom was gradually applied to the hydraulic synchronous control system, which effectively solved the control problem of the nonlinear system. However, the fuzzy rules and system design methods are completely based on experience, and it is difficult to accurately and stably control the complex system. In 1981, Canadian scholar G. Zames [10] first proposed H∞ optimal control, and then robust control was widely used in electro-hydraulic servo systems and multi-cylinder hydraulic synchronous systems. Since the 20th century, with the development of electronic components and artificial intelligence, adaptive fuzzy PID, robust control, neural network and other algorithms have been gradually applied to hydraulic synchronization systems. Because of its adjustable control parameters and strong adaptability, it can achieve high-precision synchronization accuracy and good robustness in the control of complex nonlinear hydraulic synchronization systems.

The hydraulic synchronous control system usually requires the actuators in the hydraulic transmission mechanical equipment to operate synchronously with the same displacement or speed at the same time [11]. Generally, due to the existence of nonlinear factors such as leakage, impurities, machining error, and component wear in the hydraulic synchronization control system, it is difficult to ensure the accuracy of the hydraulic synchronization system. Although the existing conventional and simple hydraulic synchronization system is low in cost, it cannot meet the requirements of high-precision synchronization equipment. In order to make the hydraulic synchronization equipment have good operation performance, some relevant scholars have optimized the hydraulic components and hydraulic circuit, but the synchronization accuracy improved by this method is limited. Other scholars have used closed-loop control to apply relevant intelligent control algorithms to the hydraulic synchronization control system, and the control accuracy and reliability and stability have been greatly improved.

Gu et al. [12] focused on the phenomenon that the synchronous open-loop hydraulic synchronous control system of two hydraulic cylinders has no feedback regulation, and there is a constant accumulation of errors and low control accuracy. Through comparing several improvement schemes, the hydraulic synchronous circuit was optimized, and finally the scheme of reasonable configuration of the one-way valve by the synchronous motor for flow compensation was adopted. Through repeated optimization of the opening of the one-way valve, the synchronous error of the two cylinders in the full stroke of 500 mm was not more than 3 mm. Chen et al. [13] proposed a fuzzy control method for piston motion trajectory synchronization aiming at the synchronization error caused by the load imbalance of dual hydraulic cylinders. Based on the fuzzy control, two fuzzy controllers were introduced to eliminate the track tracking error of hydraulic cylinders and the synchronization error between hydraulic cylinders. The experiment shows that the composite control method can control the synchronization error within ±10 mm. Yang et al. [14] proposed a single neuron PID cross-coupling control strategy for the synchronization error caused by the uneven stress of the two hydraulic cylinders of the hydraulic bending machine during the working process, and compared it with the PID control effect under master–slave and cross-coupling. The simulation results showed that the maximum synchronization error of the master–slave PID control was 0.9 mm, and the maximum synchronization error of the cross-coupling PID control was 0.79 mm. The two-cylinder system controlled by neuron cross-coupling PID had a faster response, and the maximum synchronization error was 0.27 mm.

Traditional PID control needs to rely on accurate mathematical models because its parameters are not adjustable. It can achieve a good control effect in a linear system. For complex nonlinear hydraulic systems, some scholars combine PID control with fuzzy control and adaptive control to achieve high synchronization control accuracy. Zhang et al. [15] proposed a synchronization control strategy based on fuzzy PID control, which uses fuzzy rules to realize the real-time adjustment of PID control parameters. Taking the piston displacement of a hydraulic cylinder as an indicator, the joint simulation of AMESim and Simulink shows that the synchronization accuracy of double hydraulic cylinders under this control strategy is between 0 and 0.66 mm, which meets the operation requirements of the shear-type synchronous lifting mechanism. Liu et al. [16] investigated the hydraulic pin type lifting system of the jack up wind power installation ship, a two-stage hydraulic synchronous control system. The speed tracking and displacement coupling synchronization control strategy was adopted for single pile legs to improve the synchronization accuracy, and the fuzzy PID control algorithm was adopted to improve the robustness of the system. The joint simulation of AMESim and Simulink software showed that the maximum synchronization error of the sensor was still less than 0.3 mm when it was disturbed, the system had a higher synchronization accuracy, and had the advantages of stable operation, no overshoot and higher reliability. Li et al. [17] designed a fuzzy adaptive PID controller to be used in the steel plate hydraulic synchronization technology. The joint simulation of AMEsim and Simulink software showed that the displacement synchronization error of the two hydraulic cylinders was always less than 0.1 mm, and the proposed control strategy had a faster response speed, stronger stability, and a higher control accuracy. Zhao et al. [18] proposed a fuzzy PID controller based on particle swarm optimization to solve the problem of synchronization error caused by friction and uneven stress of a new type of hydraulic fan barring drive cylinder group. The particle swarm optimization algorithm was used to iteratively optimize the quantization factors Ke, Kec, and scale factor Ku of the fuzzy controller. Simulation analysis by Simulink software showed that the hydraulic synchronization system under particle swarm composite PID control had a higher synchronization accuracy than traditional PID control, and that fuzzy PID control had the advantages of smaller overshoot, shorter adjustment time, and stronger adaptive ability. Wang et al. [19] proposed a control method combining adaptive robust error sign integration control and extended state observation (ESO) control and combined them with cross-coupling control. The simulation results showed that the control strategy can achieve good synchronous tracking performance in the double hydraulic cylinder actuator system. Fang et al. [20] introduced fuzzy PID control into the hydraulic servo system of a C32 friction welding machine, which improved the closed-loop control characteristics of the welding process. However, due to the small design tonnage, the difficulty of multi-cylinder synchronous control was not considered. In order to solve the difficulty of multi-cylinder synchronization control accuracy, Wang et al. [21] proposed a synchronization controller based on a neural network to solve the synchronization error caused by the difficulty of the coordinated operation of the hydraulic multi-actuator synchronization system. Through the backstepping technology, the synchronization accuracy of multiple hydraulic cylinders was greatly improved. Finally, the effectiveness of the control strategy was verified by simulation and experimentation.

According to the above research, closed-loop control is often used in the hydraulic synchronization control system to improve the accuracy of the hydraulic synchronization control system. This is because there is no feedback signal regulation in the hydraulic synchronous control system under open-loop control, and its control accuracy depends entirely on the accuracy of the hydraulic components to control the actuator. In addition, the synchronization error of the hydraulic system will continue to accumulate under long-term operation, so its synchronization accuracy and anti-interference are poor. The hydraulic synchronous control system under closed-loop control adds online detection and feedback of the actuator displacement and speed between the controller and the controlled object, and the feedback value is compared with the preset control parameters. The controller outputs the signal adjustment amount to compensate the synchronization error of the hydraulic system, which can significantly improve the synchronization accuracy and stability of the hydraulic synchronous control system [22–25].

This paper analyzes the development process and research significance of the hydraulic synchronous control system, studies three control modes and related intelligent control algorithms of the closed-loop hydraulic synchronous control system of the hydraulic synchronous control system, summarizes the causes of synchronization errors, and puts forward some corresponding solutions to improve the control accuracy of the hydraulic synchronous system, Finally, based on the above research, the development trend of hydraulic synchronous control system is predicted and prospects are proposed. With guaranteed control accuracy of the hydraulic synchronization system, the research in this paper can provide some measures to reduce the synchronization error, and has certain reference significance for the selection of control methods and control algorithms.

2. Research Significance of the Hydraulic Synchronous Control System and Development History of Control Method

2.1. Research Significance of the Hydraulic Synchronous Control System

Hydraulic synchronous control systems are a branch of hydraulic systems, and their research direction mainly includes hydraulic synchronous control methods and control algorithms. The hydraulic synchronous system has the advantages of high control accuracy, fast response speed, and easy application in high-power equipment, so it has a pivotal position in synchronous control systems [26], and its research significance has the following aspects.

(1) It can improve the synchronization control accuracy and maintain the good performance of mechanical equipment.
(2) It can reduce the torsional force and friction between the actuators due to the synchronization error, reduce the wear and torsional deformation of the mechanism, and extend the service life of the hydraulic equipment.
(3) It can increase the efficiency of the hydraulic system, thus saving more energy.

2.2. Development History of Hydraulic Synchronous Control Method

With the development of modern control technology, all kinds of hydraulic equipment require a higher and higher accuracy of the hydraulic synchronous system. According to the development process of the synchronization control method, it can be divided into the first, second, and third generations of the synchronization control method [27].

The first generation of the hydraulic synchronization control method mainly adopts open-loop control to regulate the output of the actuator, using hydraulic components to change the pressure change in the hydraulic system to control the output of the hydraulic source and change the flow and pressure of the fluid in the hydraulic line, so as to control the synchronization movement of the actuator. The first-generation control method is low cost, simple, and practical, but the synchronization accuracy of the control system is not high and is generally used for industrial equipment with low control accuracy requirements.

With the social demand for high-precision machining and the improvement of sensor and control technology, the requirements for hydraulic synchronous equipment have gradually increased. The second-generation hydraulic synchronization control method widely uses the electro-hydraulic control method, introducing a signal (speed, load, displacement, etc.) feedback link, detecting feedback on the output quantity of the controlled object, comparing the feedback signal value with the ideal target value, and outputting the regulation electric signal to the electro-hydraulic valve by the controller, thus regulating the actuating element to reduce the synchronization error. The second generation of hydraulic synchronization control method adopts closed-loop control, mostly used for hydraulic system control using an electronic component system-hydraulic pilot control-hydraulic actuator, which regulates the synchronization error and increases the synchronization accuracy of hydraulic actuator.

With the further improvement of automation technology and control technology, higher requirements for hydraulic synchronization accuracy have been put forward. The third-generation hydraulic synchronization control method mainly uses a microprocessor combined with different control algorithms, introduces sensors (force sensors, displacement sensors, angle sensors, speed sensors, etc.) to compare the actual control accuracy and ideal control accuracy of the system, uses a microprocessor as the control unit, all kinds of sensors to detect and feedback the hydraulic synchronization control system, and combines different pairs of control algorithms to reduce synchronization interference and error.

The third generation hydraulic synchronization control method has higher automatic intelligence, but its regulating actuators have a certain lag. In the future, the hydraulic synchronization control system can be digitized and intelligently controlled through the construction of a hydraulic characteristic information network, the collection of flow and pressure at the important nodes of the hydraulic circuit, and the active real-time regulation of the hydraulic synchronization control system by making full use of intelligent algorithms and information-variable technology.

3. Analysis of Synchronization Error of Hydraulic Synchronous Control System

To achieve synchronous control of the hydraulic system, ideally it is only necessary to allow equal pressure and equal amount of hydraulic oil to flow through the actuator with identical structural parameters. Generally, the hydraulic synchronization system has synchronization errors due to oil pollution and leakage in the hydraulic synchronization system, uneven friction and load of the actuators, asymmetrical arrangement, back pressure of the actuators, oil pressure fluctuation, manufacturing and installation errors of components, and other factors. Even if closed-loop control is adopted, the hydraulic synchronization system under actual working conditions is complex and nonlinear due to its control hysteresis. It is difficult to establish its accurate mathematical model to achieve high-precision control, so the generation of synchronization error is inevitable [28–33].

At present, the hydraulic synchronous control system mainly uses the error feedback correction method to reduce the error accumulation of the hydraulic actuator, so as to improve its synchronization accuracy, but the method has a certain lag. In the future, with the continuous development of artificial intelligence and big data, active feedback adjustment is expected to solve this problem. There are many reasons for synchronization error of the hydraulic synchronization control system, which are mainly summarized in the following 10 aspects.

(1) There is a fluid leak in the hydraulic synchronous control system. Due to the different manufacturing accuracy of hydraulic components and the clearance caused by long-time movement friction, the connection of hydraulic pipelines becomes loose, resulting in the leakage of hydraulic oil, thus affecting the synchronization accuracy of hydraulic systems.

(2) The friction of hydraulic actuators is uneven. The uneven friction of the hydraulic actuator will lead to the uneven force in the movement process, which will affect the synchronization accuracy of the hydraulic synchronization control system.

(3) Uneven load of the hydraulic synchronous system. When the load on the same hydraulic actuator end is different, synchronization error will occur, and the hydraulic cylinder or hydraulic motor with a large load has a small oil intake and a slow movement speed.

(4) Asymmetrical arrangement of the hydraulic synchronization control system. The asymmetric arrangement of hydraulic components and pipelines will make the hydraulic power and hydraulic impact of each branch in the hydraulic synchronization control system different, so that the force is different in the start stop phase, which will cause certain errors to the synchronization accuracy. Symmetrical arrangement can be adopted to make the hydraulic resistance of each hydraulic branch equal, thus improving the synchronization accuracy of the hydraulic system.

(5) Back pressure of actuators. When the hydraulic oil flows along the pipeline in a closed container, due to the obstruction of the actuating element, the flow direction of the oil will be rapidly changed, resulting in a pressure opposite to the movement direction. If the back pressure valve is installed on the actuators, due to the influence of various factors, the back pressure value on each branch is often different, which affects the accuracy of the hydraulic synchronous control system. The oil pipes at the oil return port of each actuator can be collected first, and then flow through the back pressure valve, thereby reducing the influence of back pressure on hydraulic synchronization accuracy.

(6)   There are manufacturing and installation errors in hydraulic components. Manufacturing and installation errors can cause initial unsynchronized errors in the hydraulic system, resulting in different friction resistance and oil resistance of hydraulic components. To reduce the initial asynchronous error, the following measures can be taken to reduce the manufacturing and installation errors of hydraulic components, and compensate the synchronization error of the hydraulic synchronization control system.

(7)   Pressure fluctuation of hydraulic oil. The source of pressure fluctuation is mainly the pressure fluctuation generated by the hydraulic source and the pressure fluctuation generated by the throttle valve (hydraulic resistance) due to the load change, which can affect the dynamic characteristics of the hydraulic branch and the synchronization accuracy of the hydraulic system. The following measures can be taken to reduce the pressure fluctuation of hydraulic oil: (1) choose a hydraulic source with a small pressure fluctuation or add an accumulator at the outlet of the hydraulic source; (2) add a pressure compensation (load sensitive) device at the proper position in the hydraulic system to effectively control the pressure fluctuation in the hydraulic synchronous control system.

(8)   Insufficient rigidity. Mechanical equipment with insufficient rigidity can produce obvious stress and deformation after being stressed; insufficient rigidity of the hydraulic medium can cause the fluctuation of the oil in the hydraulic pipeline, and the ambient temperature will affect the viscosity and compressibility of the fluid, which can result in insufficient rigidity of the hydraulic system and hydraulic synchronization error, even out of tolerance. The following measures can be taken to reduce the insufficient rigidity: (1) use mechanical equipment with better rigidity or optimize the structure of mechanical equipment to increase its resistance to stress and deformation; (2) optimize the connection form of the components and reasonably simplify the structure to increase its rigidity; (3) select a hydraulic medium with strong rigidity to reduce the influence of fluid viscosity and compressibility on hydraulic synchronization accuracy; (4) choose a working environment with relatively stable ambient temperature, or add equipment to keep the hydraulic oil temperature constant, so as to stabilize the viscosity and compressibility of the oil.

(9)   The oil is mixed with gas. Gas in the hydraulic oil can affect the compression ratio of the oil and the dynamic performance of the hydraulic system, thus affecting the normal oil supply required by the hydraulic system, resulting in synchronous accuracy errors of the actuating elements. The following measures can be taken to reduce the gas in the oil: (1) use reliable seals to ensure good sealing at the oil suction; (2) add an inclined metal mesh to the tank to float up and release the air in the oil; (3) add an automatic air release device to the hydraulic synchronous system.

(10) The oil is mixed with impurities. Impurities in the oil cause a flow error at the flow regulating valve port, which affects the synchronization accuracy. The following measures can be taken to reduce the impurities in the oil: (1) sealing the oil tank with a more reliable device to prevent small impurities from entering the hydraulic system; (2) cleaning the filter regularly to reduce impurities in the oil.

4. Research on Hydraulic Synchronous Control System

The hydraulic synchronization control system is widely used because of its advantages such as easy control, simple structure, and being able to adapt to high power and complex environments. At the same time, the continuous improvement and optimization of the hydraulic synchronization system has always been a research hotspot in the hydraulic field [34]. According to the different application fields, control modes and tasks, the hydraulic synchronization system can be classified into several categories [35–37].

According to whether there is feedback in the hydraulic synchronization control system, it can be divided into "open-loop synchronization" or "closed-loop synchronization". Open-loop synchronization is on a simple scale, has low control accuracy, and poor robustness. Closed-loop synchronization can be widely used in various hydraulic equipment; because of the feedback regulation of its intelligent controller, its synchronization accuracy is significantly improved, its complexity and processing scale become larger, and the robustness of the system is better. In addition, different intelligent control algorithms can be combined in closed-loop control to select the best control scheme, but the closed-loop control system under complex machine learning and a neural network often has the phenomenon of over-learning, which increases the control cost of the system. In addition, the closed-loop hydraulic synchronization control system can be subdivided into "equivalent synchronization", "master–slave synchronization" and "cross-coupling synchronization", according to the different control modes.

According to whether the hydraulic synchronous control system has compensation, it can be divided into "uncompensated synchronization" and "compensated synchronization". The uncompensated hydraulic synchronous system is simple, has low control accuracy and poor robustness, while the compensated hydraulic synchronous control system timely compensates the synchronization error, so its control accuracy is high, and its robustness is strong. In addition, the compensated hydraulic synchronization system can be divided into "pre-valve compensation hydraulic synchronization" and "post-valve compensation hydraulic synchronization" according to the different compensation positions.

According to the different control elements, the hydraulic synchronization control system can be divided into "pump-controlled synchronization" and "valve-controlled synchronization". The pump-controlled synchronization efficiency is high, the response speed is slow, the control accuracy under constant load condition is high, and the valve-controlled synchronization efficiency is low, but the robustness is good, the response speed is fast, and the synchronization accuracy under dynamic load condition is high. In addition, according to the different types of control valves, the valve-controlled hydraulic synchronization control system is divided into "flow synchronization", "electro-hydraulic proportional synchronization", and "electro-hydraulic servo synchronization".

In addition, other hydraulic synchronization systems are divided into "hydraulic cylinder synchronization" and "hydraulic motor synchronization" according to the different actuating elements. According to different control tasks, it can be divided into "position synchronization" and "speed synchronization". According to the number of actuators, it can be divided into "double actuator synchronization" and "multiple actuator synchronization". According to different installation methods, it can be divided into "horizontal synchronization" and "vertical synchronization". The classification of hydraulic synchronization control systems is shown in the following Figure 1.

This paper will carry out a detailed study on the hydraulic synchronization system according to the relevant classification in Figure 1 above.

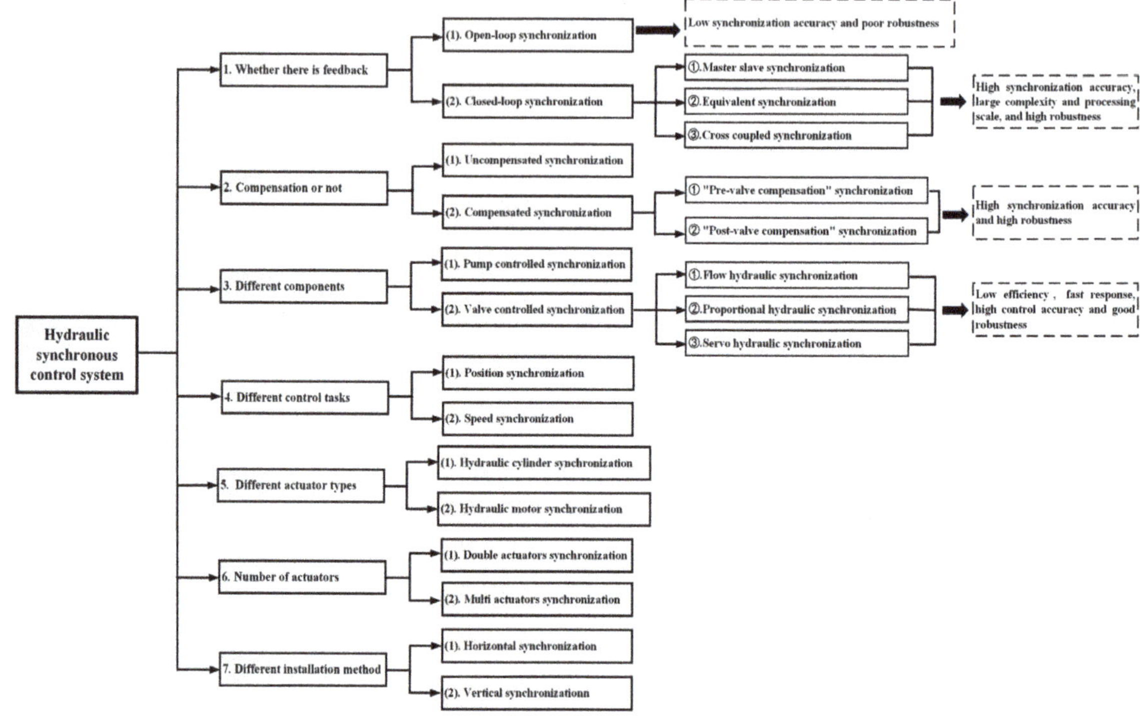

Figure 1. Classification diagram of hydraulic synchronous control system.

4.1. Open-Loop and Closed-Loop Hydraulic Synchronization Control System

The hydraulic synchronous control system can be divided into open-loop hydraulic synchronous control systems and closed-loop hydraulic synchronous control systems according to whether there is feedback signal adjustment [38].

4.1.1. Open-Loop Hydraulic Synchronous Control System

There is no feedback signal regulation in the hydraulic synchronization system of open-loop control, and its synchronization control accuracy depends entirely on the accuracy of the hydraulic components (such as step valve, throttle valve, and speed regulating valve) to control the actuator. The output signal of the actuator is not detected and fed back in the open-loop control hydraulic synchronization system, so the accuracy and anti-interference of the open-loop control are poor. Common open-loop synchronous control circuits include a mechanical rigid synchronous circuit, flow control valve synchronous circuit, series cylinder synchronous circuit, synchronous cylinder synchronous circuit, parallel motor, or parallel pump synchronous circuit. Generally speaking, because of its low cost and simple structure, the open-loop control hydraulic synchronization control system is widely used in situations with small load, low synchronization accuracy requirements, simple control loop, and close synchronization distance. The open-loop control of the hydraulic synchronization system, due to the absence of feedback compensation, will make the synchronization error gradually accumulate and increase, and cannot eliminate the errors caused by hydraulic oil leakage, load angle bias, and the manufacture and installation of hydraulic components.

To make the hydraulic synchronization control system under open-loop control realize accurate synchronization, Ding et al. [39] proposed a load sensing synchronization control method consisting of a load sensing unit and a synchronization valve. The load sensing unit consists of a load sensing pump and a load sensing valve. The load sensing pump provides the pressure and flow required by the system through the pressure-closed-loop control. The load sensing valve can improve the ability to resist offset load through pressure compensation. The synchronizing valve is located between the load sensing pump and the load sensing valve to achieve an even distribution of the flow supplied by the pump. The experimental results show that compared with the traditional synchronous valve control, this control mode has stronger control accuracy, efficiency, and synchronous speed regulation capability.

4.1.2. Closed-Loop Hydraulic Synchronous Control System

With the development of modern control theory and control technology, various closed-loop hydraulic synchronous control systems are widely used in high-precision hydraulic synchronous drive equipment [22,40,41].

The hydraulic synchronous control system under closed-loop control has a high synchronization accuracy because of the feedback regulation. The controller can conduct online feedback comparison between the output value and the ideal setting value and reduce the control deviation after regulation. The control modes of the hydraulic synchronous control system can be divided into master–slave control, equivalent control, and cross-coupling control according to the different feedback control modes of output values [42,43].

Lorenz et al. [44] proposed two synchronous control modes: equivalent control and master–slave control. Equivalent control [45–47] is a parallel structure, in which multiple hydraulic actuators form a closed loop independently, tracking the target output value signal set by the control system, so that each actuator can achieve synchronous control. Since each actuator in the equivalent control synchronizes the feedback signal independently and tracks the motion error, the synchronous tracking error of each actuator determines the error of the equivalent control. The equivalent control block diagram of the hydraulic synchronization control system is shown in Figure 2.

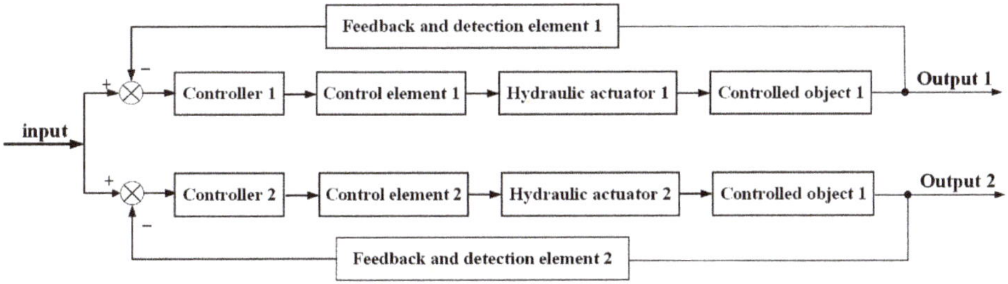

Figure 2. Equivalent control block diagram of hydraulic synchronous control system.

Equivalent control requires that each hydraulic component and electrical component of the hydraulic synchronization system must have a high matching consistency, and if the consistency of each component is good, the hydraulic system with equivalent control has a high synchronization accuracy.

The master–slave control [48–50] is a series structure. First, the output of an actuator is selected as the ideal output signal in the master–slave controlled hydraulic synchronization system; that is, one actuator is the main hydraulic oil circuit, and the other actuator is the slave hydraulic circuit. Then, the error feedback signal in the main circuit is input to the other slave hydraulic circuits through the feedback detection elements in the main hydraulic circuit. The master–slave control has a delay feedback that means the slave hydraulic circuit follows the signal of the main hydraulic circuit; especially in the start stop phase of the hydraulic synchronous control system, this delay is more significant. The slave actuating element has a large tracking lag to the master actuating element, which will cause the master–slave control error. The master–slave control block diagram of the hydraulic synchronization control system is shown in Figure 3.

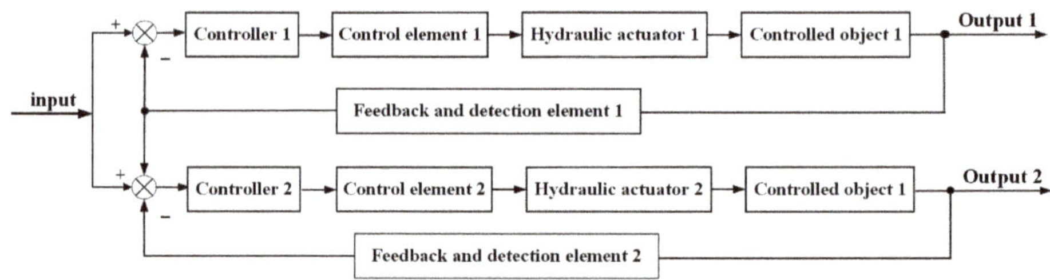

Figure 3. Master–slave control block diagram of hydraulic synchronous control system.

There is a certain delay error in the master–slave-controlled hydraulic synchronization system, and to reduce the tracking delay error of the master–slave-controlled hydraulic circuit, a front feedback device needs to be added to feed back the synchronization error signal of the system to the control link [51].

There is no coupling relationship between the master–slave control and the hydraulic branches of the same control, so the controller is relatively simple, and the synchronization accuracy is relatively general. Therefore, the cross-coupling control is gradually studied and applied in the hydraulic synchronization control system by scholars. Cross-coupling control [52,53] establishes a coupling relationship between hydraulic branches, detects the output deviation values of multiple hydraulic branches, and feeds them back to the controller. When the synchronous deviation exceeds a certain range, the synchronous controller immediately compensates for the coupling of two hydraulic actuators, thereby effectively reducing the synchronous error. The cross-coupling control combines the advantages of master–slave control and equivalent control, and the synchronization accuracy is further improved. The cross-coupling control block diagram of hydraulic synchronization control system is shown in Figure 4.

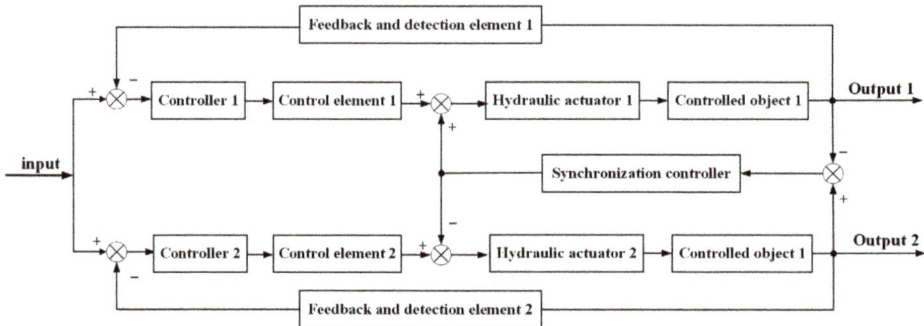

Figure 4. Cross-coupling control block diagram of hydraulic synchronous control system.

However, the cross-coupling control has certain limitations for any hydraulic system with more than two hydraulic branches. When there are many controlled hydraulic actuators, the control mode will become too complex, and the robustness, stability, and synchronization of the hydraulic synchronization control system will become worse. To solve the synchronization accuracy of multiple actuators, adjacent cross-coupling is gradually being studied by more and more scholars. Adjacent cross-coupling hydraulic synchronization control [54,55] is to locally synchronize the actuating elements. When the controller synchronizes each hydraulic branch, it outputs synchronization error feedback for its own hydraulic branch and outputs synchronization error feedback to its adjacent hydraulic branch, to control the output error compensation of adjacent hydraulic branches and achieve synchronization control of each hydraulic branch. The adjacent cross-coupling control block diagram of the hydraulic synchronization control system is shown in Figure 5.

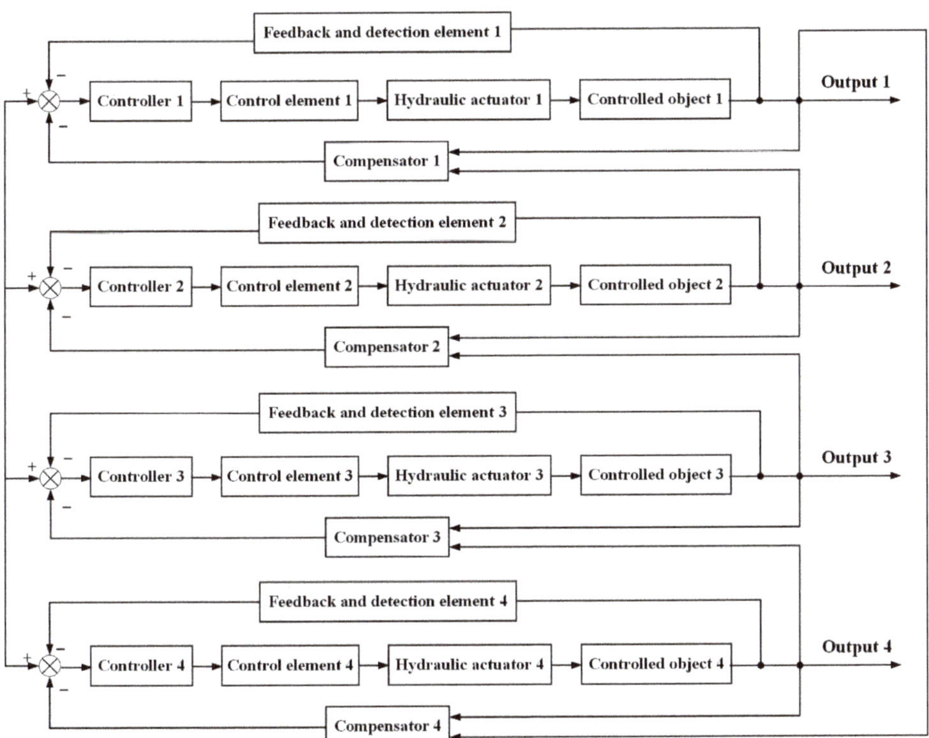

Figure 5. Adjacent cross-coupling control block diagram of the hydraulic synchronization control system.

Wu et al. [56], based on the adjacent cross-coupling synchronization control, designed the adjacent cross-coupling fuzzy self-tuning integral separation PID synchronization controller by combining fuzzy control with integral separation PID control, combined with three hydraulic synchronous closed-loop control modes (master–slave, equivalent, and adjacent cross-coupling). Through the joint simulation with AMESim and Simulink software, the three closed-loop synchronous control methods were compared and analyzed. The results show that in the four cylinders synchronous control system, the overshoot under the master–slave synchronous control mode is 6%, and the maximum synchronous error is 0.26 mm; the overshoot under the same synchronous control mode is 4.2%, and the maximum synchronous error is 0.15 mm; the overshoot of adjacent cross-coupling synchronization control mode is 2.8%, and the maximum synchronization error is 0.12 mm. The reason for the above phenomenon is that the master–slave synchronous control first sets a main hydraulic branch, and the output of the other slave hydraulic branches should follow the set main hydraulic branch, so there will be some hysteresis in the control process. All controllers of equivalent synchronous control follow the same set ideal output value, and the actuating elements of each hydraulic branch are not regulated by synchronous controller. The adjacent cross-coupling synchronous control can not only track and control the given ideal output value, but also compensate the synchronous error of the actuating elements of different adjacent hydraulic branches. Therefore, the control accuracy of adjacent cross-coupling synchronous control under closed-loop control is higher than that of master–slave control and equivalent control.

Compared with master–slave control and equivalent control, cross-coupling control can obtain a higher accuracy in the hydraulic synchronous system. Therefore, more and more scholars are combining cross-coupling control with other control methods to further develop the field of cross-coupling control. Sun et al. [57] proposed an adaptive robust cross-coupling control strategy by combining the cross-coupling control mode with the adaptive robust control, and its maximum synchronization error is not more than ±0.1 mm, which further improves the synchronization accuracy of the hydraulic press slide leveling electro-hydraulic control system. Liu et al. [58] proposed a control method based on fuzzy adjacent coupling for synchronous control of multiple hydraulic cylinders to be leveled, designed an adjacent coupling controller, fed back the position synchronization error of two adjacent hydraulic cylinders to the synchronous controller, and used the fuzzy controller to obtain the parameters that are difficult to obtain. The simulation and experimental results show that the fuzzy adjacent coupling control can quickly reduce the synchronization error of the hydraulic cylinder in the acceleration and deceleration phase.

Through the above research on the equivalent, master–slave, cross-coupling, and adjacent cross-coupling control of closed-loop hydraulic synchronous control systems, it can be concluded that their control effects and characteristics are shown in Table 1 below:

Table 1. Comparison of the effects of four control methods for closed-loop hydraulic synchronization systems.

Control Mode	Control and Control Effect	Characteristic
equivalent	Zhang [45] adopted PID equivalent control, with good synchronization control effect; Tian [46] adopted PID equivalence and master–slave control, which reduces the maximum synchronization error by 80.649% and steady-state error by 30.306% compared to master–slave control.	Each actuator independently synchronizes the feedback signal without coupling, and the synchronization accuracy is relatively high. However, it requires that each element have a high matching ability.
master–slave	Li [49] adopted a master–slave synchronous control scheme of one master and three slaves to synchronously control four hydraulic cylinders. The average displacement fluctuation of the hydraulic cylinder is 0.189 mm, and the maximum fluctuation is 0.273 mm	There is a tracking lag error between the main hydraulic circuit and the slave hydraulic circuit, especially during the startup and shutdown stages; there is no coupling and the synchronization accuracy is average.
cross-coupling	Koren [52] used cross-coupling to synchronize the dual axis hydraulic system. The results show that cross-coupling improves synchronization accuracy, but the response speed of each axis will be slightly lower.	The compensation hydraulic circuit can be coupled, combining the advantages and disadvantages of master–slave and equivalent control, with high synchronization accuracy, robustness, and stability. However, for complex multi-hydraulic circuit-synchronous systems, there are limitations, and the control accuracy is general.
adjacent cross-coupling control	Wu [56] adopted a PID controller with fuzzy self-tuning integral separation under adjacent cross-coupling for the synchronous control of four hydraulic cylinders. The results show that the overshoot of adjacent cross-coupling synchronization control is 2.8%, and the maximum synchronization error is 0.12 mm. Sun [57] adopted a robust adaptive cross-coupling strategy, with the maximum synchronization error not exceeding ±0.1 mm.	It can solve the high-precision control problem of complex multiple hydraulic circuits, with high robustness and reliability.

4.2. Hydraulic Synchronous Control System without Compensation and with Compensation

The hydraulic synchronization control system can be divided into the "compensated hydraulic synchronization control system" and "non-compensated hydraulic synchronization control system", according to whether there is compensation in the hydraulic circuit.

4.2.1. Hydraulic Synchronous Control System without Compensation

In the hydraulic synchronous control system without compensation, the actuator is not compensated, and the accuracy of the hydraulic synchronous control system is determined by the manufacturing and installation errors of the hydraulic components. Generally speaking, the control accuracy of the hydraulic synchronous control system without compensation is relatively low, and it is mostly used in situations where the load is not large, and the synchronous accuracy is not required to be too high.

4.2.2. Compensated Hydraulic Synchronous Control System

In the hydraulic synchronous control system with compensation, when the load changes, the hydraulic oil flow from the hydraulic pump to the actuator via the hydraulic valve will change, affecting the control accuracy of the hydraulic synchronous system. The compensated hydraulic synchronous system introduces the compensation valve to regulate the flow, or through the sensor to feedback the error value to the control signal. Through the corresponding control algorithm, the signal to be compensated and adjusted is input to the hydraulic pump/hydraulic motor and control valve to compensate and regulate the flow and pressure in the hydraulic circuit. The control system belongs to closed-loop control. Because it can adjust the flow and pressure of the synchronous control system, the control accuracy of the hydraulic synchronous control system is significantly improved. According to different positions of the compensation valve and control valve, it can be divided into either a "pre-valve" compensation or "post-valve" compensation hydraulic synchronous control system [59,60].

The pre-valve compensation hydraulic synchronization control system is to set the pressure compensation valve at the front of the control valve port. The spring chamber of the pressure compensation valve receives the outlet pressure of the control valve port, and the non-spring chamber is connected to the inlet pressure of the control valve port. The differential pressure before and after the two control valves for valve front compensation is equal and both are constant. The flow that can be achieved is only related to the opening of the control valve and is not affected by the load change.

The post-valve compensation hydraulic synchronization control system is to arrange the pressure compensation valve behind the throttle valve. The spring chamber of the pressure compensation valve receives the maximum load pressure selected by the shuttle valve, and the non-spring chamber is connected to control the pressure behind the controllable throttle valve. The differential pressure before and after the two control valves in the post-valve compensation is equal but not constant; when the flow provided by the pump cannot meet the requirements, the flow of the two branches will be reduced in equal proportion. The actuator with the large load will not stop working, and the function of anti-flow saturation can be realized. The schematic diagram of the hydraulic synchronous control system for "pre-valve" compensation and "post-valve" compensation is shown in Figure 6 [59].

With the development of electro-hydraulic control and hydraulic energy-saving technology, more and more scholars have studied the compensated hydraulic synchronous control system and made a series of achievements. Li [61] established an integrated platform for the synchronous test of sunken ships, analyzed the synchronous hydraulic system of active compensation and passive compensation of sunken ships, further designed the overall hydraulic system of the hydraulic heave compensation test platform, and verified the accuracy of platform synchronous motion and the effectiveness of active compensation through tests. Zheng [62] studied the hydraulic synchronization system for lifting and fishing operations, designed a semi-active wave hydraulic cylinder synchronization compensation system with hydraulic cylinders in series–controlled by using double closed-loop PID and compared and analyzed through Matlab software simulation: master–slave PID synchronization control, parallel PID control and cross-coupling PID synchronization control. The results show that the error decay rate of master–slave PID synchronization control is 47.26%, the error decay rate of parallel PID synchronization control is 16.81%, and the error decay rate of cross-coupling PID control is 69.56%. It further proves that the cross-coupling PID synchronization control can quickly reduce the synchronization error and achieve stable conditions, and its control effect is the best.

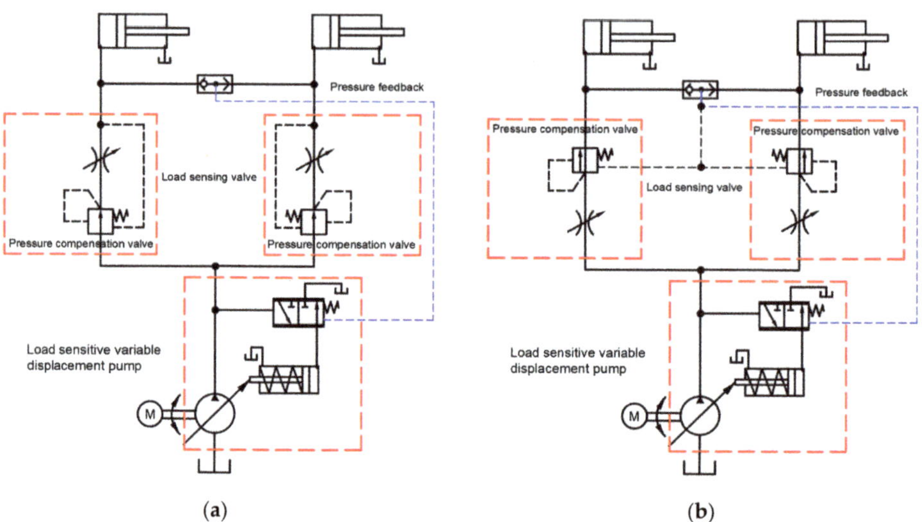

Figure 6. Schematic diagram of compensated hydraulic synchronous control system. (**a**) Schematic diagram of pre-valve compensation hydraulic synchronization control system; (**b**) Schematic diagram of post-valve compensation hydraulic synchronization control system.

Zhou et al. [63] studied the valve-controlled asymmetric hydraulic cylinder synchronization system, proposed a two-layer fuzzy controller to improve the response speed of the hydraulic cylinder, and designed a feedforward compensation control mode under cross-coupling to reduce the synchronization error. The simulation results show that the tracking accuracy and synchronization accuracy of the hydraulic cylinder with dual compensation under uneven load have been greatly improved.

4.3. Pump-Controlled and Valve-Controlled Hydraulic Synchronous Control System

Hydraulic synchronization control systems can be divided into "pump-controlled hydraulic synchronization control systems" and "valve-controlled hydraulic synchronization control systems", according to different control modes and control valve components [64].

4.3.1. Pump-Controlled Hydraulic Synchronous Control System

Pump controlled is also called volume controlled [65]. The pump-controlled hydraulic synchronous control system controls the synchronous movement of the actuating elements by changing the displacement of the hydraulic pump to make each actuator input the same amount of oil. Compared with the valve-controlled hydraulic synchronization system, the pump-controlled hydraulic synchronization system adopts the way of pump controlling the actuating elements, which reduces the number of hydraulic valves and the layout of hydraulic pipelines, significantly reduces the overflow throttling loss of the hydraulic system, the leakage of hydraulic oil, the friction heat of the system and other energy loss, and improves the efficiency of the system. However, under the same conditions, the natural frequency of the pump control system is much lower than that of the valve control system, which makes the response speed of the pump control system not high. The volumetric efficiency of the pump-controlled hydraulic synchronization system is high, and a low synchronization error can be obtained under the condition of constant load or small load change.

Peng et al. [66] designed a closed-loop digital PID control algorithm to control the speed of the hydraulic pump-controlled motor due to the synchronization error problem during its operation, and verified the good stability and rapidity of the system by optimizing the PID control parameters through simulation. Zhang et al. [67] put forward a loader lifting device with closed pump-controlled three chamber hydraulic cylinder, which adds a potential energy recovery chamber connected with the accumulator, and changed the original asymmetric two chamber hydraulic cylinder of the boom into a symmetric hydraulic cylinder. The simulation proved that the valve control system can not only significantly reduce energy consumption, but also significantly improve the response speed and control accuracy of the boom.

4.3.2. Valve-Controlled Hydraulic Synchronous Control System

The valve-controlled hydraulic synchronization control system realizes synchronization by controlling hydraulic valves (proportional valves, servo valves, pressure compensation valves, synchronization valves, diverter valves, etc.) and specially designed valve blocks. The valve control system usually selects proportional and servo valves as its control components, which can be known from the basic formulas of pump control system and valve control system [68]: in the case of the same controlled object, the natural frequency of the valve control system is greater than that of the pump control system. The valve control system has faster response and better dynamic characteristics. Therefore, the valve control system is mostly used under the working condition of dynamic load. The components of the valve-controlled hydraulic synchronous control system are arranged separately, which does not affect the performance of the system and is more convenient for maintenance. Due to the throttling speed regulation of the throttle port and overflow valve and the centralized oil supply of the system, the valve-controlled hydraulic synchronous system has the disadvantages of large throttling loss, large pressure loss, large hydraulic oil leakage, and low system efficiency, resulting in large system power loss.

The valve-controlled hydraulic synchronization control system can be divided into "flow hydraulic synchronization control systems", "electro-hydraulic proportional hydraulic synchronization control systems" and "electro-hydraulic servo hydraulic synchronization control systems" according to different types of control valves.

The flow hydraulic synchronization control system [69,70] regulates the flow in the hydraulic circuit through the flow control valve to make the flow into and out of the hydraulic cylinder/hydraulic motor equal, thus realizing the synchronization of the hydraulic system. The flow hydraulic synchronous system often uses a speed-regulating valve, diversion and collection valve, and throttle valve to control the hydraulic circuit. The synchronization accuracy of the hydraulic synchronization system of the speed-regulating valve is generally less than 4–5%, which is often used in situations where the flow is not very large, the speed change is small, and the synchronization accuracy is required to be high. The synchronization accuracy of the hydraulic synchronization system of the diverter and collector valve is generally 1–3%, its deviation correction capability is large, its use is simple, and its cost is low. It is often used in situations where the hydraulic circuit is simple and the flow and load change range are not large. The hydraulic synchronizing system of the throttle valve has low synchronization accuracy, but its cost is low, adjustment is simple, and multi cylinder synchronization can be achieved. It is often used in situations where the synchronization accuracy is not high, the flow is small, the load is not large, and the load is stable.

The electro-hydraulic proportional hydraulic synchronization control system [71,72] is controlled by the electro-hydraulic proportional valve. The electro-hydraulic proportional valve is a new type of electro-hydraulic control element, which is mainly composed of a proportional electromagnet and a working valve core, which can change the displacement of the working valve core according to the size of the electrical signal of the system, to adjust the flow and pressure at both ends of the electro-hydraulic proportional valve. Compared with the traditional hydraulic synchronization control system, the electro-hydraulic proportional hydraulic synchronization system has greatly improved the response speed, low cost, strong anti pollution ability, and high control accuracy. The electro-hydraulic proportional hydraulic synchronous control system is generally composed of a controller, electro-hydraulic proportional unit, actuator, controlled object, etc. Its control block diagram is shown in the following Figure 7.

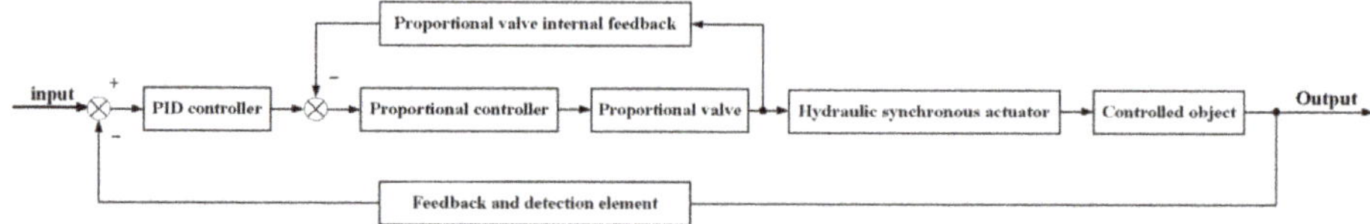

Figure 7. Block diagram of electro-hydraulic proportional hydraulic synchronization control system.

The electro-hydraulic servo hydraulic synchronization control system [73,74] is controlled by the electro-hydraulic servo valve. The electro-hydraulic servo valve has the advantages of small dead zone, zero covering, fast dynamic response, and high oil cleanliness. The electro-hydraulic servo hydraulic synchronous control system can adjust the flow into the hydraulic actuator at any time, with good dynamic characteristics and high control accuracy [75]. The electro-hydraulic servo hydraulic system has the characteristics of intrinsic nonlinearity and parameter nonlinearity. Intrinsic nonlinearity mainly refers to the dead zone characteristics of control elements and the coupling characteristics of flow in the system, while parameter nonlinearity mainly refers to the uncertainty of temperature change, load change, and oil leakage. The synchronization error caused by the above nonlinear factors can be reduced by selecting an effective control algorithm. The electro-hydraulic servo hydraulic synchronization control system is generally composed of a servo amplifier, controller, electro-hydraulic servo valve, and actuating element. Its control block diagram is shown in Figure 8.

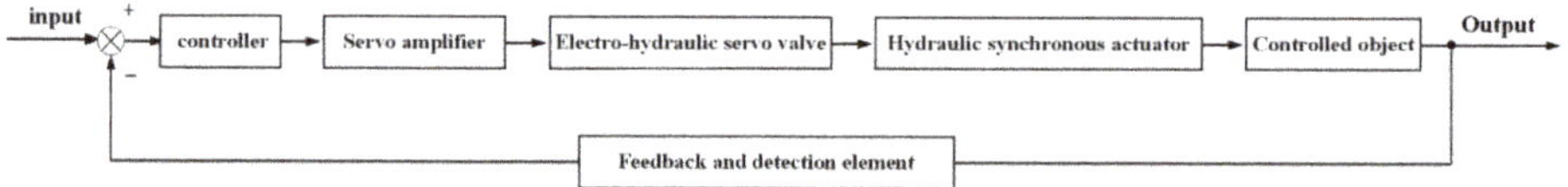

Figure 8. Block diagram of electro-hydraulic servo hydraulic synchronization control system.

With the development of modern control, more and more scholars have carried out research related to hydraulic synchronous control systems under valve control and pump control. Prabhakar et al. [76] compared the dynamic performance of pump control and valve control for a hydraulic motor, and the results show that the valve-controlled hydraulic motor system is more sensitive. Zhao et al. [77] established the control model for the four-motor hydraulic synchronous control system controlled by an electro-hydraulic proportional valve based on a fuzzy self-tuning PID control strategy. AMESim/Simulink software simulation analysis showed that the electro-hydraulic proportional valve-controlled four-motor hydraulic synchronous control system has a fast response and high synchronization accuracy.

In addition, some scholars combine valve control with pump control. The new synchronous control system with pump control as the main and valve control as the auxiliary can integrate the respective advantages of valve control and pump control, making the system have the characteristics of both control accuracy and response speed. Dong et al. [78] applied valve control and pump control to the synchronous system of a hydraulic excavator to improve the response speed and control accuracy of the system, designed the flow pressure to match the independent control of the inlet and outlet, applied it to the hydraulic excavator system to reduce the pressure fluctuation of the hydraulic excavator boom during operation from 6.9 Mpa to 1.7 Mpa, and at the same time, the energy consumption of one working cycle of the boom was reduced by 15%. The control strategy significantly reduces the working pressure difference at the valve port and improves the system stability and energy utilization.

4.4. Other Hydraulic Synchronization Control Systems

According to the different actuators [79], the hydraulic synchronization control system can be divided into the "hydraulic cylinder synchronization" and "hydraulic motor synchronization" systems. The hydraulic motor synchronization control system can be divided into "gear motor synchronization system" and "plunger motor hydraulic synchronization system" according to the type of hydraulic motor. Ideally, by comparison, the synchronization accuracy of the plunger hydraulic synchronous motor can reach more than 99%, and the synchronization accuracy of the gear hydraulic synchronous motor is about 95%, and the external load has a great impact on it.

According to different motion states [80], the hydraulic synchronization control system can be divided into "position hydraulic synchronization control system" and "speed hydraulic synchronization control system". Position synchronization requires that different hydraulic actuators have the same position at the same time during movement or when static. Speed synchronization requires that different hydraulic actuators have the same speed.

According to the number of actuators [81], the hydraulic synchronous control system can be divided into "double actuator" and "multi-actuator" hydraulic synchronous control systems. The hydraulic synchronous system with double actuators has less synchronous error accumulation due to fewer actuating elements. The synchronous system of multiple actuators has many components in the circuit, and there are manufacturing, installation, and leakage errors among the hydraulic components, which makes the accumulation of synchronous errors among its mechanisms relatively large. To improve the synchronization accuracy of multiple actuators, closed-loop control and flow compensation are often used.

According to different installation position [82], the hydraulic synchronization control system can be divided into "horizontal" and "vertical" hydraulic synchronization control system. The actuating element of the horizontal hydraulic synchronous system is fixed horizontally, as the hydraulic cylinder synchronous control system is not affected by its own gravity, and the change in the hydraulic cylinder's own gravity will not affect the synchronization accuracy. In comparison, the vertical hydraulic cylinder synchronization system has its own load under the vertical installation of the hydraulic cylinder, and the uneven change of the oil quality in the hydraulic cylinder also has a certain impact on the synchronization accuracy of the system.

5. Research Status of Control Algorithms of Hydraulic Synchronous Control System

With the development of modern control methods, hydraulic synchronous systems are more widely used in the engineering field, and at the same time, people also put forward higher requirements for hydraulic synchronization accuracy, anti-interference, stability, and response speed. The hydraulic synchronous system mostly adopts closed-loop control, and the simple closed-loop control can no longer meet the requirements of high-precision synchronous control due to the hysteresis of error feedback and the influence of nonlinear factors. Therefore, scholars at home and abroad combine different control algorithms with the hydraulic synchronous system to further improve the accuracy, stability, and anti-interference ability of the hydraulic synchronous system. The control algorithms of the hydraulic synchronous control system mainly include fuzzy control, PID control, fuzzy PID control, auto disturbance rejection control, sliding mode control, robust control, adaptive robust control, machine learning control, neural network control, and networked control, etc.

5.1. Fuzzy Control

Fuzzy control makes the control variables of the hydraulic synchronous system fuzzy, uses fuzzy rules, and finally defuzzles to simulate human reasoning and decision-making, to realize the control of the hydraulic synchronous system. The fuzzy control block diagram of the hydraulic synchronization control system is shown in Figure 9.

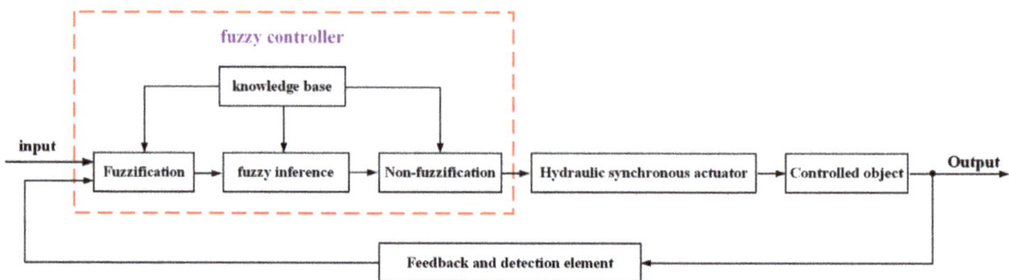

Figure 9. Fuzzy control block diagram of hydraulic synchronization control system.

Fuzzy control is mostly applied to hydraulic synchronous systems that are difficult to establish accurate mathematical models. However, this control algorithm will have a certain steady-state error, which requires researchers to summarize a lot of experience. The accuracy of the control system needs to be guaranteed by establishing complete fuzzy control rules. Liu et al. [83] established a fuzzy feedforward controller for symmetric double hydraulic cylinders, and simulation results show that the synchronization error between the two cylinders is within 15 mm, which verifies the feasibility of the dual hydraulic cylinder synchronization control system under fuzzy control. Chen et al. [84] designed a dual-integrated fuzzy controller based on fuzzy control, in which the fuzzy coordinator is used to process the synchronization error signal of the dual hydraulic cylinders, and the fuzzy tracking controller is used to detect the tracking error of the tracking dual hydraulic cylinders. The test results show that the maximum synchronization error of the two hydraulic cylinders is stable within ±4 mm under the working conditions of unbalanced load, system uncertainty, and large change of system interference.

The synchronization error of the hydraulic synchronous system under the fuzzy control is relatively large, and to make it have a higher synchronization accuracy, it needs to combine other algorithms and control methods to optimize. Li et al. [85] proposed a cross-coupled hydraulic fuzzy synchronization control strategy based on the fuzzy control principle and applied it to the two plug-in main valves of the plug-in electro-hydraulic servo valve, and the simulation results show that the control strategy can effectively compensate the synchronization error. Jia [86] designed a fuzzy-PID controller for the multi-stage hydraulic cylinder synchronization control system by combining fuzzy control and PID control in Simulink. The joint simulation of AMESim and Simulink shows that the synchronization accuracy of the multi-stage hydraulic cylinder synchronization system under fuzzy PID control is more stable and higher than that under fuzzy control.

5.2. PID Control

The PID algorithm [45,46,87,88] is widely used in hydraulic synchronous systems in recent years due to its simplicity, easy implementation, simple algorithm, and good robustness. Compared with fuzzy control, PID control has a better control effect in linear systems. PID control is composed of a proportional factor P, integral factor I, and differential factor D, and when these three parameters are adjusted properly, the hydraulic synchronous system can obtain a good control effect. The block diagram of PID control of the hydraulic synchronization control system is shown in Figure 10.

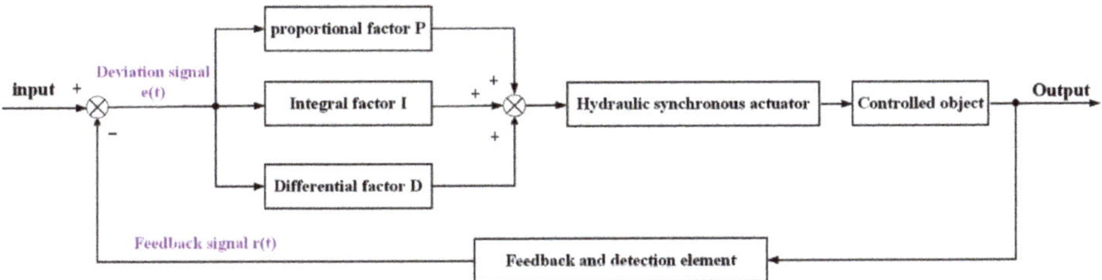

Figure 10. PID control block diagram of hydraulic synchronization control system.

Wei et al. [89] studied the step response of the four-cylinder synchronous platform under different parameters based on PLC control and the PID algorithm, and the results showed that the top mold hydraulic system under the PID control algorithm had a faster response speed, stable operation, and no overshoot. Li et al. [90] compared and analyzed the dynamic control effects of conventional PID and fractional PID on the motion of hydraulic cylinders based on the synchronous control strategy and method of double-cylinder four-column hydraulic press, and the simulation results show that the fractional PID controller has better anti-interference ability and robustness.

PID control has certain limitations in the control of hydraulic synchronous system because its parameters are not adjustable, and it depends on accurate mathematical models. Therefore, some scholars combine PID control with fuzzy control, adaptive control, neural network control, and other algorithms to further improve the control accuracy of the hydraulic synchronous system. Huang et al. [91], based on the control mode of deviation coupling, used an algorithm to optimize the PID control parameters, and studied the synchronization performance of multi-channel hydraulic cylinders with different parameters. The simulation results show that the control strategy has the characteristics of rapid system response, no overshoot, and a high steady-state accuracy. Yu et al. [92] combined the fuzzy adaptive control and PID control to realize the real-time adjustment of PID control parameters, and adopted the master–slave control mode for the hydraulic lifting synchronization system. The simulation shows that the improved control strategy can make the system respond faster and become more stable when the synchronization error is within the allowable range.

5.3. Fuzzy PID Control

Some scholars have combined fuzzy control and PID control to apply them to the hydraulic synchronous control system. First, the fuzzy controller performs fuzzy inference on the input control quantity, and then adjusts the parameters of the PID controller according to the working conditions of the hydraulic synchronous system. Then, the PID controller regulates the hydraulic synchronization system based on the deviation signal, thereby controlling the hydraulic actuator to synchronize the movement. The fuzzy PID control block diagram of the hydraulic synchronization control system is shown in Figure 11.

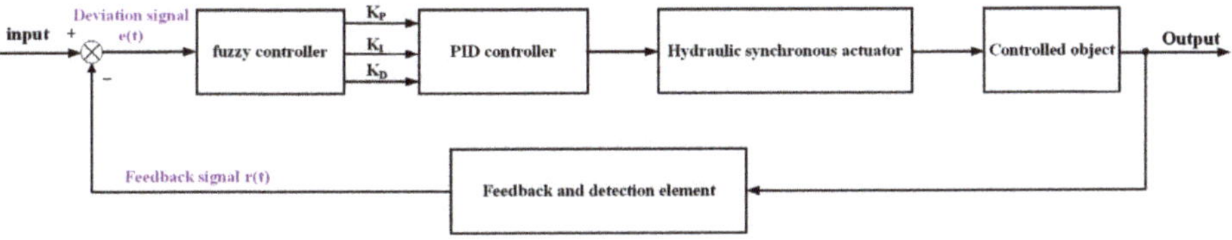

Figure 11. Fuzzy PID control block diagram of hydraulic synchronization control system.

Compared with traditional PID control, fuzzy PID control [93] combines the advantages of fuzzy control and PID control. It can not only realize the real-time adjustment of PID control parameters, but also apply to nonlinear systems. Moreover, the hydraulic synchronous system under this control strategy has a higher control accuracy and better control effect. Liu et al. [94] used the fuzzy PID control strategy to control the two-way electro-hydraulic dual-cylinder synchronous drive system. The PID control parameters can be automatically adjusted according to the change in the working condition error during the control process; the simulation results show that there is a small following error between the slave cylinder and the active cylinder under the fuzzy PID control, and the synchronization accuracy is high.

To make the synchronization control effect better, some scholars have proposed multiple fuzzy PID control strategies. Based on the dual fuzzy PID control with compensation factors, Mu et al. [95] proposed an adaptive fuzzy controller based on a model reference to control the hydraulic circuit of three hydraulic cylinders synchronization control, which reduced the impact of the hydraulic synchronization system due to the changes in the nonlinear friction, load, and operating parameters.

The combination of fuzzy PID control, adaptive control, neural network control, and other algorithms can further improve the control effect of the hydraulic synchronous system [96]. Mu et al. [97] designed a fuzzy parameter adaptive PID synchronization control strategy based on master–slave synchronization control for the hydraulic synchronization system of the inclined gating machine and added an accumulator pressure maintaining circuit at the outlet of the hydraulic pump of the hydraulic synchronization system. The simulation results of Simulink and AMESim software show that the control strategy has better synchronization accuracy and was more energy saving than the traditional PID hydraulic synchronization control strategy. Wu et al. [56] compared the synchronization control accuracy of the fuzzy PID controller and the fuzzy self tuning integral separation PID controller. The test showed that the overshoot amount of the fuzzy PID control algorithm was 4.8%, and the maximum synchronization error was 0.19 mm; the overshoot amount of fuzzy self-tuning integral separation PID control algorithm was 2.8%, and the maximum synchronization error was 0.12 mm. The reason for this phenomenon is that the fuzzy self-tuning integral separation PID controller integrated with the integral separation strategy can close the integral link when the synchronization error is large to prevent the control flow from supersaturation, so its control accuracy is higher than the fuzzy PID control.

5.4. Auto Disturbance Rejection Control

Active disturbance rejection control (ADRC) [98,99] can effectively solve the problem of system disturbance; the algorithm can estimate and compensate the disturbance to the hydraulic synchronous control system in real time, and the system response is more rapid and accurate. Compared with PID control, ADRC is more convenient to adjust parameters and has better control accuracy.

Ren et al. [100] designed an auto disturbance rejection controller to counteract the internal and external disturbances to the speed system, and added PID control in the feedback loop, which enhanced the dynamic adjustment ability and anti-interference ability of the system under disturbances. This significantly improved the synchronization accuracy of the dual electro-hydraulic position servo hydraulic synchronization control system. Lu et al. [101] introduced the auto-disturbance rejection control method into the hydraulic suspension multi-cylinder synchronization system based on the adjacent cross-coupling control method, and compared it with the PID control strategy based on the adjacent cross-coupling. The simulation shows that when the adjacent cross-coupling and auto-disturbance rejection control method is used, the synchronization error of the suspension system group is within ±0.2 mm. When the adjacent cross-coupling and PID control method is used, the synchronization error range of the suspension system group is −2~4 mm, Thus, it is verified that the designed control strategy has strong robustness and adaptability, and has high synchronization accuracy.

5.5. Sliding Mode Control

In the hydraulic synchronous control system, the sliding mode control sets the corresponding sliding plane according to the dynamic characteristics of the system, and then makes the system gradually converge to the sliding plane through the control function. When the state of the controlled object reaches the sliding plane, the system will move along the plane under the control of the equivalent function. Sliding mode control has the advantages of simple structure, insensitivity to parameter disturbances and changes, and is a nonlinear control strategy based on switching characteristics. Synovial membrane control has higher accuracy and robustness than PID control [102,103].

Guo et al. [104] designed an adaptive sliding film synchronous position control strategy with a feedforward compensator based on the improved sliding film approach law, and through the joint simulation of the dual hydraulic cylinders of the hydraulic support used in mining by Matlab and AMESim, and the comparative analysis with PI control and fuzzy PID control, the superiority of the control strategy was further verified. Zhang et al. [105] studied the performance of master–slave synchronous control of hydraulic support and proposed an adaptive sliding film control method based on sliding film control and fuzzy control. The simulation results show that compared with the fuzzy PID control, the adaptive sliding film control has better stability, synchronization, and higher synchronization accuracy, and can better overcome the problem of large fuzzy PID steady-state error.

5.6. Robust Control

Robust control [106] is based on the system state space model, which can solve nonlinear control problems without an accurate mathematical model. Robust control is applied to hydraulic synchronization equipment because it can ensure stability after the model and parameters of hydraulic synchronization system change, and has high robustness.

Zhou et al. [107] designed a robust controller composed of a stabilizing compensator and a servo compensator to solve the problem of small suppression and poor robustness of the multi-cylinder synchronous drive control system of a giant die-forging hydraulic press to system parameter changes. The simulation results show that the multi-cylinder hydraulic synchronous system still has good synchronization accuracy and response speed even in the case of large parameter changes. Liu et al. [108] established the mathematical model of the synchronization system of the giant die-forging hydraulic press, and designed a robust controller for the synchronization control of the giant die-forging hydraulic press. The simulation results show that the system synchronization error increases with the increase of the eccentric load, but the steady state error is basically close to 0, which proves that the robust controller has a good inhibition effect on the nonlinear fluctuation of the synchronization system, and can effectively solve the problem of multi-cylinder synchronization control.

5.7. Adaptive Robust Control

Robust control is applicable to hydraulic synchronous control systems with a large variation range of uncertain factors and a small stability margin. Because it generally does not work in the optimal state, the steady-state control accuracy of the system is poor. Therefore, in order to make the system work in the optimal state, some scholars have proposed adaptive robust control. Adaptive robust control [109,110] combines the advantages of robust control and adaptive control. Based on robust control, parameter adaptive control is added to the feedforward link of the hydraulic synchronous control system to enable the system to work in the optimal state and ensure that the hydraulic actuator has high-precision motion tracking performance and high synchronization accuracy when moving.

Dou [111] studied the synchronization of unsymmetrical hydraulic cylinders, designed a robust adaptive tracking controller based on the nonlinear coupling model, selected the adaptive backstepping design method, and combined it with the parameter adaptive law to solve the uncertainty of some parameters of the system. The simulation in Amesim software shows that the maximum synchronization error of the two hydraulic cylinders is about 1.4 mm, and this control method has good synchronization. Liu et al. [112] designed an adaptive robust controller based on the new integrated actuator electro-hydraulic composite cylinder to compensate the continuous friction it suffered, introduced the robust integral of the error signal into the controller to compensate the approximate error caused by the external disturbance, and the simulation verified the effectiveness of the control method. Li et al. [113] designed an adaptive robust H∞ control and proposed an adaptive controller to update the parameters to control the influence of uncertain parameters and external disturbances during the movement of double hydraulic cylinders. The good control ability of the control system was verified through simulation, which provided a reference for the synchronous control of multiple hydraulic cylinders.

5.8. Machine Learning Control

Machine learning [114,115], as an artificial intelligence algorithm with multiple fields, is dedicated to studying how computers simulate and realize human learning behavior and gradually improve and perfect their own performance through autonomous learning. The process of machine learning modeling is to divide the data into training sets and test sets. The test set is the independent data of the training set. It does not participate in the training at all, but is only used for the final evaluation of the model. With the continuous development of artificial intelligence, machine learning is now gradually being applied to the hydraulic synchronous control system.

Compared with traditional control, machine learning control is more stable and reliable, with higher control accuracy. Fu et al. [116] proposed a control method combining fuzzy PID control and feedforward control based on machine learning to solve nonlinear problems such as a sluggish and dead zone in the electro-hydraulic system of the boom of a mainstream pump truck, and fitted the machine learning and fuzzy PID to obtain the control algorithm that can be applied in PLC. The test results show that the control algorithm can greatly improve the tracking control synchronization accuracy of the boom. Wei [117] studied the control method based on the fusion of machine vision, machine learning and the fuzzy model and established the prediction model of slab bending based on machine learning. Subsequently, the author introduced the differential evolution algorithm and Bayesian optimization algorithm to select the optimal parameters of the machine learning algorithm in order to realize the high-precision detection and prediction of slab bending, and the synchronous and accurate compensation of the roll gap tilt compensation value for the hot strip rolling process. The experiment shows that the accuracy of the stochastic forest prediction model based on differential optimization is 96.3% in the range of ± 3 mm, and the accuracy of the prediction model based on Bayesian optimization support vector is 99.25% in the range of ± 2 mm. The machine learning model can meet the accuracy requirements.

5.9. Neural Network Control

Neural network is based on simulating the structure and function of the biological brain, and quickly approximates the complex model of nonlinear multi-input and multi-output. Its control parameters can be adjusted continuously with the changes in environment and load conditions, so it can achieve good control accuracy. However, neural networks are prone to over-learning, and their control cost is high. With the development of modern control theory, more and more scholars have applied neural network control to hydraulic synchronous systems and achieved good control results. The relevant neural network control algorithm can automatically adjust the control parameters online, which gives the hydraulic synchronization system a strong anti-interference ability. The neural networks commonly used in the hydraulic synchronization system include the BP neural network and immune neural network.

The BP neural network [118,119] (error return neural network) is widely used in hydraulic neural network synchronization control. The BP neural network uses a nonlinear continuous transformation function to make neurons in the hidden layer have a learning ability. Its core idea is to use error to reverse correct the control system, so that the system has a strong learning ability, memory ability, and nonlinear mapping ability. Liu et al. [120] proposed the master–slave synchronization control strategy of the BP neural network for the four-cylinder pod thruster installation platform that needs synchronization control, and designed a PID controller based on the BP neural network in combination with the PID controller. The simulation verified that the synchronization error of the smooth operation of the four hydraulic cylinders was within 0.1 mm, and the PID control principle under the BP neural network was shown in Figure 12.

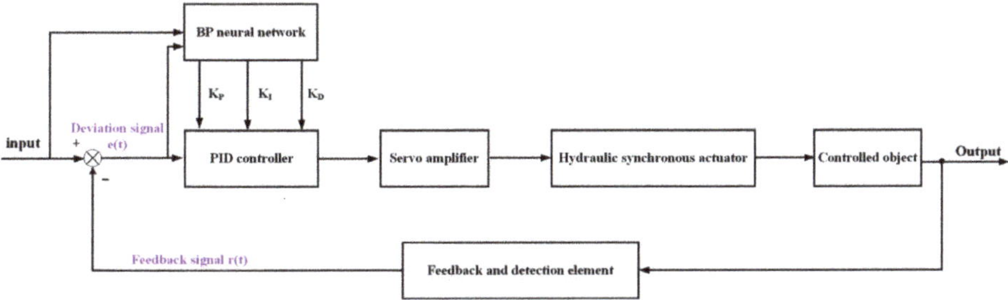

Figure 12. PID control principle under BP neural network.

The immune neural network [121] is a control algorithm formed by studying the immune algorithm based on the working principle of the biological immune feedback mechanism and applying it to a neural network. Its key is the approximation of antibody suppression function. Based on the property that a neural network can approach any nonlinear function, Liu [122] used the BP neural network to approach the antibody suppression regulation function, and combined it with a PID controller to design an immune neural network control method. This uses the output of the actuator to approach the antibody suppression regulation function in real time, thus realizing the self-adjustment of proportional integral differential parameters, and further improving the control effect of the dual hydraulic cylinders. The simulation results show that when the system damping is 0.1, the maximum synchronization error of two hydraulic cylinders is 0.014 mm, which further verifies that the control algorithm can meet the control requirements of a high precision double-cylinder forging machine.

Neural network control is often combined with fuzzy control, PID control, and adaptive control because of its strong robustness, high control accuracy, and the ability to solve the control of complex, uncertain, nonlinear, and time-varying systems, which further significantly improves the control accuracy of hydraulic synchronous system. Xu et al. [123] combined a neural network and fuzzy control to form a fuzzy neural network synchronization control algorithm, applied it to a double-cylinder forging hydraulic press, and established a mathematical model of the double-cylinder synchronization control. The experiment shows that the fuzzy neural network control algorithm has good synchronous control accuracy.

5.10. Networked Control

With the continuous development of Internet technology, the hydraulic synchronization control system is gradually developing to wireless network control. Some scholars have used sensors and a wireless network signal transceiver to achieve hydraulic synchronization control, so that it has a faster response speed and higher synchronization accuracy.

To solve the problem of a long-distance hydraulic pipeline and signal line in hoisting a hydraulic lifting system, Dong et al. [124] proposed an intelligent hydraulic synchronous lifting scheme based on a wireless network communication module, and designed a distributed iterative learning controller, which can effectively overcome the impact of communication time delay and saturation. AMESim simulation verified the effectiveness of the control strategy, and the six hydraulic-cylinder system can achieve accurate synchronization control in a short time. Lu et al. [125], based on the robust control theory, set the sensor and actuator nodes as time-driven, set the controller as event-driven, used the wireless networked synchronization system to connect the distributed subsystems that needed synchronous output, and established the error model of the hydraulic synchronization system that could reduce the network delay function. The simulation shows that the displacement of the active cylinder and the slave cylinder can be gradually synchronized, and the experiment shows that the displacement synchronization error of the driving cylinder and the slave cylinder is within ±1 mm.

After the above research on hydraulic synchronization control algorithms, their control effects and characteristics can be summarized as shown in Table 2.

Table 2. Comparison of Different Control Algorithms for Hydraulic Synchronization System.

Control Algorithm	Control and Control Effect	Characteristic
fuzzy control	Liu [83] used a fuzzy feedforward controller to synchronize symmetric dual hydraulic cylinders, and the synchronization error between the two cylinders is within 15 mm; Chen [84] used a dual integrated fuzzy controller to synchronize the dual hydraulic cylinders, and the maximum synchronization error of the hydraulic cylinders did not exceed ±4 mm.	Researchers need to summarize and establish complete fuzzy control rules to ensure synchronization accuracy, commonly used in hydraulic synchronization systems that are difficult to establish accurate mathematical models. Synchronization control accuracy is poor and is often used in combination with other control algorithms.
PID control	Wei [89] used a PID controller to synchronously control four hydraulic cylinders. The results show that the system has fast response speed, stable operation, and no overshoot, but the control accuracy is average.	The control is simple, easy to implement, has good robustness, and has high control accuracy in linear systems. However, the control parameters are not adjustable, the adaptability is weak, and the synchronization accuracy is general. Often used in conjunction with other algorithms.
Fuzzy PID control	Mu [97] used a fuzzy parameter adaptive PID synchronization controller to synchronize the tilting pouring machine. The results show that this control strategy has higher synchronization accuracy and more energy saving compared to PID control; Wu [56] used a fuzzy PID controller to synchronously control 4 cylinders drive hydraulic press. The experiment shows that the overshoot of the fuzzy PID control algorithm is 4.8%, and the maximum synchronization error Is 0.19 mm.	Combining the advantages of fuzzy control and PID control, it can not only adjust PID control parameters in real time, but also be applicable to nonlinear systems. High control accuracy and good control effect.
auto disturbance rejection control	Lu [101] adopted an active disturbance rejection control and adjacent cross-coupling control method to synchronize the suspension hydraulic cylinders. Simulation results show that the synchronization error of the suspension system group is within ±0.2 mm.	It can estimate and compensate the disturbance of hydraulic synchronous control system in real time, and improve the response speed of the system. High synchronization control accuracy, high robustness, and strong adaptability.
sliding mode control	Zhang [105] proposed an adaptive sliding film control method based on sliding film control and fuzzy control for synchronous control of hydraulic supports. Compared with fuzzy PID control, adaptive sliding film control has better steady-state performance and higher synchronization accuracy.	Simple structure, insensitive to parameter disturbances and changes, and higher control accuracy and robustness compared to PID control.

Table 2. *Cont.*

Control Algorithm	Control and Control Effect	Characteristic
robust control	Liu [108] designed a robust controller for synchronization control of giant die forging hydraulic presses. The simulation results show that the system synchronization error increases with the increase of eccentric load.	Nonlinear control problems can be solved without the need for accurate mathematical models, but hydraulic systems typically do not work in optimal conditions and have poor control accuracy.
adaptive robust control	Dou [111] designed a robust adaptive tracking controller based on a nonlinear coupled model to synchronize asymmetric hydraulic cylinders. The results show that the maximum synchronization error of the two hydraulic cylinders is about 1.4 mm.	Combining the advantages of robust control, it can also make the hydraulic synchronous system work in the optimal state, with strong robustness and high steady-state control accuracy.
machine learning control	Wei [117] used a control method that combines machine vision, machine learning, and fuzzy models to synchronously and accurately compensate the roll gap tilt compensation value. Experiments show that the accuracy of the prediction model reaches 99.25% within the range of ± 2 mm, meeting the requirements of high-precision synchronous control.	It is more stable and reliable than traditional control, and can self-tune control parameters, resulting in higher synchronization control accuracy.
neural network control	Liu [120] used a PID controller based on BP neural network, simulation shows that the synchronization error of the four hydraulic cylinders is less than 0.1 mm during smooth operation. Liu [122] adopted an immune neural network control method combined with a PID controller, and when the system damping is 0.1, the maximum synchronization error of the two hydraulic cylinders is 0.014 mm.	Synchronous control has high accuracy, strong anti-interference ability, and can self-tune control parameters, but it is prone to excessive learning and high control cost.
networked control	Lu [125] used a wireless networked synchronization system based on robust control to connect distributed subsystems that require synchronous output, with the synchronization error of the active cylinder and the slave cylinder being less than ± 1 mm.	Fast response speed and average synchronization accuracy.

6. Discussion

The hydraulic synchronization control system has the characteristic of nonlinearity and is time-varying, and the adoption of an appropriate control algorithm is an important research direction of the hydraulic synchronization system. Common advanced control algorithms include fuzzy control, fuzzy PID control, adaptive robustness control, machine learning control, and neural network control, etc., but they have their own advantages and disadvantages. For a nonlinear, easily disturbed, time-varying hydraulic synchronous control system, a single control algorithm often cannot achieve a good control effect, so some scholars have organically combined multiple control algorithms and thus achieved a good control effect.

With the development of hydraulic synchronization control systems towards automation, low energy consumption, low leakage, low noise, high response, anti-interference, intelligence, and high accuracy, higher requirements are put forward for the control strategy and hydraulic components. The designed hydraulic synchronous system first must meet the requirements of high precision, needs to have fast response, overshooting, strong anti-interference ability and other characteristics, and the selected control strategy can ensure that the system has a good control effect when it is subjected to external disturbance. The prediction and prospect of the development trend of the hydraulic synchronous control system [22,126–133] can be summarized as follows.

(1) Reliable and safe. On the premise of ensuring the accuracy of hydraulic synchronization, with the improvement of control technology, new technology, and processing technology, the hydraulic synchronization system will develop towards more safety and reliability in the future.

(2) Energy conservation and environmental protection. An intelligent control algorithm and hydraulic components with new structure will be adopted to improve the efficiency and response rate of the hydraulic system, thus reducing the system energy consumption and oil leakage and further improving the energy utilization rate.

(3) Automation. The new intelligent control strategy will be adopted to enable automatic regulation and further improve the control effect of the hydraulic synchronization control system.

(4) Digitalization. Rational use of mechanical and computer controllers will further improve the hydraulic system, will be further combined with new control technology to form the integration of a "mechanical-electrical-hydraulic-controller", and then realize digital control. A discrete digital signal will be used to control the discrete hydraulic fluid, and then the digital control of the hydraulic synchronous control system will be realized.

(5) Integration. The integrated hydraulic components can make the structure of the hydraulic synchronization control system more compact, without the connection of hydraulic pipe fittings, thus effectively eliminating the vibration and oil leakage at the oil pipes and joints.

(6) Intelligence. In the future, intelligent new hydraulic pumps, intelligent electro-hydraulic (digital) valves, intelligent electronically controlled synchronous cylinders, etc., can make the accuracy of hydraulic synchronization control system higher.

(7) Neural network. The use of a neural network and other related control algorithms will become the development trend of hydraulic synchronization control systems in the future due to its advantages of continuous self-learning, fast approaching to nonlinear complex models, and high control accuracy.

(8) Innovation. With the continuous development of science and technology, the new pressure compensation synchronous control systems, the new valveless hydraulic synchronous control systems and the new hydraulic synchronous control systems with wireless monitoring function will continue to enrich the hydraulic field.

(9) Initiative. With the continuous development of artificial intelligence and big data, the future hydraulic synchronous control system can slowly realize the functions of active prediction and active flow distribution, so as to self-diagnose and control the hydraulic system, which will further improve the system response speed and synchronization accuracy.

(10) Networking. The wireless network transceiver will be added to the hydraulic synchronization system, and the distributed sub-hydraulic branch system that needs synchronous operation will be connected through the Internet, which can realize data acquisition, communication, and control of electro-hydraulic valve at low cost, making the hydraulic synchronization system respond faster, and effectively solving the control problem of the hydraulic synchronization control system at a long distance.

7. Conclusions

The hydraulic synchronous control system is the core of hydraulic synchronous motion machinery and equipment, which determines the function and technical performance of these machines and equipment. The main problems of the hydraulic synchronization system under the current closed-loop control are: there is a feedback adjustment lag, it is difficult to combine between different control algorithms, it is difficult to achieve accurate synchronization under eccentric load and other problems, and with the improvement of control accuracy, the requirements and costs of hydraulic components are getting higher and higher. It is believed that with the continuous development of artificial intelligence and control algorithms in the future, the above problems can be gradually solved. With the continuous development of control theory, the improvement of computer level and the continuous improvement of new intelligent manufacturing methods and processes, the cost of the hydraulic synchronization system will be gradually reduced, and its combination with intelligent control algorithms will also be closer. Digitalization, integration, intelligence, neural networking, innovation and networking are gradually becoming the future development trend of the hydraulic synchronous control system, among which the more promising development is neural network and networking, which can significantly improve the control accuracy of the hydraulic synchronous system and reduce the time of the feedback adjustment. At the same time, more scholars are also required to carry out relevant research on the hydraulic synchronous control system, so as to continuously promote the progress and development of the hydraulic field.

Author Contributions: Conceptualization, R.L. and W.Y.; methodology, X.D.; software, J.X.; validation, W.Y., Q.S. and Y.Z.; formal analysis, R.L.; investigation, X.D.; resources, W.Y.; data curation, X.D.; writing—original draft preparation, W.Y.; writing—review and editing, R.L.; visualization, J.X.; supervision, X.D.; project administration, Q.S.; funding acquisition, R.L. All authors have read and agreed to the published version of the manuscript.

Funding: 1. Key R&D plan of Shandong Province, China, grant number 2021CXGC010207; 2. Creation Project of Key Components Of High Performance Sowing and Harvesting and Intelligent Work Tools, grant number 2021CXGC010813; 3. General project of Shandong Natural Science Foundation, grant number ZR2021ME116; 4. Key R&D plan of Shandong Province, China, grant number 2020CXGC011005.

Acknowledgments: We would like to thank our tutor, Ruichuan Li, for all his support and guidance. I would like to thank my colleagues for their care and help in my daily work.

Conflicts of Interest: The authors declare no conflict of interest.

References

1. Wen, D.S.; Zhen, X.S.; Chen, F.; Jiang, F.; Shang, X.D. A comparison analysis of hydraulic synchronous multi-motor and traditional synchronous motor. *J. Harbin Inst. Technol.* **2017**, *49*, 173–177.
2. Zhang, L.; Cong, D.; Yang, Z.; Zhang, Y.; Han, J. Optimal Design and Hybrid Control for the Electro-Hydraulic Dual-Shaking Table System. *Appl. Sci.* **2016**, *6*, 220. [CrossRef]
3. Zhao, L.Z. Hydraulic System Design and Control Performance Research of Wind Turbine Turning. Master's Thesis, Shenyang University of Technology, Shenyang, China, 2022.
4. Zhang, X.H.; Liu, P.Y.; Wei, W.J.; Tian, G.P.; Li, H.T. Design and characteristic simulation of variable gear synchronous shunt. *Trans. Chin. Soc. Agric. Eng.* **2017**, *33*, 63–69.
5. Tian, Y. Study on Synchronization System of Hydraulic Cylinders with Opposite Movement. Master's Thesis, Changchun University of Technology, Changchun, China, 2009.
6. Zhao, C.Y.; Duan, W.H.; Jiang, M.C.; Wang, Y.H. Self-synchronization Theory of Pile Sinking System Driven by Dual Hydraulic Motors. *J. Northeast. Univ.* **2022**, *43*, 1446–1452+1483.
7. Sun, X. Research on Force Synchronization Control Strategy of Double Hydraulic Cylinder. Master's Thesis, Harbin University of Technology, Harbin, China, 2020.
8. Liu, F. Research on the Synchronization Control for the Proportional Valve Controlled Hydraulic Cylinder. Master's Thesis, Beijing Institute of Technology, Beijing, China, 2016.
9. Dickinson, H.W. Joseph Bramah and his inventions. *Trans. Newcom. Soc.* **1941**, *22*, 169–186. [CrossRef]
10. Zames, G. Feedback and optimal sensitivity: Model reference transformations, multiplicative seminorms, and approximate inverses. *IEEE Trans. Autom. Control* **1981**, *26*, 301–320. [CrossRef]
11. Li, H.J.; Wang, L.; Kang, B.B. Research on Synchronous Control of Dual-hydraulic Cylinder Based on Grey Prediction Theory. *Mach. Tool Hydraul.* **2020**, *48*, 97–100.
12. Gu, H.T.; Sun, C.X.; Zhu, W.J. Analysis and Improvement of Open Cycle Synchronous Control Hydraulic System for Two Cylinder in a Wind Tunnel. *Mach. Tool Hydraul.* **2022**, *50*, 124–128.
13. Chen, T.H.; Li, H.X. Research on Synchronous Fuzzy Control of Duplex Hydraulic Piston Trajectory. *Mach. Tool Hydraul.* **2017**, *45*, 146–148. [CrossRef]
14. Yang, L.; Cen, Y.W.; Ye, X.H.; Huang, J.Z. Synchronous Control Research for Hydraulic Bending Machine Based on Single Neuron PID Strategy. *Mach. Tool Hydraul.* **2017**, *45*, 119–123.

15. Zhang, L.D.; Li, Y.S. Synchronous Control of Double Hydraulic Cylinders of Scissors Aerial Work Platform Based on Fuzzy PID. In Proceedings of the 2020 5th International Conference on Electromechanical Control Technology and Transportation (ICECTT), Nanchang, China, 15–17 May 2020.
16. Liu, Z.L.; Liu, M.Q.; Wang, J.M.; Wang, Z.M. Design and simulation of pile leg synchronous control system for jack up wind turbine installation ship. *Mach. Tool Hydraul.* **2022**, *50*, 113–117.
17. Li, D.B. Research on Hydraulic Synchronous Lifting Technology of Steel Formwork. Master's Thesis, Changchun University of Technology, Changchun, China, 2022.
18. Zhao, L.Z.; Su, D.H.; Zuo, W.; Zou, H.T. Synchronous control system of hydraulic cylinder of fan coil car controlled by particle swarm fuzzy PID. *J. Mech. Electr. Eng.* **2022**, *39*, 961–966.
19. Wang, H.; Chen, J.H.; Yao, J.Y. RISE-Based Cross-Coupling Synchronization Control for a Dual Redundant Hydraulic Actuation System with Extended State Observer. In Proceedings of the 2021 China Automation Congress (CAC), Beijing, China, 22–24 October 2021.
20. Fang, Z.; Huang, C.L.; Xu, Y.S.; Zhang, C.M. Research on Closed-loop Control System for Friction Welding Machine Moving Velocity Based on PID. *Mach. Tool Hydraul.* **2020**, *48*, 29–32+36.
21. Wang, Y.F.; Ding, H.G.; Zhao, J.Y.; Li, R. Neural network-based output synchronization control for multi-actuator system. *Int. J. Adapt. Control* **2022**, *36*, 1155–1171. [CrossRef]
22. Cao, Y.; Li, Q.M.; Wu, G.Q.; Qiu, K.W.; Yu, R.C. The research progress of hydraulic synchronization system. *Mod. Manuf. Eng.* **2014**, *410*, 136–140.
23. Liao, L.; Xiang, G.F.; Zheng, X.J.; Zhu, Y.Q.; Xiao, Q.; Dian, S.Y. Synchronous Control of Parallel-Hydraulic System Based on Linear Active Disturbance Rejection. *Modular Mach. Tool Autom. Manuf. Tec.* **2023**, *588*, 60–63+68.
24. Xia, Y.M.; Shi, Y.P.; Yuan, Y.; Zhang, Y.M.; Yao, Z.W. Analyzing of influencing factors on dynamic response characteristics of double closed-loop control digital hydraulic cylinder. *J. Adv. Mech. Des. Syst* **2019**, *13*, 59–76. [CrossRef]
25. Chen, Z.J.; Ding, Z.; He, H.N. Design analysis and improvement of hydraulic synchronous system of HSM coil lifting table. *Metal. Ind. Autom.* **2020**, *44*, 74–78.
26. Hu, M.Q. Research on Hydraulic Synchronous and Positioning Control System for Shield Machine Support. Master's Thesis, Shenyang University of Technology, Shenyang, China, 2019.
27. Wang, G.J.; Gao, Y.; Zhang, F.B.; Sun, J.; Li, J.P. Simulation Research of Hydraulic Multi-cylinder Synchronous Lifting System Based on Gear-motor Flow-divider. *Mach. Tool Hydraul.* **2022**, *50*, 148–152.
28. Tang, Y.H.; Liu, Z.W.; Qing, X.Q. Analysis of the factors affecting the performance of the synchronization system for giant forging hydraulic press. *Forg. Stamp. Technol.* **2014**, *39*, 77–83+103.
29. He, M. Research on Water Hydraulic Synchronization Systems. Master's Thesis, Kunming University of Technology, Kunming, China, 2011.
30. Jin, Y.; Xia, Y.M.; Lan, H.; Kang, H.M. Design of Hydraulic Synchronization Systems Eliminating Bias Load Effect. *Mach. Tool Hydraul.* **2014**, *42*, 107–110.
31. Liu, X.L.; Feng, Z.H.; Shi, X.; Yang, H.L. Application of Hydraulic Synchronous Motor for Two-cylinder Lifting Equipment. *Hydraul. Pneum. Seals* **2020**, *4*, 44–46.
32. Bing, C.S.; Zhang, Y.Y.; Fu, Y. Design of Hydraulic Synchronization System for Super Low Profile Alignment Scissor Lift. *Hydraul. Pneum. Seals* **2020**, *40*, 53–57.
33. Liu, J.S.; Ding, H.G.; Li, H.N.; Zhao, J.Y. Hydraulic Synchronous System Design for Temporary Support Pushing Cylinders. *Chin. Hydraul. Pneum.* **2017**, *307*, 73–76.
34. Meng, Y.P. The Research of Working Principle in the Multi-Cylinder Electro-Hydraulic Proportional Synchronous Control System. Master's Thesis, Inner Mongolia University of Science and Technology, Baotou, China, 2008.
35. Liu, X.F. Research on Synchronous Control Technology of Dual motor in Crawler Crane Lifting System. Ph.D. Thesis, Jilin University, Jilin, China, 2012.
36. Ren, D.L. Study of Hydraulic Syuchronous System in Practice. Master's Thesis, Shenyang University of Technology, Shenyang, China, 2009.
37. Li, Z. Research on Control System of Double Hanging Point Hydraulic Hoist Based on Synchronous Drive Technology. Master's Thesis, Harbin Institute of Technology, Harbin, China, 2019.
38. Shi, G.L.; Shi, W.X.; Li, T.S. Hydraulic synchronous closed-loop control and its application. *Mach. Tool Hydraul.* **1997**, *4*, 3–7+2.
39. Ding, H.G.; Liu, Y.Z.; Zhao, Y.B. A new hydraulic synchronous scheme in open-loop control: Load-sensing synchronous control. *Meas. Control* **2020**, *53*, 119–125. [CrossRef]
40. Pan, M.T.; Zhang, B.; Zhao, Q.; Zhou, M.M.; Zuo, S.Y. Research on Control Strategy of Multi Cylinder Synchronization Based on Heave Compensation Platform. *Mach. Tool Hydraul.* **2022**, *50*, 123–128.
41. Sheng, T. Structure Design of Umbilical Fatigue Test Machine and Research on Its Hydraulic synchronous Control System. Master's Thesis, Harbin Engineering University, Harbin, China, 2019.
42. Zhang, X.M. Study on Synchronous Control of the Hot Platen Lifting System in Continuous Flat Press Based on Predicative Control. Ph.D. Thesis, Chinese Academy of Forestry Sciences, Beijing, China, 2018.
43. Deng, Y.; Tang, B.; Pan, Y.; Chen, J.H. Research on Key Technologies of Synchronization Control for Multi-Axis Hydraulic System. *Mach. Des. Manuf.* **2022**, *377*, 173–178+182.

44. Lorenz, R.D.; Schmidt, P.B. Synchronized Motion Control for Process Automation. In Proceedings of the Conference Record of the IEEE Industry Applications Society Annual Meeting, San Diego, CA, USA, 1–5 October 1989.
45. Zhang, T.J. Analysis and Study on Electro-Hydraulic Synchronization System of the Swing Thru of Container Self-Loading Trailers. Master's Thesis, Chang'an University, Xi'an, China, 2014.
46. Tian, Y.; She, Y.; Wang, X.B. "Series Parallel" Synchronous Control Structure of Four Column Hydraulic Press. *Chin. Hydraul. Pneum.* **2021**, *353*, 20–26.
47. Wang, C.L.; Guo, Z.F.; Dong, Y.L.; Zhao, K.D. Research of equal status synchronizing controller for the dual hydraulic driven motors based on QFT. *Chin. Hydrau. Pneum.* **2011**, *244*, 18–23.
48. Liu, F. Simulation and Research of Double Hydraulic Cylinder Synchronization Control System on Four-Column Hydraulic press. Master's Thesis, Shanghai University of Engineering Science, Shanghai, China, 2016.
49. Li, L.Y.; Jia, Y.Z.; Wang, X.G. Study on Hydraulic Cylinder of Roller Leveller to Master-Slave Control Precision of Press Down Synchronously. *Mach. Tool Hydraul.* **2020**, *48*, 128–132.
50. Xu, Y.H.; Han, H.Y.; Wei, C.M. Position-pressure Master-slave-control Method of Electro-hydraulic Servo System. *Chin. Hydraul. Pneum.* **2017**, *30*, 15–20.
51. Li, X.X. Research on Synchronization Control of Hydraulic Motor for Tipping Table of Positioning Machine. Master's Thesis, Yanshan University, Qinhuangdao, China, 2021.
52. Koren, Y. Cross-coupled biaxial computer control for manufacturing systems. *J. Dyn. Syst. Meas. Control* **1980**, *102*, 265–272. [CrossRef]
53. Song, Y.Y. Research of Dual Hydraulic Cylinders Synchronous Control Based on Single Neuron PID. *Hydraul. Pneum. Seals* **2014**, *34*, 45–47.
54. Zhou, G.Y. Research on Double Fuzzy Four-Cylinder Synchronous Control Based on Adjacent Cross Coupling. Master's Thesis, Lanzhou University of Technology, Lanzhou, China, 2020.
55. Lin, J.Q. Design and Research of Synchronous Loading System of the Instrumented Wheel-Set Calibration Test Bench. Master's Thesis, Changchun University of Technology, Changchun, China, 2016.
56. Wu, C.H.; Hao, X. Fuzzy self-tuning integral separation PID synchronous control on multi-cylinder hydraulic press. *Forg. Stamp. Technol.* **2022**, *47*, 146–153.
57. Sun, C.G.; Yuan, R.B. Adaptive robust cross-coupling position synchronization control of a hydraulic press slider-leveling. *Sci. Prog.* **2021**, *104*, 1–19. [CrossRef]
58. Liu, R.X.; Chen, K. Research on Fuzzy Adjacent Coupling Synchronization Control in Multi-cylinder Leveling System. *China Mech. Eng.* **2016**, *27*, 2316–2321.
59. Liu, Y.Z. Study on Load Sensitive Variable Speed Flow Dividing Principle and Synchronous Drive Characteristics. Master's Thesis, China University of Mining and Technology, Xuzhou, China, 2021.
60. Ding, H.G.; Liu, Y.Z.; Zhao, J.Y. Load Sensitive Variable Speed Synchronous Drive Principle and Characteristics. *Chin. Hydraul. Pneum.* **2021**, *45*, 77–81.
61. Li, S. Design of Hydraulic Heave Compensation Test Platform for Wreck Synchronous Lifting. Master's Thesis, Dalian Maritime University, Dalian, China, 2018.
62. Zheng, S.L. Research on Semi-Active Wave Compensation Technology for Hydraulic Synchronous Lifting and Salvage. Master's Thesis, Harbin Engineering University, Harbin, China, 2020.
63. Zhou, Q.; Zhang, D.M.; Wang, L.; Lv, L.L. Synchronous Control of Hydraulic Cylinder Position with Double Compensation. *Chin. Hydraul. Pneum.* **2021**, *45*, 158–164.
64. Gao, J.; Yan, G.Y.; Peng, G.Y.; Xi, G.N. Experiment Study on Performances of Servo Pump Used in Hydraulic Press Machine. *Adv. Mater. Res.* **2012**, *516*, 892–895. [CrossRef]
65. Shi, Y.F. Study on Continuous Jack-Up System of Large Structures Based on Multi-Point Volume Synchronization. Master's Thesis, Tianjin University, Tianjin, China, 2015.
66. Peng, Z.X.; Yuan, S.H. A Closed Loop Speed Control System for Hydraulic Pump-Motor. *Trans. Beijing Inst. Technol.* **2009**, *29*, 205–208+232.
67. Zhang, X.G.; Wang, X.Y.; Zhang, H.J.; Quan, L. Characteristics of Wheel Loader Lifting Device Based on Closed Pump-controlled Three-chamber Hydraulic Cylinder. *Trans. Chin. Soc. Agric. Mach.* **2019**, *50*, 410–418.
68. Yin, X.Y. Characteristic comparing between pump-control and valve-control electricity-hydraulic servo system and their rational select principle during the design of mining-hydraulic equipments. *Coal Mine Mach.* **2002**, *11*, 37–39.
69. Ning, C.X.; Zhang, X.S. Study of the Hydraulic Synchronous Circuit and Synchronous Control of the Hydraulic Hoist. *Appl. Mech. Mater.* **2013**, *2212*, 275–277. [CrossRef]
70. Teng, L.; Wang, S.; Wang, X.B. Design of High Pressure Water Jet Decontamination Device with Pressure and Flow Synchronous Control Function. *Nucl. Power Eng.* **2020**, *41*, 153–157.
71. Hui, S. The Mechanism and Application Research of Electro-Hydraulic Proportional Synchronous Control Based on Graph Theory. Master's Thesis, Kunming University of Technology, Kunming, China, 2015.
72. Yang, W.B.; Hu, J.K.; Wang, Z.P. Dynamic characteristics simulation of two-stage bi direction hydraulic synchronization control system. *J. Zhejiang Univ.* **2014**, *48*, 1107–1113.
73. Zhang, S.Y. Research on Two-Cylinder Synchronous Control System Based on Two-Dimensional (2D) Servo Valve. Master's Thesis, Zhejiang University of Technology, Hangzhou, China, 2020.

74. Chen, G.; Zhu, S.S.; Wang, Q.X.; Wu, H.H. Development and Application of Electro-hydraulic Control Technology. *Mach. Tool Hydraul.* **2006**, *4*, 1–3+11.
75. Li, X.C. Research on Synchronous Control of Multiple Electro-Hydraulic Servo Actuators. Master's Thesis, University of Electronic Science and Technology of China, Chengdu, China, 2020.
76. Prabhakar, k.; Dasgupta, k.; Ghoshal, S.K. A comparative analysis of the pump controlled, valve controlled, and prime mover controlled hydromotor drive to attain constant speed for varying load. *ISA Trans.* **2022**, *120*, 305–317.
77. Zhao, P.Z.; Wang, Y.J.; Yao, K. Co-simulation on Motors Synchronous Control of Hydrostatic Driving System for Wheel Amphibious Vehicle Based on AMESim/Simulink. *Hydraul. Pneum. Seals* **2014**, *34*, 31–34.
78. Dong, Z.X.; Huang, W.N.; Ge, L.; Quan, L.; Huang, J.H.; Yang, J. Research on the Performance of Hydraulic Excavator with Pump and Valve Combined Separate Meter in and Meter Out Circuits. *J. Mech. Eng.* **2016**, *52*, 173–180. [CrossRef]
79. Cao, Y. Analysis and Improvement for Synchronous Motor Control System of Multi-cylinder Synchronization Hydraulic System. *Mach. Tool Hydraul.* **2017**, *45*, 99–101.
80. Li, C. Application of Hydraulic Synchronous Control Circuit in Hydraulic Control of Coal Mine Machinery. *Intern. Combust. Eng. Parts* **2021**, *327*, 90–91.
81. Sui, G.D. Research on a New Type of Hydraulic Synchronous Multi-Motor. Master's Thesis, Yanshan University, Qinhuangdao, China, 2019.
82. Zhang, S.L. Research of Self-Propelled Corn Harvest and Hydraulic Control System. Master's Thesis, Hebei University of Science and Technology, Shijiazhuang, China, 2016.
83. Liu, A.L.; Li, S.L.; Yu, H.L. Research on Synchronous Control of Two-cylinder Hydraulic System Based on Fuzzy Theory. *Chin. Hydraul. Pneum.* **2016**, *296*, 44–47.
84. Chen, C.Y.; Liu, L.Q.; Cheng, C.C.; George, T.C. Fuzzy controller design for synchronous motion in a dual-cylinder electro-hydraulic system. *Control Eng. Pract.* **2015**, *16*, 658–673. [CrossRef]
85. Li, B.R.; Li, F.; Shi, B.K.; Deng, H.F. Research on cross-coupled synchronization fuzzy control of double valve. In Proceedings of the 2015 International Conference on Fluid Power and Mechatronics (FPM), Harbin, China, 5–7 August 2015.
86. Jia, S.B. Research on Four-Cylinder Synchronous Control System of Multi-Stage Hydraulic Cylinder. Master's Thesis, Taiyuan University of Science and Technology, Taiyuan, China, 2012.
87. Wu, N.; Yuan, M.W. Modeling and simulation of double cylinder hydraulic synchronous control system for forging press. *Forg. Stamp. Technol.* **2020**, *45*, 144–150.
88. Shi, L.; Wang, M.; Liu, J.Y.; Sun, T.W.; Wang, S.H. Research on motion control of EDM spindle based on feedforward fuzzy PID. *Mod. Manuf. Eng.* **2021**, *495*, 7–12+119.
89. Wei, L.J.; Gu, Q.Q.; Xin, Y.L.; Lu, Y.F. Design on Hydraulic Control System for Multi-cylinder Synchronous Lifting of Jack-up Formwork System. *Mach. Tool Hydraul.* **2021**, *49*, 65–69.
90. Li, H.L.; Pang, B. Research on Synchronization Control System of Double Hydraulic Cylinder Four-column Hydraulic Press. *Mach. Tool Hydraul.* **2019**, *47*, 82–85+91.
91. Huang, Z.H.; Wang, G.R.; Liu, F.X. Synchronization Design of Multi-channel Hydraulic Servo System. *Chin. Hydraul. Pneum.* **2012**, *255*, 29–31.
92. Yang, Y.; Jin, X. Simulation analysis of cooperative motion fuzzy control of distributed lifting units. In Proceedings of the 2021 International Conference on Energy, Power and Electrical Engineering (EPEE2021), Jilin, China, 24–26 September 2021.
93. Jiao, D.Q.; Wang, D.H. Research on Double Hanging Point Hydraulic Synchronous Control System Based on Fuzzy PID. In Proceedings of the 2021 2nd International Conference on Artificial Intelligence and Information Systems (ICAIIS 2021), Chongqing, China, 28–30 May 2021.
94. Liu, Z.; Chen, J.; Zhang, K. New Hydraulic Synchronization System Based on Fuzzy PID Control Strategy. *Adv. Mater. Res.* **2013**, *2717*, 229–233. [CrossRef]
95. Mu, T.C.; Zhang, D.M.; Zhou, W.M. Research on control system of synchronous flexible stamping device based on fuzzy PID algorithm. *Manuf. Technol. Mach. Tool* **2022**, *715*, 94–97.
96. Zhou, P.Q. Research on Synchronous Control Technology for Double Hoist Based on Fuzzy Neural Network. *Mach. Des. Manuf.* **2016**, *307*, 64–68.
97. Mu, T.C.; Zhou, Q.N.; Zhang, D.M.; Lv, L.L. Study on Synchronization System of Tilting Pouring Machine Based on Fuzzy PID Control. In Proceedings of the 2021 6th International Conference on Intelligent Informatics and Biomedical Sciences (ICIIBMS), Oita, Japan, 25–27 November 2021.
98. Wang, L.X.; Zhao, D.X.; Liu, F.C.; Liu, Q.; Meng, F.L. Linear active disturbance rejection control for electro-hydraulic proportional position synchronous. *Control Theory Appl.* **2018**, *35*, 1618–1625.
99. Wang, C.W.; Quan, L.; Zhang, S.J.; Meng, H.J.; Lan, Y. Reduced-order model based active disturbance rejection control of hydraulic servo system with singular value perturbation theory. *ISA Trans.* **2017**, *67*, 455–465. [CrossRef]
100. Ren, X.; Zhao, Y.Y. Electro-hydraulic position synchronization control system based on auto-disturbances rejection and feedback. In Proceedings of the World Automation Congress 2012, Puerto Vallarta, Mexico, 24–28 June 2012.
101. Lu, Z.S.; Cai, W.; Zhao, J.Y.; Qin, Y.L.; Ren, W.B.; Li, Z.B. Active Disturbance Rejection Control for Adjacent Cross Coupling Synchronization of Hydraulic System Group. *Chin. Hydraul. Pneum.* **2020**, *348*, 29–34.

102. Zhang, B.Z. Analysis and Experimental Study on Vehicle Stability Control of a New Four-Wheel Steering. Ph.D. Thesis, Hunan University, Changsha, China, 2016.
103. Qin, T.; Quan, L.; Li, Y.H.; Ge, L. A Dual-Power Coordinated Control for Swing System of Hydraulic-Electric Hybrid Excavator. In Proceedings of the 2021 IEEE/ASME International Conference on Advanced Intelligent Mechatronics (AIM), Delft, The Netherlands, 12–16 July 2021.
104. Guo, Y.N.; Zhang, Z.; Liu, Q.Y.; Nie, Z.; Gong, D.W. Decoupling-based adaptive sliding-mode synchro-position control for a dual-cylinder driven hydraulic support with different pipelines. *ISA Trans.* **2022**, *123*, 357–371. [CrossRef] [PubMed]
105. Zhang, Y.; Zhao, J.H. Synchronizing Control Characteristics Analysis of Hydraulic Support Test-bed Based on Adaptive Sliding Mode Control. *Mach. Tool Hydraul.* **2021**, *49*, 176–179.
106. Hu, T.; Shen, L.Q.; Fu, J.; Fan, T.X. Robust control and electromechanical-hydraulic joint simulation of spacecraft servo mechanism. *Syst. Eng. Electron.* **2023**, *45*, 1–10.
107. Zhou, Y.C.; Liu, S.J.; Huang, M.H.; Deng, Y.; Duan, J. A Synchronous Drive Control System for a Huge Scale Press Based on a Robust Regulator. *Mech. Sci. Technol. Aerosp. Eng.* **2011**, *30*, 501–506.
108. Liu, Z.W.; Huang, M.H.; Liu, X.L. Research on robust control of synchronous control system for giant die forging hydraulic press. *Forg. Stamp. Technol.* **2011**, *36*, 88–93.
109. He, C.Y.; Shi, G.L.; Guo, Q.Y.; Wang, D.M. Adaptive Robust Control Strategy for Valve Controlled Asymmetric Hydraulic Cylinder Position Control System. *J. Shanghai Jiaotong Univ.* **2019**, *53*, 209–216.
110. Lyu, L.T.; Chen, Z.; Yao, B. Energy saving motion control of independent metering valves and pump combined hydraulic system. *IEEE/ASME Trans. Mech.* **2019**, *24*, 1909–1920. [CrossRef]
111. Dou, H.B. Adaptive Tracking Control for Synchronization Motion of Lift System with Two Asymmetric Hydraulic Cylinders. *Chin. Hydraul. Pneum.* **2017**, *316*, 82–89.
112. Liu, H.S.; Shen, W. Research on Adaptive Robust Control of Electro-hydraulic Composite Cylinder Under Complex Load Conditions. In Proceedings of the 2020 Chinese Control and Decision Conference (CCDC), Hefei, China, 22–24 August 2020.
113. Li, S.J.; Wang, W. Adaptive robust H∞ control for double support balance systems. *Inf. Sci.* **2020**, *513*, 565–580. [CrossRef]
114. Zhang, Y. Research on the Position Control of Loader Working Device Based on Machine Learning. Master's Thesis, Jilin University, Changchun, China, 2020.
115. Zuo, H.; Wang, Y.L.; Jin, R.; Huang, H.G. Energy consumption modeling method of pump unti based on machine learning. *J. Hefei Univ. Technol.* **2022**, *45*, 582–588+619.
116. Fu, W.J.; Wang, L.; Yin, J.; Wu, L.; Li, M.J.; Huang, X.; Liang, P.S.; Xiong, S. Research on position control technology of boom of pumptruck based on fuzzy PID of machine learning and feedforward compensation. *Chin. J. Constr. Mach.* **2022**, *20*, 140–145.
117. Wei, Z.P. Research on Vision Based Compensation Control of Hot Continuous Rolling Process. Master's Thesis, Inner Mongolia University of Science and Technology, Baotou, China, 2022.
118. Xiao, Z. Research on Synchronization Control System of Giant Forging Hydraulic Press Based on BP Neural Network. Master's Thesis, Central South University, Changsha, China, 2012.
119. Li, B.; Zhang, Y.D.; Yuan, L.P.; Zhu, G.Q.; Dai, X.S.; Su, W.H. Predictive control of Plantar Force and Motion Stability of Hydraulic Quadruped Robot. *China Mech. Eng.* **2021**, *32*, 523–532.
120. Liu, F.H.; Zhang, J.J.; Li, X.; Xiang, W.B. Hydraulic Synchronization Control Strategy of Pod Thruster Installation Platform. *Ship Eng.* **2021**, *43*, 105–110.
121. LI, S.Z.; Li, P.D.; Li, C. Double-Cylinder Hydraulic Press Synchronous PID Control Based on Immune Neutral Network Algorithm. *Mech. Eng.* **2019**, *332*, 139–142.
122. Liu, Y. Study on the Synchronization Method of Double-Cylinder Forging Hydraulic Press Based on Immune Neural Network PID Algorithm. Master's Thesis, Hefei Polytechnic University, Hefei, China, 2013.
123. Xu, X.D.; Bai, Z.F.; Shao, Y.Y. Synchronization Control Algorithm of Double-Cylinder Forging Hydraulic Press Based on Fuzzy Neural Network. *Algorithms* **2019**, *12*, 63. [CrossRef]
124. Dong, L.J.; Qiu, M.F.; Nguang, S.K. Design and Advanced Control of Intelligent Large-Scale Hydraulic Synchronization Lifting Systems. *J. Control Sci. Eng.* **2019**, *8*, 1–10. [CrossRef]
125. Lu, F.P.; Xue, W.; Zhu, X.H. Research on Networked Hydraulic Synchronous Control System. *J. Phys.* **2020**, *1621*, 1–11. [CrossRef]
126. Zhai, S.; Yu, Z.; Jin, B. Research Status and Development Trend of Hydraulic Control System for Multi-legged Walking Robot. *Robot* **2018**, *40*, 958–968.
127. Liu, Z.W. Research on Synchronous Control System for Giant Forging Hydraulic Press and Its Robust Control. Ph.D. Thesis, Central South University, Changsha, China, 2011.
128. Liu, H.J.; Zhi, S.Y. High Precision Synchronous Mechanical Motion Hydraulic Compensation Method. *Appl. Mech. Mater.* **2015**, *3744*, 810–813. [CrossRef]
129. Shi, H.; Mei, X.S. New Features and Development Prospect of Modern Hydraulics Drive Techniques. *Mach. Tool Hydraul.* **2017**, *45*, 158–166.
130. Huang, H. Synchronization Control Strategy of Electro-Hydraulic Multiple Cylinder Position Servo System. Master's Thesis, Jiangsu University, Zhenjiang, China, 2022.

131. Sun, Y.L. Research on the Synchronization and Fault-Tolerance Control Strategy of Hydraulic Press Based on Multi-Objective Optimization. Master's Thesis, Tianjin University of Technology, Tianjin, China, 2022.
132. Zou, H.T. Research on Control Performance of Hydraulically-Assisted Universal Vehicle Based on Electro-Hydraulic Servo Control. Master's Thesis, Shenyang University of Technology, Shenyang, China, 2022.
133. Ye, H.; Ni, X.; Chen, H.; Li, D.; Pan, W. Constant Speed Control of Hydraulic Travel System Based on Neural Network Algorithm. *Processes* **2022**, *10*, 944. [CrossRef]

 *processes* 

Review

A Review of Automobile Brake-by-Wire Control Technology

Xuehui Hua [1], Jinbin Zeng [2], Haoxin Li [2], Jingkai Huang [2], Maolin Luo [2], Xiaoming Feng [3], Huiyuan Xiong [4] and Weibin Wu [2,*]

[1]   College of Automotive Engineering, Foshan Polytechnic, Foshan 528100, China
[2]   College of Engineering, South China Agricultural University, Guangzhou 510642, China
[3]   School of Mechanical Automotive Engineering, South China University of Technology, Guangzhou 510642, China
[4]   School of Intelligent Systems Engineering, Sun Yat-sen University, Shenzhen 518107, China
*   Correspondence: wuweibin@scau.edu.cn

Abstract: Brake-by-wire (BBW) technology is crucial in driverless cars. The BBW technology, which has a faster reaction time and greater stability, can improve passenger safety in driverless cars. BBW technology refers to the removal of some complicated mechanical and hydraulic components from the traditional braking system in favor of using wires to transmit braking signals, which improves braking performance. Firstly, this paper summarized BBW's development history as well as its structure, classification, and operating principles. Subsequently, various control strategies of the BBW system were analyzed, and the development trend and research status of the motor brake-control strategy and wheel-cylinder pressure-control strategy in the braking force-distribution strategy were analyzed respectively, and the brake fault-tolerance technology and regenerative-braking technology were also analyzed and summarized. Finally, this paper summarized the various technologies of BBW, taking the electromechanical brake (EMB) in the braking system as an example to discuss the current challenges and the way forward.

Keywords: brake-by-wire technology; motor brake-control strategy; wheel cylinder pressure-control strategy; brake fault-tolerance technology; regenerative-braking technology

Citation: Hua, X.; Zeng, J.; Li, H.; Huang, J.; Luo, M.; Feng, X.; Xiong, H.; Wu, W. A Review of Automobile Brake-by-Wire Control Technology. *Processes* **2023**, *11*, 994. https://doi.org/10.3390/pr11040994

Academic Editors: Francisco Ronay López-Estrada and Guillermo Valencia-Palomo

Received: 22 February 2023
Revised: 19 March 2023
Accepted: 21 March 2023
Published: 24 March 2023

1. Introduction

More effective and energy-saving drive-by-wire technology has evolved with the advent of intelligent and networked automobiles [1,2]. Drive-by-wire technology was initially utilized in the aerospace industry as a crucial component of the braking system. The quick development of intelligent networked vehicles has drawn a lot of attention. For intelligent networked vehicles, the quick advancement of wire-control technology offers not only a reliable control basis but also guidance for developing unmanned vehicles [3].

The most significant and technically complex aspect of chassis technology is the brake-by-wire (BBW) system [4]. The engine, in the case of an electric vehicle, was eliminated, therefore it is unable to supply a vacuum source to the vacuum booster for engine-based braking assistance [5]. As a result, an improvement to the braking system is inevitable. Moreover, more cooperative, intelligent, and responsive motion actuators are needed for the autonomous driving of automobiles. To facilitate braking and effectively recover braking energy to extend the range of electric vehicles, brake-by-wire technology regulates the motor. At the same time, it can execute intelligent driving commands with pinpoint accuracy. The brake-by-wire system will undoubtedly be the best option for braking system improvement.

The brake-by-wire system, as an important safeguard mechanism in automobile active safety, has the advantages of environmental protection, accurate pressure regulation, and fast response time. The BBW systems can readily combine the anti-lock braking system (ABS), electronic stability controller system (ESC), and regenerative braking system (RBS) through the implementation of synergistic control techniques. Some benefits of these pairings include enhanced brake stability and energy recovery [6,7].

At the time when the BBW system had more mature technology and loading cases, some famous auto parts manufacturers in developed countries, such as Bosch, Siemens, Continental Teves in Germany, Delphi, TRW in the United States, Hyundai, Mobis, Mando in South Korea, Haldex and SKF in Sweden, had carried out research on EMB and developed their own products. Mobis and Mando in Korea, and Haldex and SKF in Sweden, had all conducted EMB research and developed their own products. A number of Chinese universities and automobile companies, including Jilin University, Tongji University, Tsinghua University, Beijing University of Technology, and Geely Automobile, developed preliminary programs [8–11].

2. Development History

Several types of braking systems have evolved since Wilhelm Maybach invented the drum brake in 1900. Their evolution can be loosely divided into the four stages listed below.

The first stage is ABS, anti-lock braking system. To achieve the best braking effect, the system may automatically alter the wheel brake force while braking. The wheels can prevent skidding during braking by regulating and managing the brake line pressure. To optimize braking performance, keep the wheels at a 15–25% slip rate of motion while rolling [12]. Proportion Integral Derivative (PID) control [13], neural network control [14], fuzzy control [15], fuzzy PID control [16], and logic threshold control [17–20] are the primary control systems utilized for automobile ABS. ABS is standard equipment in almost every car in the globe. The majority of them are based on Bosch's logic threshold ABS control system. The logical threshold control method is straightforward, easy to apply, and extremely adaptable. It maintains the slip rate and angular acceleration in the best range and is mature for the application. Today, improving control techniques is the main focus of ABS optimization. Li made improvements to the automotive ABS system and put out a plan to use the peak wheel speed linkage to address the vehicle speed and standard slip rate. To solve the control problem of an anti-lock braking system (ABS) in an automobile under many challenging driving circumstances. He created a hierarchical intelligent control system that includes organizational coordination, parameter correction, and human control [21,22].

The second stage is ESC/ESP. The electronic stability controller system (ESC) is a new type of vehicle active safety system. It is an extension of the vehicle's anti-lock braking system (ABS) and traction control system (TCS). On this basis, a transverse swing rate (yaw rate) sensor, a lateral acceleration sensor, and a steering wheel angle sensor were added to the vehicle steering when driving. Through ECU control, ensure the vehicle's stability and safety while driving, braking, and steering. The ECU electronic control ensures the vehicle's stability and safety while driving, braking, and steering. The vehicle dynamic stability control system, as shown in Figure 1, provides internal feedback control via ESP. The performance of the wheel will be unstable and non-linear at the limit of adhesion between the tire and the road. In this case, the vehicle dynamics control system will assist the car in maintaining control [23–25]. The sensors provide the ECU with information on the vehicle's current status [26]. Two new pressure limiting valves and two suction valves were added to the ABS's eight solenoid valves, totaling 12 solenoid valves capable of switching the system's active boost circuit and return circuit. To ensure the stability of the vehicle driving, solenoid valves, plunger pumps, and other key components have different parameters of the series of products, which can be diversified with the trial of different models.

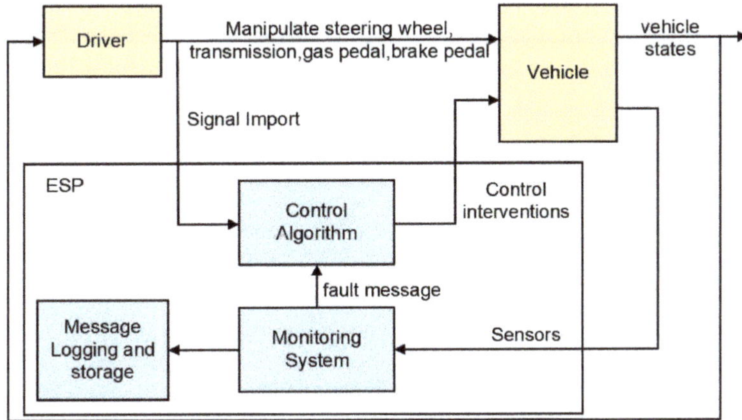

Figure 1. Components of vehicle dynamic stability control systems.

The third stage is IPB+RBU. For L3 and L4 of autopilot, Booster proposed a brake-by-wire technology, which is IPB+RBU. Integrated Power Brake (IPB) integrates brake vacuum pump, vacuum brake booster, ESP and other important actuators. In the IPB, the highly integrated power brake is complemented by a redundant brake unit. This has ensured that even in the event of an integrated power brake failure, the vehicle can stop safely and reliably without driver intervention with the following advantages:

(a) Low weight and simple arrangement;
(b) Low costs. There are no vacuum pumps and hydraulic pipes, thus reducing production costs;
(c) Short braking distances for automatic emergency braking systems thanks to the highest-pressure build-up dynamics;
(d) Long ranges. Powered by an efficient brake regeneration system, it can greatly improve the range of hybrid and electric vehicles.

By analyzing the requirements of the braking system, it is necessary to consider the design of redundant braking. In addition to the primary brake control unit, a secondary brake control unit is required, and the system should also be equipped with extended functions such as system status detection, redundancy control, and back-up vehicle stability control. Booster's solution is to use the IPB as the primary braking system to perform the braking request in most cases. In the case of IPB failure, RBU (redundant brake unit) can be used as the redundant brake.

The fourth stage is EMB. It is electromechanical and can be easily integrated with other electronic control systems of the vehicle to realize more functions, such as braking, ABS, EBS, ESP, automatic driving, and optimized energy recovery. Based on all these advantages, EMB technology is bound to develop vigorously and present serious challenges to the hydraulic braking system in the future.

EMB is a new concept of braking systems. At the stage of EMB, load, cost, reliability and other constraints have blocked the industrialization. However, as a parking brake it is used in industry. However, thanks to the advantages of fast response, energy saving and environmental protection, it has become one of the development directions of the future vehicle braking system.

3. Structure of Brake-by-Wire System

Electronic components had largely taken the role of mechanical and hydraulic components in the brake-by-wire system. The brake pedal is no longer mechanically connected to the braking wheel cylinders. Electronic brake pedal sensors monitor the driver's braking activity and translate it into electrical signals for the electronic control unit (ECU) to analyze. Then the ECU sends the proper instructions to the electric drive units, including the high-pressure accumulator and the motor. In order to produce braking power, the motor or high-pressure brake wheel cylinder eventually presses the push rod.

The brake-by-wire system can be classified into two categories: hydraulic and mechanical, depending on the various actuators. The hydraulically controlled actuation system, which includes the electro-hydraulic brake system (EHB), is structurally similar to the conventional hydraulic brake system. It only kept the original hydraulic brake system and swapped out the vacuum assist for the motor assist [27,28]. But, as an intermediate of the brake system with wire control. Major automakers continue to favor EHB due to its outstanding compatibility and simplicity of retrofitting to the original brake system for adoption. The electromechanical brake is a braking system that employs an electric signal to regulate the mechanical structure, eliminating the transmission mechanism such as hydraulic pressure, and using a motor as the power source. Being a mechanically driven system, it is more environmentally friendly and sensitive, and eliminates the brake fluid delay effect. It can also actively implement the longitudinal dynamics integration of EMB and other control systems via algorithmic advancements to achieve a better braking effect. As intelligent car technology evolves, EHB will be replaced by EMB, which is more in line with the needs of future braking systems Because of its distinct features, the EMB system is currently attracting considerable interest from Chinese and global automotive manufacturers. It will become a research center for future automobile braking.

3.1. Electro-Hydraulic Brake System (EHB)

As a transitional product between the traditional brake system and the brake-by-wire system, the operating mechanism of the EHB has replaced the traditional hydraulic brake pedal with an electronic brake pedal and used a hydraulic device instead of a bulky vacuum booster. The hydraulic control unit in the EHB automatically adjusts the braking pressure of the wheels according to different driving conditions to provide sufficient braking force. EHB has eliminated the coupling between the brake pedal and the driver without a vacuum booster, with advantages of compact structure, convenient and reliable control, low braking noise, and good braking comfort. The composition of EHB system can be seen in Figure 2 [29,30]. EHB can also control the braking force of each wheel individually to easily maximize the braking energy recovery. Therefore, EHB has special application value for electric vehicles. Dependent on the hydraulic drive unit, EHB can be classified as a high-pressure accumulator type or an electric pump type. Table 1 displays the hydraulic control system for the electro-hydraulic braking system.

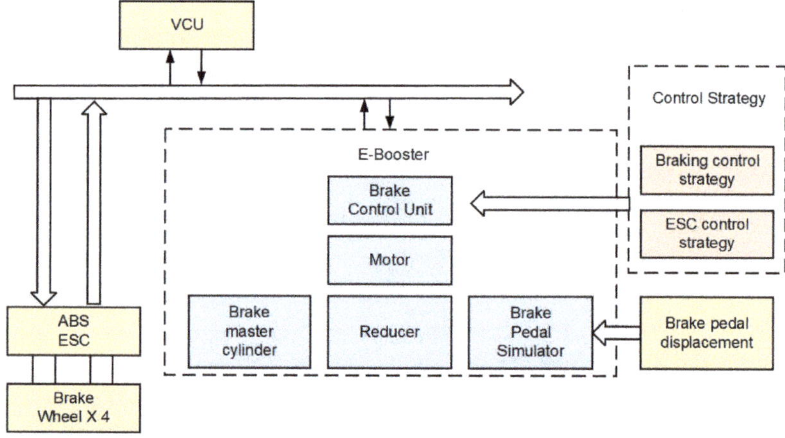

Figure 2. The composition of EHB system.

Table 1. Hydraulic pressure-control scheme of master cylinder of electro-hydraulic braking system.

Hydraulic Drive Unit Form	Control Variable	Control Algorithm	Reference
Motor and Electric hydraulic pump	Main cylinder hydraulic pressure	Feedback control	[31–33]
		PID control	[34]
		Taguchi method and Segmentation PI control	[34]
		Gain scheduling PI control	[35]
		Feedforward and feedback control	[36]
		Friction compensation and PID control	[37]
		Friction compensation and adaptive robust control	[37]
		Friction compensation and sliding mode variable structure control	[38]
	Main cylinder piston push rod displacement	PID control	[39,40]
		Sliding mode variable structure control	[40]
		Prediction based on hydraulic model	[39]
	Motor rotation angle	Feedback control	[41]
	Main cylinder hydraulic pressure and main cylinder piston push rod displacement	Switching control	[42]
		Cascade control	[43]
		Cascade anti-winding control	[44]
	Main cylinder hydraulic pressure and motor current	Cascade control	[45]
Hydraulic pump and high pressure accumulator	Main cylinder hydraulic pressure	Switch control	[46]
		Pulse width modulation control	[41]
		PID control	[47]
		Sliding mode variable structure control	[48]

3.1.1. High Pressure Accumulator Type

Among them, the hydraulic brake system of high-pressure accumulator type provides the master cylinder hydraulic or wheel cylinder braking force through the high-pressure accumulator high-pressure energy. Thus, the dynamic regulation of braking force is achieved. After obtaining the driver's intention through the brake analog pedal, ECU sends a command to the vehicle controller. The high pressure accumulator, solenoid valve and pump are controlled to produce appropriate hydraulic pressure. When pressure in the high-pressure accumulator is insufficient, the hydraulic pump will pressurize the high-pressure accumulator.

In the 1990s, when braking systems were still in their infancy, a high-pressure accumulator kind of hydraulic braking system was devised to overcome the problem of the brake pump's delayed response and low flow rate. In which high pressure is generated by an electric motor plunger pump and stored in advance in a high-pressure accumulator. The high-pressure accumulator received braking fluid at a faster pace, which shortened the brake's response time. When the brake was activated, the high pressure was swiftly released [49]. In 1994, the first high-pressure accumulator brake system was created. It's created using analogy and Saber modeling. Toyota then introduced the first direct-drive wheel cylinder actuator-equipped EHB into production. It has two linear solenoid valves installed in the hydraulic adjustment unit, allowing for accurate control of the pressure in each wheel cylinder. In the system failure backup braking circuit, the front and rear chambers of the brake master cylinder were connected to each of the 4-wheel cylinders via switching solenoid valves. Also, the hydraulic adjustment unit's high-pressure accumulator, hydraulic pump, and motor were all detached separately. To implement the hydraulic power-assist function, the high-pressure accumulator outlet was connected to the brake master cylinder [50]. Bosch improved the design of the Sensotronic Brake Control (SBC), which was based on Toyota and shared many structural and functional similarities with the EHB. With the evident exception that the hydraulic adjustment unit of SBC was outfitted with two separate pistons for the front axle brake circuit. In the event of a system failure, the SBC system can only brake on the front axle wheels. The high-pressure nitrogen leakage from the high-pressure accumulator can be efficiently eliminated by the separating pistons on the front axle brake circuit [8]. The Mercedes R230 SL received the design for the first time in 2001. The high-pressure accumulator received high pressure from the electric motor, which was used to immediately supply braking pressure to the wheel cylinders [51]. They were made by Bosch, Siemens, Continental Teves, and Toyota [52–54]. These pioneering producers and researchers gave EHB a direction for future study. For the earlier concepts, Li proposed an electro-hydraulic braking system and its braking control method using a high-pressure accumulator. The high-pressure accumulator enabled the brake system to be realized with quick and precise pressure control and provided the driver with good pedal feedback.

3.1.2. Electric Pump Type

In the 1990s, Bosch designed and built the first EHB system based on the more mature ESP system and conducted real-world tests with good results [55]. The electric pump hydraulic braking system is driven by an actuator that directly drives the master cylinder for braking. The design retains the traditional brake anti-lock system ABS (Anti-lock Brake System), and also has the advantages of high control reliability and small changes to the traditional structure, which is now one of the mainstream development trends of the brake-by-wire system. Sun proposed a pump-controlled direct-drive the brake-by-wire unit, which uses pump-controlled direct-drive volumetric servo technology, and directly drives the bidirectional gear pump, to achieve control and rapid adjustment of the brake wheel cylinder pressure, eliminate the throttling loss of the valve control system, and effectively improve the system efficiency. The rotary motor output shaft directly drives the hydraulic pump scheme, and the hydraulic pump forward outlet port is connected to the brake wheel cylinder, that is, the hydraulic pump directly drives the wheel cylinder piston, and the forward inlet port is connected to the low-pressure accumulator to realize the function of controlling the wheel cylinder pressure [56].

Gu presented a passenger automobile brake-by-wire technology. A primary pressure supply unit was directly powered by an electric motor, and an auxiliary pressure supply unit with a high-pressure accumulator was also part of the system. The synergistic action of the two units results in accurate pressure regulation [57]. Li suggested a direct-actuated valve-based quick reaction braking system. It reduced the number of solenoid valves that needed to be configured by driving the valve spool directly with a solenoid linear actuator. And the sliding film variable structure control and adaptive robust control [58] swiftly modify the oil pressure of the brake wheel cylinder; Gong created a novel electromagnetic linear actuator-based braking unit for the all-electric driving characteristics of electric cars. By driving the electro-hydraulic brake unit directly to improve braking performance [59]. As a result, some manufacturers created electronic wedge brake (EWB). This is a self-excitation function linear actuation system [60]. Jo researched EWB modeling and control. A new single-motor EWB has been created. Controlling electronic actuators and self-exciting wedge mechanisms provided braking power. Because EWB did not require vacuum boosters or master cylinders, they were straightforward to configure in the cabin [61]. Streli and Balogh created a mechanical model that takes frictional factors into account and an EWB controller. A sliding mode controller was developed based on a dynamic equation model of the electronic wedge brake. Based on a simplified electronic wedge brake model, Kwangjin Han accomplished clamping force prediction and provided a contact point identification technique. Simulation and experimentation with a prototype EHB were used to confirm this controller's effectiveness [60]. The findings demonstrate that, in comparison to the pre-optimized EHB system, the optimized EHB system can provide the same vehicle stabilization with less boost torque during braking [62].

3.2. Electromechanical Braking System (EMB)

EMB stands for electronically managed mechanical braking. It entirely gets rid of the hydraulic system. It offers features like an efficient structure and environmental protection, among others. Due to the absence of the braking fluid's delay effect, it is also incredibly sensitive. The management of braking pressure is more exact with active braking function. EMB can simultaneously disconnect the brake pedal force and guarantee pedal feeling in the vehicle. To increase the rate of energy recovery when used in conjunction with the higher controller without altering the system's organizational structure. Simple adjustments to the control algorithm level can integrate EMB with the longitudinal dynamics of other systems. Decoupled composite braking systems can be implemented with wire-controlled actuation. EHB and EMB are two types of linear control technology currently in use. On the basis of Conventional Hydraulic Brake (CHB), EHB adds brake pedal sensor, pedal stroke simulator, master cylinder pressure sensor, pressure regulator, and pressure controller, and each wheel's braking force can be independently controlled. The EHB is based on the hydraulic circuit of the conventional automotive braking system and has a number of inherent flaws, including numerous hydraulic lines, a sizable vacuum booster, a challenging layout and assembly, a lengthy response time to braking, a high level of pedal vibration, and high manufacturing and maintenance costs. Instead of a hydraulic circuit, EMB is an electromechanical system composed primarily of pedal simulators, EMB actuators, and controllers that control the clamping force of the actuator to achieve independent control of each wheel braking force. EMB has numerous advantages over EHB, as shown in Table 2 [6,63].

Table 2. Performance comparison of various brake.

Type	Advantages	Disadvantages
CHB	(1) Large brake power (2) Proven technology	(1) Non-continuous control of braking force (2) Brake pedal jitter (3) Long braking force response time (4) Requiring complex electronics (5) Complex hydraulic circuit (6) Pollution of the environment
EHB	(1) High power (2) Braking power four-wheel independent control (3) Can be coupled with regenerative braking	(1) Non-continuous control of braking force (2) Brake pedal jitter (3) Long braking force response time (4) Requiring complex electronics (5) Pollution of the environment
EMB	(1) Braking power four-wheel independent control (2) No pedal jitter (3) Fast response of braking force control (4) Facilitates coupling with other active safety control systems (5) Light weight and small size (6) Avoids brake drag	(1) Need for redundant design (2) Requiring a power supply system of 42 V or more

The EMB is made up of a brake actuator on the brake wheel (which includes a torque motor, gear reducer, differential mechanism, motion converter, and brake caliper), and a central controller, the structure of which is represented in Figure 3. The motor serves as the primary driving source, with the reducer and torsion mechanism amplifying the spinning torque. The rotating motion of the motor is converted into the linear motion of the brake caliper by the motion conversion mechanism (ball screw structure or bevel gear structure), which then acts on the brake disc in the blocking condition.

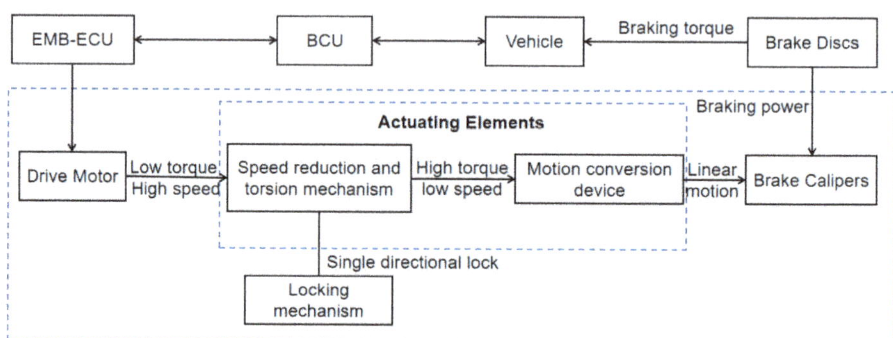

Figure 3. Electromechanical brake system structure composition.

As shown in Figure 4, The EMB actuator is designed using the vehicle's braking performance requirements and installation space limitations as input, the actuator's size limitation, the target maximum braking clamping force and its response time, the braking gap and its gap elimination time, and other performance objectives [64]. The performance criteria for electrical components, mechanical transmission components, and sensors are then obtained individually. Finally, the performance parameters of the specific motor type, operating range, rotational inertia, mass, operating voltage, and operating current, etc., are added to the design of the motor and its drive controller, the design of transmission parts, and the selection of sensors, and the detailed design of the actuator is realized.

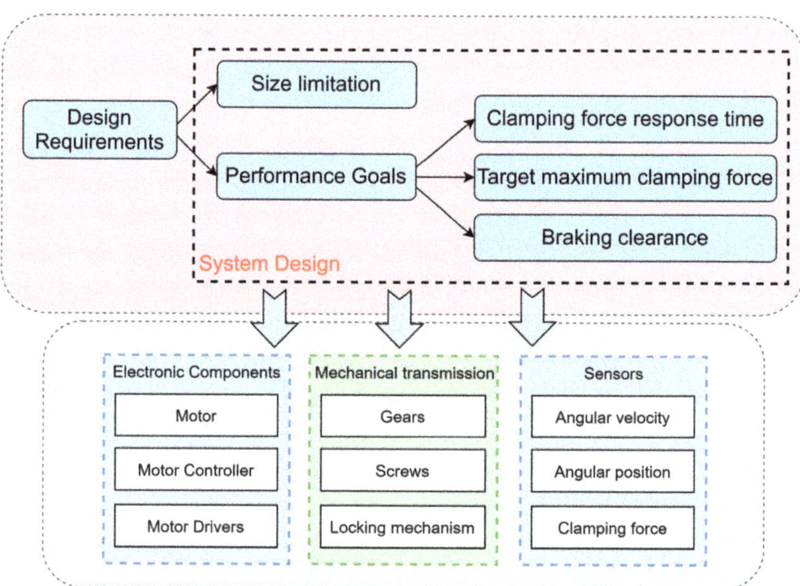

Figure 4. The flowchart of EMB actuator design.

Electromechanical brake systems were first applied in aircraft, such as the U.S. F-15 fighter jets. A new generation of aircraft will use brake-by-wire technology, according to a 2004 statement from Boeing. Numerous automotive companies and research institutions have become interested in EMB as a result of its successful use in the aviation industry. And it gradually moved on to the field of automotive braking [65,66].

For example, Bosch, Siemens, and Continental Teves in Germany, Delphi and TRW in the United States, Hyundai Mobis and Mando in Korea, Inc., have all demonstrated a great interest in researching EMB actuators since the 1990s. In Sweden, Haldex, SKF had investigated EMB, created their own EMB actuators, and conducted a number of studies on them [67,68].

The EMB designed by Bosch adopts the structure of external motor rotor [67,68]. The motor drove the internal planetary wheel system, and then the rotational motion exported from the planetary wheel system was converted into linear motion through the bevel gear structure. The gears would push the friction blocks to compress the brake discs to achieve the deceleration effect. This structure is more compact and more complex. The planetary gear reducer with a ball screw mechanism was chosen as the answer by the majority of the businesses represented by Continental Teves. The parking brake gear construction was likewise fixedly attached to the rotor. The solenoid's magnetism was dissipated while braking. To fix the rotor, the pawl made contact with the gear. The braking force already applied before halting was maintained in order to achieve the parking brake function, locking the entire transmission mechanism.

Nowadays, EMB research has been conducted in China, with universities serving as the primary research subjects. Early Chinese research on the EMB system dates back to a 2005 patent application by Tsinghua University. Tsinghua University's Zhao examined and compiled data on the structural designs of EMB actuators used both domestically and overseas. A structural plan with a torque motor, planetary gear reduction, and ball screw was his suggestion. The EMB controller without a pressure sensor was created by Han from Jilin University, who also finished the off-line simulation and debugging. Li from Chongqing University finished designing the braking control algorithm for the EMB technology after analyzing the response characteristics of the actuator. The stiffness of the brake caliper and ball screw, which impact the stability of the EMB, were looked into modeling studies by Yu of Shenyang University of Technology and colleagues, who also adjusted the EMB's structural properties. In the design of the EMB actuator, Yun of Zhejiang University converted the rotor to a hollow type and positioned the motion translation mechanism in this area [27,28,69–72]. Based on the existing commercial vehicle disc brake, Song proposed a commercial vehicle driving and parking brake electromechanical actuator based on a dual motor and self-locking mechanism. Sun proposed a dual-planetary EMB system based on the front-drive electric vehicle. Compared with the ordinary EMB structure, the structure of the dual-planetary EMB is optimized, which effectively reduces the motor power and improves the actuator performance. Aiming at the problems of large volume and low braking performance of commercial bus brakes, Shen proposed an auxiliary braking actuator-double stator eddy current brake and main braking actuator-electro-mechanical brake. Hye-Yeon Ryu proposed an EMB system using a motor controller without a large braking force. Zhou proposed an EMB system based on slip rate control. It is concluded that the system can improve the braking effect. Giseo proposed a new EMB clamping force controller, which can control the additional cost of existing sensor installation and response delay. It can be seen that with continuous research and development, electromechanical braking is expected to be popularized in the braking system of future automobiles due to its many advantages.

Research on EMB started late in China. So far, research on EMB systems in domestic universities and research institutes is still based on prototypes and has not yet been commercially applied. In general, there is still a large gap compared with foreign technologies, and the unremitting efforts of relevant scientific and technical workers are needed to narrow the technology gap and catch up with imported advanced levels [67]. Hence, as intelligent technologies and electrification continue to advance. The EMB will almost certainly be more up to date. Because of its distinct advantages, the EMB system is currently attracting considerable interest from domestic and foreign automobile manufacturers. It has become a research priority in the field of automotive braking. However, due to the high safety risk of motor power supply failure, the EMB system still has a long way to go before it can enter mass production without an additional backup mechanism for brake failure [6,64].

4. The Control Strategy of BBW System

4.1. Motor Braking Force Distribution Strategy

In the process of braking force distribution control of the brake-by-wire system, there are mainly electric machine power control and wheel cylinder pressure control. Reasonable distribution of braking force of motor and hydraulic system can improve the braking stability of BBW system, shorten the braking distance and increase the braking comfort. The traditional braking force distribution strategy is usually set up in the form of fixed ratio function of axle load or vehicle deceleration. As the fixed ratio function of the brake force distribution system has a simple and poor adaptability, it cannot better perform the performance of the braking system. Therefore, it is necessary to optimize the design of the braking force distribution control strategy to continuously improve the vehicle braking performance.

The braking force distribution method based on the slip rate is a more reliable distribution method to realize the electric mechanism power distribution. Li at Shanghai Jiaotong University used the control distribution algorithm to achieve the optimal distribution of the tire force based on the symbol-holding quadratic programming method [73]. Peng proposed an optimal braking force distribution strategy along with the constrained optimization problem based on the slip rate as a real-time solution method. The proposed optimal braking force distribution strategy can always ensure that the front wheel slip rate is higher than the rear wheel slip rate under different braking conditions. As a result, it not only improves the stability of vehicle braking, but also shortens the braking distance [74]. For the braking force distribution, H Fennel proposed that the slip rate of the rear wheels should track that of the front wheels in a fixed ratio [75]. In contrast, the ratio is divided into four segments for separate control in the literature [76]. Zhang proposed a hierarchical control system to solve the problems of poor anti-interference performance, poor electromechanical coordination performance and large target tracking errors when autonomous vehicles perform braking. In the upper-level controller, a feedback controller is built based on the desired speed sequence of the unmanned system. The desired braking deceleration is used as the front-end input to compensate for the target braking torque, and the speed error used as the feedback input to correct the target torque difference. In the lower controller, a coordinated braking force distribution algorithm based on fuzzy control is established by considering the characteristics of mechanical braking and motor braking [77].

4.2. Wheel Cylinder Pressure-Control strategy

The introduction of the brake master cylinder has caused a response delay in the linear hydraulic brake system. Frictional nonlinearity and initial pressure difference have different effects on pressure control. A digitally controlled proportional valve is the most effective and straightforward method in achieving efficient brake pressure modulation. It can achieve the continuous control of hydraulic fluid flow, and thus go on to achieve linear control of hydraulic pressure. To improve the control accuracy of wheel cylinder hydraulics and the performance of the hydraulic braking system, researchers at home and abroad had conducted a comprehensive study on the design and control methods of hydraulic actuators. Li proposed a pressure-control strategy. Based on the open-loop test analysis of the electric master cylinder, the integrated in-line hydraulic brake (IEHB) system was designed. Combined with the pressure segmentation control architecture, a feed forward and feedback PID method based on the auxiliary boost coefficient compensation was used to regulate the electric master cylinder. And a logic valving method was used to control the booster valve, pressure reducing valve and electric pump. Thus, the design of a pressure controller based on this IEHB system was realized [78]. Chen Lv proposed a novel sliding mode control method based on high precision hydraulic feedback modulation. A hydraulic brake system dynamics model including valve dynamics was established. An open-loop load pressure control method based on the linear relationship between pressure drop and coil current in the critical open-loop equilibrium state of the valve was proposed, and the effectiveness and superior performance of the proposed closed-loop modulation method verified [79].

4.3. Stability Control Strategy Technology

For conventional cars, the electronic stability controller (ESC) system is controlled by direct transverse moment. As the wire-controlled actuation technology increasingly develops, electrical signal transmission has replaced the conventional automotive drive line. It can make the most of the advantages such as the rapid electrical signal transmission to improve the handling stability of the vehicle [80,81]. Therefore, it is urgent to research the dynamics stability control system of brake-by-wire.

Vehicle stability control parameters include the lateral stability control of the vehicle. Factors affecting the lateral stability of the vehicle mainly include the yaw rate and the sideslip angle of the mass center. For the vehicle stability control, there are four main control methods directly transverse moment control, active front wheel steering, active suspension and active anti-roll stabilizer [82,83].

The transverse moment stabilization control system mainly adopts a hierarchical control structure. The computation time is thus reduced by establishing a parallel controller. Its good real-time performance can meet the requirements of the braking system. In response to the above issues, Qin proposed a structure of the new stable hierarchical control system. Based on the feedforward joint control, the upper layer is an additional transverse moment controller, which is able to resist the influence of changing road adhesion conditions. The lower-level torque distribution controller adopts an optimal distribution control algorithm to achieve independent and optimal distribution of braking force to all four wheels of the vehicle [84].

The structure divides the whole control system into three layers as follows:

(1) The layer of vehicle state observation: Real-time monitoring of vehicle parameters;

(2) The layer of motion control target tracking decision: When the vehicle has a tendency to destabilize, the magnitude of the transverse swing torque to prevent transverse swing instability will be calculated according to the driver's intention and the actual state of the vehicle;

(3) The layer of torque distribution control: Optimally distributes the additional transverse sway torque to the brake actuators of the four wheels when the vehicle dynamics constraints are satisfied.

Currently, the dynamics stability control system is mainly through tracking control of the yaw rate and the centroid sideslip angle. Tracking control of yaw rate is commonly used in the regular vehicle driving, which can ensure the vehicle handling stability during normal driving, but it may fail under extreme working conditions. Tracking control of centroid sideslip angle is a good complement to this defect, which can quickly adjust the vehicle state when it loses stability control and ensure the normal vehicle driving. Therefore, most of the research on vehicle stability control focus on coordinating tracking control of the yaw rate and the centroid sideslip angle, and the two cooperate to ensure the handling stability and improve driving comfort and safety.

Mirzaeinejad H et al. proposed a vehicle stability control system as shown in Figure 5. The designed control system is divided into two layers. The upper layer calculates the front outer yaw moment and the corrected lateral force referring to the yaw rate and proposes an algorithm for the distribution of tire braking force. In the lower layer, a new nonlinear multivariable controller is optimally designed. The weighted combination of the front outer yaw moment, corrected lateral force and the tracking error are used to define the vehicle stability performance index. The defined performance index is minimized to obtain a closed form multivariate control law. Through the proposed fuzzy scheduling method, the designed integrated controller is automatically tuned so that the weighting factors can change softly under different driving conditions, thus realizing the adaptive regulation of vehicle stability control [85].

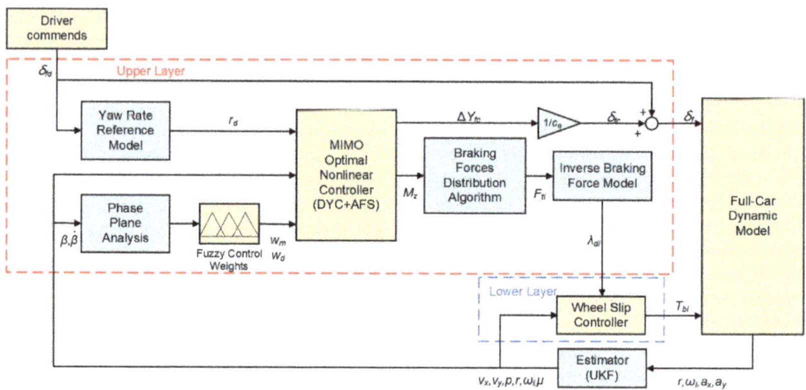

Figure 5. Overall structure of the proposed integrated control system.

Besides, the research on vehicle stability control focuses on the tracking control of yaw rate and center point sideslip angle under extreme operating conditions. Various control algorithms have been used, such as sliding mode control [86,87], fuzzy logic control [88], control allocation methods [89], optimal control [90,91], model predictive control strategies [92], and robust control [93]. In the literature, an extra direct yaw moment control was obtained in real time by tracking and controlling the angular velocity of the transverse swing and the lateral deflection angle of mass center of a robust controller [94]. The literature proposed a vehicle stability control algorithm based on brake-by-wire, which adds a brake redundancy to the vehicle stability control to improve the braking safety [95]. It also proposed a cooperative control of vehicle lateral stability using active front-wheel steering and differential braking, which softly constrains the yaw rate velocity by a model prediction algorithm and controls the centroid sideslip angle to always stay within a certain range so as to maintain the stability and comfort during driving [96].

4.4. Brake Fault-Tolerant Technology

The brake-by-wire system is expected to completely replace conventional braking systems because of advantages of fast response and high accuracy. However, all mechanical connections between the brake pedal and the brake actuator have been eliminated in the brake-by-wire system. While providing better braking performance, this also poses significant safety challenges to the in-line braking systems. A fault-tolerant and fail-safe system architecture is therefore needed. Given such issues, experts have put forth some solutions. Based on the analysis of transient fault propagation characteristics of BBW systems, Shuang Huang proposed a hierarchical transient fault-tolerance scheme with embedded intelligence and resilient coordination. In this scheme, most transient faults can be quickly handled at the node level by a feature-code-based detection method. The remaining are those that cannot be detected directly at the node level. Transient faults that degrade the system performance through fault propagation and evolution can be detected and recovered at the system level by functional and structural models. A sliding mode control algorithm and task reassignment strategy were designed [97]. A comprehensive review of fault-tolerant brake-by-wire system was presented in the literature [98], which discussed fault detection methods for low-cost components, analyzes fault-tolerant design principles between sensors, actuators, communications and provided comprehensive considerations. Liu at Tsinghua University conducted a redundant braking design for the ABS module in the brake-by-wire and designed a redundant ABS control system architecture. By recognizing the driver's driving signals such as braking, steering and driving, estimating and monitoring each state parameter during vehicle braking, the redundant ABS control system can determine the intervention and withdrawal timing of the redundant braking function through driving intention and vehicle parameters. Among them, pressure controls used variable to gain PID control. When the redundant ABS control system works, the SMC (sliding mode control) algorithm will be used to control the slip rate of the wheels, thus realizing the redundant ABS control function of the vehicle and improving driving safety [99].

4.5. Regenerative Braking Technology

The concept of energy conservation and environmental protection takes root. People pursue green and renewable environmental energy such as automobile fuel and pay more and more attention to the energy utilization of automobiles [100]. Therefore, the research on automotive energy recovery system is particularly important. Automotive braking energy recovery [101], also known as regenerative braking technology, is a process of vehicle deceleration and braking. Through the energy conversion device, it can convert part of the braking kinetic energy into other forms of energy. While maintaining a steady deceleration of the vehicle, the energy will be saved to a device such as a battery, capacitor or high-speed flywheel for reuse in the vehicle [102,103]. The schematic of the comprehensive vehicle energy recovery system can be seen in Figure 6. Fuel economy can be effectively improved by efficient utilization of regenerative braking and improvement of regenerative braking energy. Current research on regenerative braking technology focuses on the synergistic control of regenerative braking and hydraulic braking. It is necessary to ensure that the total braking force matches the braking force required by the driver [104,105]. This paper analyzes the structure and characteristics of various researchers on braking regeneration strategies in a comparative manner. The following Table 3 describes the relevant researches:

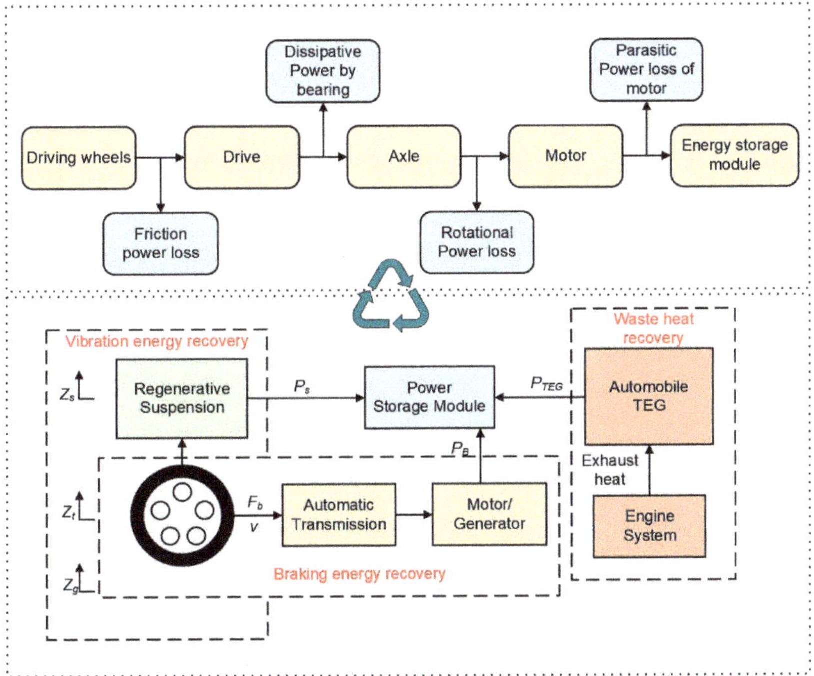

Figure 6. Schematic of the comprehensive vehicle energy recovery system.

Table 3. List of relevant studies on braking.

Topics	Goals	Reference
A refined energy management strategy.	(1) Modeling the vehicle energy recovery system; (2) An assistant power balance strategy(APBS); (3) Bench tests;	[106]
Electro-hydraulic proportional control for brake energy regeneration.	(1) Maximized the recovery of braking energy; (2) Driver's braking feel;	[107]
Development of hydraulic servo brake system for cooperative control with regenerative brake.	(1) To adjusting the stroke simulator mechanism; (2) Sliding mode control used to feedback control solenoid valves;	[48]
A fuzzy-logic-based regenerative braking strategy (RBS) integrated with series regenerative braking.	(1) According to the distribution law to calculate the braking force; (2) According to the fuzzy logic controller to determine the regenerative braking force;	[108]
A control strategy combining the logic threshold method and the key parameter optimization algorithm.	(1) To combine the merits of both batteries; (2) Proposed a logical threshold method; (3) To real-time control the torque distribute;	[109]
The integration of sensor-based wire-controlled dynamic fault-tolerant drive	(1) The different sensor measurement transitions, drive input drive profiles and redundancy management algorithm; (2) Improve the conversion performance of the sensor;	[110]

5. Challenges and Future Directions of EMB

The EMB system provides a way for the development of intelligent driving. However, up to now, the brake-by-wire system still cannot be put into use on a large scale, indicating that there are still many issues to be solved.

(1) Safety issues. In the process of automobile development, automobile safety is the first factor to be considered in the design of automobile structure. Automobile safety can be divided into active safety and passive safety. Among them, the braking system belongs to the category of active safety. In order to ensure the safety of drivers and passengers, the design scheme of braking redundancy should be fully considered when designing the electromechanical braking system. The biggest application scenario of EMB system is commercial vehicles. For commercial vehicles, they need to face many complex road conditions, such as high temperature, plateau cold road, muddy road and so on. As the installation form of EMB is similar to that of hub motor, a large number of vibration frequencies will be transmitted to the system, causing the socket to become loose and the actuator to break. In addition, the severe environment will lead to EMB system sensor detection failure due to high temperature or cold. Therefore, the brake redundancy design of EMB will become a significant obstacle to further growth;

(2) Stable response of EMB. EMB is transmitted by electrical signals. The sensor collects the braking signal of the brake simulation pedal and transmits it to the actuator, and the braking time is very short. Therefore, the sensitivity of the sensor will greatly affect the response time of EMB. However, at this point, the sensor's delayed reaction is an issue;

(3) Energy demand of braking force. The traditional vehicle power supply is generally 12 V, and the driving motor of the electromechanical braking system adopts a voltage of 42 V, because it is beneficial to improve the performance of the actuator. However, the high voltage brought by the 42 V power supply system will bring issues such as line insulation, withstand voltage and electromagnetic interference, which also poses a threat to the longevity of the vehicle's complete circuit system;

(4) Accurate control of cars with different parameters. Due to the uncertainty of the vehicle's driving conditions, cargo quality and the nonlinearity of braking, it is difficult to achieve precise control of braking force. Therefore, the next stage of EMB research will be to achieve accurate estimation of vehicle parameters and accurate control of braking force.

This requires the use of a variety of control methods to achieve the most suitable control of vehicle braking, and to continuously improve it to improve control accuracy.

EMB will be more in line with the characteristics of automobile integration, intelligence and automation, and more suitable for the future development direction of automobiles. In view of the problems existing in EMB, the development trend is as follows:

- Study the control methods suitable for various road conditions;
- Design a more compact structure with good seismic performance;
- Develop sensors with stable performance and high sensitivity.

In a word, the improvement, optimization, research and development of these technologies of EMB system have important value and role in improving the safety performance of automobile driving, and are conducive to promoting the development of vehicle intelligence and automatic driving technology.

6. Conclusions

As an emerging product of the era, the brake-by-wire system can provide more stable and faster braking performance for automotive braking. It also puts strict requirements on the control accuracy and robustness of the wire-controlled actuation system. The EHB based on the traditional valve technology has the advantages of high control reliability, low energy consumption, low price and easy modification on the traditional braking system. At present, the improvement of the brake-by-wire system mainly focuses on reducing energy consumption. The main strategies include:

(1) Use Two box or One box architecture solution;

(2) Improve braking accuracy by improving the control module of hydraulic adjustment unit or brake motor;

(3) Design the control method of high pressure accumulator to reduce the braking time and energy consumption;

(4) Design brake regeneration system to improve brake energy utilization.

At present, what is more widely deployed in the brake-by-wire system is EHB. EHB technology is more mature, and more adaptable to the market demand. EHB will be the direction of wireless control braking development at this stage. EMB, due to superior technical conditions, high cost, redundancy backup and thermal reliability technology, still needs to be improved. At present, there is still some distance from full commercialization. However, as the most advanced technology in braking system, EMB system has the advantages of high response speed, energy saving and environmental protection to which the traditional braking systems including EHB cannot be compared. It can be applied in L4 and L5 of automatic driving. For the research of EMB, we can only take this opportunity to rise to the challenges and carry out continuous research in the development of EMB system, and gradually deepen and refine our work. We believe that in the future, when various difficulties that constrain the development of EMB systems are overcome, the conditions for EMB industrialization will slowly mature. We also believe that the electromechanical brake technology has a bright future, and is bound to gain more and more people's attention and favor. Finally, this paper compared the advantages and disadvantages of the braking systems described in the previous paper and reviewed the efficiency of the discussed solutions used for the braking systems. The following Table 4 can describe the comparison of braking system solutions.

Table 4. Braking system solution comparison.

Type	EHB	EMB
Definition	Electro-hydraulic brake, retaining part of the hydraulic system	Electromechanical brake, no hydraulic system
Volume	Poor integration and large volume	Good integration and small size
Number of motors	One motor (in hydraulic pump)	4 motors (at the wheel)
Cost	Low Cost	High cost (the main cost is focused on the design and modification of the motor)
Response Speed	120 ms	90 ms
Brake Redundancy	A redundant system is available. If both the EHB and the redundant solution fail, it can still be changed to an unassisted hydraulic brake	No redundant system, it needs to ensure power stability and fault tolerance of each node in the communication system
Complexity	Simple structure and easy to control	Complex structure and difficult to control
Power density	Adopting brushless direct current motor, with higher power density	Adopt permanent magnet synchronous motor, less power density
Braking force	High braking power	Low braking force
Brake fluid safety	Brake fluid leakage will lead to short circuit failure of electronic components	No brake fluid, safe and light weight
Comfort	Poor comfort	Good comfort
Disadvantage	Still requires vacuum system, which increases energy consumption and noise	No redundant system, hub size limits the motor power and its operating environment, which is a great test for the motor's permanent magnets and semiconductor components

Author Contributions: Conceptualization, J.Z. and X.H.; methodology, J.Z.; software, H.L.; validation, J.H., M.L. and X.F.; formal analysis, H.X.; investigation, J.Z.; resources, W.W.; data curation, J.Z.; writing—original draft preparation, J.Z.; writing—review and editing, X.H.; visualization, X.H.; supervision, H.L.; project administration, J.Z.; funding acquisition, X.H. All authors have read and agreed to the published version of the manuscript.

Funding: This research was funded by the Guangdong Province Key Field R&D Program—Electric Vehicle Powertrain Design and Optimization Industrial Software (2021B0101220003) and the R&D Program in Key Fields of Guangdong Province—R&D and Application of Key Technologies for New Energy Vehicle Gearboxes with High Efficiency, High Precision, Long Life and Low Noise (2020B090926004).

Institutional Review Board Statement: Not applicable.

Informed Consent Statement: Not applicable.

Data Availability Statement: This study did not report any data.

Acknowledgments: The authors acknowledge the editors and reviewers for their constructive comments and all the supports on this work.

Conflicts of Interest: The authors declare no conflict of interest.

Abbreviations

The following abbreviations are used in this manuscript:

ABS	Anti-lock Braking System
BBW	Brake-By-Wire
CHB	Conventional Hydraulic Brake
EBS	Electronic Braking System
ECU	Electronic Control Unit
EHB	Electro-Hydraulic Brake
EMB	Electromechanical Brake
ESC	Electronic Stability Controller
ESP	Electronic Stability Program

EWB	Electronic Wedge Brake
IEHB	Integrated In-line Hydraulic Brake
IPB	Integrated Power Brake
PID	Proportion Integral Derivative
RBS	Regenerative Braking System
RBU	Redundant Brake Unit
SBC	Sensotronic Brake Control
SMC	Sliding Mode Control
TCS	Traction Control System
TEG	Thermoelectric Generators
UKF	Unscented Kalman Filter
VCU	Vehicle Control Unit

References

1. Guo, Y.; Ding, J.; Wang, C.; Wang, S.; Zhang, Y. Research status and application progress of chassis wire control technology for commercial vehicles. *Chin. J. Automot. Eng.* **2022**, *12*, 695–714.
2. Zhao, X.; Ye, Y.; Ma, J.; Shi, P.; Chen, H. Construction of electric vehicle driving cycle for studying electric vehicle energy consumption and equivalent emissions. *Environ. Sci. Pollut. Res.* **2020**, *27*, 37395–37409. [CrossRef] [PubMed]
3. Yu, L.; Liu, X.; Xie, Z.; Chen, Y. Review of brake-by-wire system used in modern passenger car. In Proceedings of the ASME 2016 International Design Engineering Technical Conferences and Computers and Information in Engineering Conference, Charlotte, NC, USA, 21–24 August 2016; Volume 50138, p. V003T01A020.
4. Zong, C.; Li, G.; Zheng, H.; He, L.; Zhang, Z. Study process and outlook of chassis control technology for X-by-wire automobile. *Chin. J. China J. Highw. Transp.* **2013**, *26*, 160–176.
5. Li, D.; Tan, C.; Ge, W.; Cui, J.; Gu, C.; Chi, X. Review of brake-by-wire system and control technology. *Actuators* **2022**, *11*, 80. [CrossRef]
6. Yu, Z.; Wei, H.; Songyun, X.; Lu, X. Review on hydraulic pressure control of electro-hydraulic brake system. *J. Mech. Eng.* **2017**, *53*, 1–15. [CrossRef]
7. Editorial Department of China Journal of Highway and Transport. Review on China's automotive engineering research progress. *China J. Highw. Transp.* **2017**, *30*, 1–197.
8. Nakamura, E.; Soga, M.; Sakai, A.; Otomo, A.; Kobayashi, T. *Development of Electronically Controlled Brake System for Hybrid Vehicle*; SAE Technical Paper No. 2002-01-0300; SAE International: Warrendale, PA, USA, 2002.
9. Chen, H.; Gong, X.; Hu, Y.; Liu, Q.; Gao, B.; Guo, H. Automotive control: The state of the art and perspective. *Acta Autom. Sin.* **2013**, *39*, 322–346. [CrossRef]
10. Zhang, J.; Lv, C.; Li, Y. Hybrid propulsion and hybrid braking technologies of electrified vehicles: Status and prospect. *J. Automot. Saf. Energy* **2014**, *5*, 209.
11. Li, L.; Wang, X.; Chen, S.; Chen, X.; Huang, C.; Ping, X.; Wei, L. Technologies of control-by-wire and dynamic domain control for automotive chassis. *J. Automot. Saf. Energy* **2020**, *11*, 143.
12. Jitesh, S. Antilock braking system (ABS). *Int. J. Mech. Eng. Robot. Res.* **2014**, *3*, 253.
13. Nie, J.; Zhou, P. The simulation study of ABS based on PID control. *Chin. J. Shanghai Univ. Technol.* **2010**, *18*, 6–9.
14. Mou, Y.; Zhu, R.; Li, S.; Shen, L. Fuzzy neural network control system for automotive ABS. *Chin. J. J. Shanghai Jiaotong Univ.* **1999**, *33*, 64–67.
15. Li, L.; Li, Z.; Chen, K. Analysis and simulation of ABS fuzzy control method for automobiles. *Chin. J. Jiangsu Univ. (Nat. Sci. Ed.)* **2003**, *24*, 49–52.
16. Zhu, W.; Chen, Y. Application and simulation study of fuzzy PID control in automotive ABS. *Chin. J. Jiangsu Univ. (Nat. Sci. Ed.)* **2004**, *25*, 310–314.
17. Du, Y.; Qin, C.; You, S.; Xia, H. Efficient coordinated control of regenerative braking with pneumatic anti-lock braking for hybrid electric vehicle. *Sci. China Technol. Sci.* **2017**, *60*, 399–411. [CrossRef]
18. Minh, V.T.; Oamen, G.; Vassiljeva, K.; Teder, L. Development of anti-lock braking system (ABS) for vehicles braking. *Open Eng.* **2016**, *6*, 554–559. [CrossRef]
19. Zhu, S.; Fan, X.; Qi, G.; Wang, P. Review of Control Algorithms of Vehicle Anti-lock Braking System. *Recent Pat. Eng.* **2023**, *17*, 30–45. [CrossRef]
20. Zhang, X.; Lou, Y.; Yang, X. Study of control logic for automobile anti-lock braking system. In Proceedings of the 2008 International Conference on Intelligent Computation Technology and Automation (ICICTA), Changsha, China, 20–22 October 2008; Volume 1, pp. 493–497.
21. Day, T.D.; Roberts, S.G. A simulation model for vehicle braking systems fitted with ABS. *SAE Trans.* **2002**, *111*, 821–839.
22. Li, R.; Zheng, T.; Li, Y.; Feng, Z.; Chen, W. Graded intelligent control of automotive anti-lock braking system. *J. Mech. Eng.* **2007**, *8*, 135–141. [CrossRef]
23. Van Zanten, A.T. *Bosch ESP Systems: 5 Years of Experience*; SAE Transactions: Warrendale, PA, USA, 2000; pp. 428–436.

24. Müller, A.; Achenbach, W.; Schindler, E.; Wohland, T.; Mohn, F. Das neue fahrsicherheitssystem elektronic stability program von mercedes-benz. *Automob. Z.* **1994**, *96*, 630.

25. Van Zanten, A.T.; Erhardt, R.; Pfaff, G. *VDC, the Vehicle Dynamics Control System of Bosch*; SAE Transactions: Warrendale, PA, USA, 1995; pp. 1419–1436.

26. Fennel, H.; Ding, E. *A Model-Based Failsafe System for the Continental TEVES Electronic-Stability-Program (ESP)*; SAE Technical Paper; SAE International: Warrendale, PA, USA, 2000.

27. Yu, J.; Ding, R.; He, Y.; Shao, W. Optimal design of structural parameters of electro-mechanical brakes. *Commer. Veh. Mag.* **2015**, *32*, 5–7, 37.

28. Fu, Y. Research on the Design of Automotive Electro-Mechanical Braking System and Its Key Technologies. Ph.D. Thesis, Zhejiang University, Zhejiang, China, 2013.

29. Jung, Y.S. Electro-Hydraulic Brake System and Method for Controlling the Same. U.S. Patent 9,738,263, 22 August 2017.

30. Kim, Y.K. Electro-Hydraulic Brake System. U.S. Patent 9,156,457, 29 April 2013.

31. Kreutz, S. Ideal regeneration with electromechanical brake booster (eBKV) in Volkswagen e-up! and Porsche 918 Spyder. In *5th International Munich Chassis Symposium 2014*; Springer Vieweg: Wiesbaden, Germany, 2014; pp. 549–558.

32. Ohtani, Y.; Innami, T.; Obata, T.; Yamaguchi, T.; Kimura, T.; Oshima, T. *Development of An Electrically-Driven Intelligent Brake Unit*; Technical report, SAE Technical Paper; SAE International: Warrendale, PA, USA, 2011.

33. Fujiki, N.; Koike, Y.; Ito, Y.; Suzuki, G.; Gotoh, S.; Ohtani, Y.; Yamagucki, T. *Development of an Electrically-Driven Intelligent Brake System for EV*; Technical report, SAE Technical Paper; SAE International: Warrendale, PA, USA, 2011.

34. Yu, Z.; Xu, S.; Xiong, L.; Guang, X. Robustness Hydraulic Pressure Control System of Integrated-electro-hydraulic Brake System. *J. Mech. Eng.* **2015**, *51*, 22–28. [CrossRef]

35. Wang, Z.; Yu, L.; Wang, Y.; You, C.; Ma, L.; Song, J. *Prototype of Distributed Electro-Hydraulic Braking System And Its Fail-Safe Control Strategy*; Technical Report, SAE Technical Paper; SAE International: Warrendale, PA, USA, 2013.

36. Ohkubo, N.; Matsushita, S.; Ueno, M.; Akamine, K.; Hatano, K. Application of electric servo brake system to plug-in hybrid vehicle. *SAE Int. J. Passeng. Cars-Electron. Electr. Syst.* **2013**, *6*, 255–260. [CrossRef]

37. De Castro, R.; Todeschini, F.; Araujo, R.E.; Savaresi, S.M.; Corno, M.; Freitas, D. Adaptive-robust friction compensation in a hybrid brake-by-wire actuator. *Proc. Inst. Mech. Eng. Part I J. Syst. Control Eng.* **2014**, *228*, 769–786. [CrossRef]

38. Yu, Z.; Wei, H.; Lu, X. Hydraulic pressure control system of integrated-electro-hydraulic brake system based on Byrnes-Isidori normalized form. *J. Mech. Eng.* **2016**, *52*, 92–100. [CrossRef]

39. Leiber, T.; Köglsperger, C.; Unterfrauner, V. Modular brake system with integrated functionalities. *ATZ Worldw. Emagazine* **2011**, *113*, 20–25. [CrossRef]

40. Yang, I.J.; Choi, K.; Huh, K. Development of an electric booster system using sliding mode control for improved braking performance. *Int. J. Automot. Technol.* **2012**, *13*, 1005–1011. [CrossRef]

41. Kirschner, M. Slip Control Boost Braking System. U.S. Patent 9,221,443, 29 December 2015.

42. Todeschini, F.; Corno, M.; Panzani, G.; Savaresi, S.M. Adaptive position–pressure control of a brake by wire actuator for sport motorcycles. *Eur. J. Control* **2014**, *20*, 79–86. [CrossRef]

43. Todeschini, F.; Corno, M.; Panzani, G.; Fiorenti, S.; Savaresi, S.M. Adaptive cascade control of a brake-by-wire actuator for sport motorcycles. *IEEE/ASME Trans. Mechatron.* **2014**, *20*, 1310–1319. [CrossRef]

44. Todeschini, F.; Formentin, S.; Panzani, G.; Corno, M.; Savaresi, S.M.; Zaccarian, L. Nonlinear pressure control for BBW systems via dead-zone and antiwindup compensation. *IEEE Trans. Control Syst. Technol.* **2015**, *24*, 1419–1431. [CrossRef]

45. Liu, Q.; Sun, Z. Research of automotive BBW system. *Machinery Electronics* **2004**, *11*, 24–26.

46. Kirschner, M. Control Unit and Operation Method Used for Hydraulic Brake System. CN Patent 102501841, 20 June 2006.

47. Pei, S. Study on characteristics of NEHB system and the vehicle performance based on NEHB. Master's Thesis, Wuhan University of Technology, Wuhan, China, 2014.

48. Aoki, Y.; Suzuki, K.; Nakano, H.; Akamine, K.; Shirase, T.; Sakai, K. *Development of Hydraulic Servo Brake System for Cooperative Control with Regenerative Brake*; SAE Technical Paper; SAE International: Warrendale, PA, USA, 2007.

49. Ren, H.; Haipeng, F. Research and development of autonomous emergency brake (AEB) technology. *J. Automot. Saf. Energy* **2019**, *10*, 1.

50. Soga, M.; Shimada, M.; Sakamoto, J.I.; Otomo, A. Development of vehicle dynamics management system for hybrid vehicles: ECB system for improved environmental and vehicle dynamic performance. *JSAE Rev.* **2002**, *23*, 459–464. [CrossRef]

51. Anwar, S.; Zheng, B. An antilock-braking algorithm for an eddy-current-based brake-by-wire system. *IEEE Trans. Veh. Technol.* **2007**, *56*, 1100–1107. [CrossRef]

52. Jonner, W.D.; Winner, H.; Dreilich, L.; Schunck, E. *Electrohydraulic Brake System—The First Approach to Brake-By-Wire Technology*; SAE Transactions: Warrendale, PA, USA, 1996; pp. 1368–1375.

53. Park, M.; Kim, S.; Yang, L.; Kim, K. *Development of the Control Logic of Electronically Controlled Hydraulic Brake System for Hybrid Vehicle*; SAE Technical Paper; SAE International: Warrendale, PA, USA, 2009.

54. Putra, M.; Nizam, M.; Tjahjana, D.; Waloyo, H. The effect of air gap on braking performance of eddy current brakes on electric vehicle braking system. In Proceedings of the 2019 6th International Conference on Electric Vehicular Technology (ICEVT), Bali, Indonesia, 18–21 November 2019; pp. 355–358.

55. Reif, K. *Brakes, Brake Control and Driver Assistance Systems*; Springer Vieweg: Wiesbaden, Germany, 2014.

56. Quan, Z.; Quan, L.; Zhang, J. Review of energy efficient direct pump controlled cylinder electro-hydraulic technology. *Renew. Sustain. Energy Rev.* **2014**, *35*, 336–346. [CrossRef]
57. Gu, H. Research on the Design and Control of Direct-Drive Wire-Control Actuation System for Passenger Car Motors. Master's Thesis, Jilin University, Changchun, China, 2019.
58. Li, B.; Li, D.-X.; Ge, W.-Q.; Ta, C.; Lu, J.-Y.; So, A.-J. Precision Control of Hydraulic Pressure in Fast-response Brake-by-wire System Based on Direct-drive Valve. *China J. Highw. Transp.* **2021**, *34*, 121–132.
59. Gong, X.; Qian, L.; Ge, W.; Wang, L. Research on the anti-disturbance control method of brake-by-wire unit for electric vehicles. *World Electr. Veh. J.* **2019**, *10*, 44. [CrossRef]
60. Han, K.; Kim, M.; Huh, K. Modeling and control of an electronic wedge brake. *Proc. Inst. Mech. Eng. Part C J. Mech. Eng. Sci.* **2012**, *226*, 2440–2455. [CrossRef]
61. Jo, C.; Lee, S.; Song, H.; Cho, Y.; Kim, I.; Hyun, D.; Kim, H. Design and control of an upper-wedge-type electronic brake. *Proc. Inst. Mech. Eng. Part D J. Automob. Eng.* **2010**, *224*, 1393–1405. [CrossRef]
62. Jung, K.H.; Kim, D.; Kim, H.; Hwang, S.H. Analysis of the regenerative braking system for a hybrid electric vehicle using electro-mechanical brakes. *Int. J. Automot. Technol.* **2009**, *10*, 229–234.
63. Roberts, M.; Chhaya, T. *An Approach to the Safety Design and Development of a Brake-by-Wire Control System*; SAE Technical Paper; SAE International: Warrendale, PA, USA, 2011.
64. Xia, L. The Study on Distributed Hybrid Braking System Based on EMB. Ph.D. Thesis, Chongqing University, Chongqing, China, 2019.
65. Ho, L.M.; Roberts, R.; Hartmann, H.; Gombert, B. *The Electronic Wedge Brake—EWB*; SAE Technical Paper; SAE International: Warrendale, PA, USA, 2006.
66. Yu, J. Haldex Corporation introduces electro-mechanical brake system based on wire control technology. *Commer. Veh. Mag.* **2007**, 124.
67. Liu, G. Experimental study on hardware-in-the-loop simulation of electro-mechanical braking system. *Automot. Eng.* **2006**, 929–932.
68. Yang, K. Research of Electromechanical Brake and Vehicle Stability Control System for Light Vehicle. Ph.D. Thesis, Jilin University, Changchun, China, 2009.
69. Shen, L. Research on the Mechanical System of Electro-Mechanical Brake (EMB). Master's Thesis, Chang'an University, Xi'an, China, 2017.
70. Zhao, Y. Research and Development of Electro-Mechanical Brake System Actuator. Ph.D. Thesis, Tsinghua University, Beijing, China, 2010.
71. Han, P. Research on Anti-Lock Braking System for Light-Duty Electro-Mechanical Braking Vehicles. Master's Thesis, Chang'an University, Xi'an, China, 2011.
72. Li, C. Research on EMB-Based Automotive Braking Control. Master's Thesis, Chongqing University, Chongqing, China, 2015.
73. Li, D. Study on Integrated Vehicle Dynamics Control Based on Optimal Tire Force Distribution. Ph.D. Thesis, Shanghai Jiaotong University, Shanghai, China, 2008.
74. Peng, X.; Lv, Y.; He, L. Research on braking force distribution strategy of linear control actuation system based on slip rate. *China Mech. Eng.* **2016**, *27*, 2407–2412.
75. Fennel, H. Method for Controlling Brake Force Distribution. U.S. Patent 6,322,169, 27 November 2001.
76. Zhang, W. Research on ABS-EBD Control Strategy Based on Slip Rate. Master's Thesis, Jilin University, Changchun, China, 2008.
77. Zhang, C.; Liu, Q.; Dong, H.; Chen, H.; Xi, J. Research on braking force distribution strategy of linear control actuation system based on slip rate. *Acta Armamentarii* **2022**, *22*, 2727–2737.
78. Lv, C.; Wang, H.; Cao, D. High-precision hydraulic pressure control based on linear pressure-drop modulation in valve critical equilibrium state. *IEEE Trans. Ind. Electron.* **2017**, *64*, 7984–7993. [CrossRef]
79. Lv, C.; Zhang, J.; Li, Y.; Sun, D.; Yuan, Y. Hardware-in-the-loop simulation of pressure-difference-limiting modulation of the hydraulic brake for regenerative braking control of electric vehicles. *Proc. Inst. Mech. Eng. Part D J. Automob. Eng.* **2014**, *228*, 649–662. [CrossRef]
80. Abe, M.; Kano, Y.; Suzuki, K.; Shibahata, Y.; Furukawa, Y. Side-slip control to stabilize vehicle lateral motion by direct yaw moment. *JSAE Rev.* **2001**, *22*, 413–419. [CrossRef]
81. Sawase, K.; Sano, Y. Application of active yaw control to vehicle dynamics by utilizing driving/breaking force. *JSAE Rev.* **1999**, *20*, 289–295. [CrossRef]
82. Li, L.; Jia, G.; Song, J.; Rang, X. Progress on vehicle dynamics stability control system. *J. Mech. Eng.* **2013**, *49*, 95–107. [CrossRef]
83. Wang, X.; Fei, Z.; Yan, H.; Xu, Y. Dynamic event-triggered fault detection via zonotopic residual evaluation and its application to vehicle lateral dynamics. *IEEE Trans. Ind. Inform.* **2020**, *16*, 6952–6961. [CrossRef]
84. Qin, H. Research on Lateral Stability Control Strategy for Vehicles with the Brake-by-Wire System. Master's Thesis, Hunan University, Changsha, China, 2008.
85. Mirzaeinejad, H.; Mirzaei, M.; Rafatnia, S. A novel technique for optimal integration of active steering and differential braking with estimation to improve vehicle directional stability. *ISA Trans.* **2018**, *80*, 513–527. [CrossRef] [PubMed]
86. Lu, S.B.; Li, Y.N.; Choi, S.B.; Zheng, L.; Seong, M.S. Integrated control on MR vehicle suspension system associated with braking and steering control. *Veh. Syst. Dyn.* **2011**, *49*, 361–380. [CrossRef]

87. Li, B.; Rakheja, S.; Feng, Y. Enhancement of vehicle stability through integration of direct yaw moment and active rear steering. *Proc. Inst. Mech. Eng. Part D J. Automob. Eng.* **2016**, *230*, 830–840. [CrossRef]
88. Boada, M.; Boada, B.; Munoz, A.; Diaz, V. Integrated control of front-wheel steering and front braking forces on the basis of fuzzy logic. *Proc. Inst. Mech. Eng. Part D J. Automob. Eng.* **2006**, *220*, 253–267. [CrossRef]
89. Tjonnas, J.; Johansen, T.A. Stabilization of automotive vehicles using active steering and adaptive brake control allocation. *IEEE Trans. Control Syst. Technol.* **2009**, *18*, 545–558. [CrossRef]
90. Acarman, T. Nonlinear optimal integrated vehicle control using individual braking torque and steering angle with on-line control allocation by using state-dependent Riccati equation technique. *Veh. Syst. Dyn.* **2009**, *47*, 155–177. [CrossRef]
91. Li, B.; Goodarzi, A.; Khajepour, A.; Chen, S.k.; Litkouhi, B. An optimal torque distribution control strategy for four-independent wheel drive electric vehicles. *Veh. Syst. Dyn.* **2015**, *53*, 1172–1189. [CrossRef]
92. Di Cairano, S.; Tseng, H.E.; Bernardini, D.; Bemporad, A. Vehicle yaw stability control by coordinated active front steering and differential braking in the tire sideslip angles domain. *IEEE Trans. Control Syst. Technol.* **2012**, *21*, 1236–1248. [CrossRef]
93. He, Z.; Ji, X. Nonlinear robust control of integrated vehicle dynamics. *Veh. Syst. Dyn.* **2012**, *50*, 247–280. [CrossRef]
94. Yin, G.; Wang, R.; Wang, J. Robust control for four wheel independently-actuated electric ground vehicles by external yaw-moment generation. *Int. J. Automot. Technol.* **2015**, *16*, 839–847. [CrossRef]
95. Xiang, W.; Richardson, P.C.; Zhao, C.; Mohammad, S. Automobile brake-by-wire control system design and analysis. *IEEE Trans. Veh. Technol.* **2008**, *57*, 138–145. [CrossRef]
96. Jalali, M.; Khosravani, S.; Khajepour, A.; Chen, S.k.; Litkouhi, B. Model predictive control of vehicle stability using coordinated active steering and differential brakes. *Mechatronics* **2017**, *48*, 30–41. [CrossRef]
97. Huang, S.; Zhou, C.; Yang, L.; Qin, Y.; Huang, X.; Hu, B. Transient fault tolerant control for vehicle brake-by-wire systems. *Reliab. Eng. Syst. Saf.* **2016**, *149*, 148–163. [CrossRef]
98. Isermann, R.; Schwarz, R.; Stolzl, S. Fault-tolerant drive-by-wire systems. *IEEE Control Syst. Mag.* **2002**, *22*, 64–81.
99. Liu, X.; Yu, L.; Zheng, S.; Lu, Z.; Song, J. Research on redundant anti-lock braking algorithm based on electronic booster. *Automot. Eng.* **2022**, *44*, 82–93.
100. Clegg, S. *A Review of Regenerative Braking Systems*; Institute of Transport Studies, University of Leeds: Leeds, UK, 1996.
101. Yu, W.; Wang, R. Development and performance evaluation of a comprehensive automotive energy recovery system with a refined energy management strategy. *Energy* **2019**, *189*, 116365. [CrossRef]
102. Rousseau, A.; Pagerit, S.; Gao, D.W. Plug-in hybrid electric vehicle control strategy parameter optimization. *J. Asian Electr. Veh.* **2008**, *6*, 1125–1133. [CrossRef]
103. Shimizu, H.; Moriguchi, Y.; Kondo, Y.; Tamura, M. Global warming and the feasibility of electric cars. *Int. J. Sol. Energy* **1994**, *14*, 169–179. [CrossRef]
104. Ogura, M. *The Honda EV Plus Regenerative Braking System*; EVS-14; Honda R&D Co. Ltd.: Wako, Japan, 1997.
105. Otomo, A.; Sakai, A.; Takasu, T.; Nakamura, K. *Development of Regenerative Braking System for Hybrid Vehicle*; SAE Paper; SAE International: Warrendale, PA, USA, 1999; Volume 10, p. 0062.
106. Song, R. Research on Design and Control Strategy of Electromechanical Braking System for Commercial Vehicle. Master's Thesis, Qingdao University of Technology, Shandong, China, 2022.
107. Sun, K. Optimization design and energy recovery strategy of double planetary row electro-mechanical braking system. Master's Thesis, Nanjing University of Aeronautics and Astronautics, Jiangsu, China, 2021.
108. Xu, G.; Li, W.; Xu, K.; Song, Z. An intelligent regenerative braking strategy for electric vehicles. *Energies* **2011**, *4*, 1461–1477. [CrossRef]
109. Sun, H.; Yang, L.; Jing, J.; Luo, Y. Control strategy of hydraulic/electric synergy system in heavy hybrid vehicles. *Energy Convers. Manag.* **2011**, *52*, 668–674.
110. Lii, N.; Sturm, S.; Coombs, T.A. A study on sensor fusion for fault tolerant brake-by-wire driver input design with dissimilar sensors. In *Safety-Critical Automotive Systems*; SAE International: Warrendale, PA, USA, 2006; pp. 67–73.

Article

Modified Dimension Reduction-Based Polynomial Chaos Expansion for Nonstandard Uncertainty Propagation and Its Application in Reliability Analysis

Jeongeun Son and Yuncheng Du *

Department of Chemical & Biomolecular Engineering, Clarkson University, Potsdam, NY 13699, USA; son@clarkson.edu

* Correspondence: ydu@clarkson.edu; Tel.: +1-315-268-2284

Citation: Son, J.; Du, Y. Modified Dimension Reduction-Based Polynomial Chaos Expansion for Nonstandard Uncertainty Propagation and Its Application in Reliability Analysis. *Processes* 2021, 9, 1856. https://doi.org/10.3390/pr9101856

Academic Editors: Francisco Ronay López-Estrada and Guillermo Valencia-Palomo

Received: 23 September 2021
Accepted: 15 October 2021
Published: 19 October 2021

Publisher's Note: MDPI stays neutral with regard to jurisdictional claims in published maps and institutional affiliations.

Abstract: This paper presents an algorithm for efficient uncertainty quantification (UQ) in the presence of many uncertainties that follow a nonstandard distribution (e.g., lognormal). Using the polynomial chaos expansion (PCE), the algorithm builds surrogate models of uncertainty as functions of a standard distribution (e.g., Gaussian variables). The key to build these surrogate models is to calculate PCE coefficients of model outputs, which is computationally challenging, especially when dealing with models defined by complex functions (e.g., nonpolynomial terms) under many uncertainties. To address this issue, an algorithm that integrates the PCE with the generalized dimension reduction method (gDRM) is utilized to convert the high-dimensional integrals, required to calculate the PCE coefficients of model predictions, into several lower-dimensional ones that can be rapidly solved with quadrature rules. The accuracy of the algorithm is validated with four examples in structural reliability analysis and compared to other existing techniques, such as Monte Carlo simulations and the least angle regression-based PCE. Our results show our algorithm provides accurate UQ results and is computationally efficient when dealing with many uncertainties, thus laying the foundation to address UQ in complex control systems.

Keywords: dimension reduction; polynomial chaos expansion; uncertainty analysis; nonstandard distribution; statistical moments

1. Introduction

Uncertainty, originating from the inherent randomness of a complex system, is common in first-principle models widely used in various engineering problems. Uncertainty quantification (UQ), which quantitatively studies the impact of uncertainty on a system's performance, is important since uncertainty in model parameters (e.g., external loads, geometry, and material properties in a complex system) can degrade the accuracy of model predictions, thus affecting prediction-based control design and analysis. Many UQ approaches have been developed and applied to engineering problems to improve a system's performance [1–8]. For example, reliability assessment has been explored for the controlled structures of viscoelastic dampers [3], and the stochastic responses of bridge structures under uncertainty have been studied [4]. Further, reliability-based design optimization (RBDO), used to consider uncertainty in process design, has been developed for structural analysis, including the sequential approximate optimization to deal with multimodal random variables [6] and the adaptive surrogate-based algorithm to replace expensive-to-evaluate models [7]. A hybrid reliability-based optimization method was also proposed for thermal structure design [9]. However, several challenges remain unsolved. For example, a major challenge of existing UQ tools is how to quantify the effect of a large number of uncertainties on model responses in an accurate and computationally efficient manner.

There are several popular UQ methods, including Monte Carlo (MC) simulations [10,11], Taylor expansion-based methods [12,13], and polynomial chaos expansion (PCE) [14,15].

Among these, MC has been widely used since it is easy to implement. It only requires random generation of samples from the distribution of uncertainty and repetitive execution of the first principle models with each sample. Despite the apparent simplicity, MC and its associated approaches are insufficient because they often require a large number of samples to ensure UQ accuracy, thus increasing the computational cost. To reduce the computational cost, different variants of MC were developed, which include an active learning approach to combine Kriging and MC simulation (i.e., AK-MCS) [8]. Further, the Taylor expansion-based method is another way for UQ, which is effective for less complicated models [16,17]. This includes the first-order reliability method (FORM) and second-order reliability method (SORM) [12,18] in structural reliability. However, when models involve complex nonpolynomial terms, the UQ accuracy of the Taylor expansion-based method cannot be guaranteed [12,13,18].

Recently, PCE has been used in different areas including the communities of structural reliability and control theory [4,19]. This approach was proposed by Wiener [14] to quantify uncertainty that follows a normal distribution and later extended to uncertainty with several standard distributions (e.g., uniform) by Xiu [20,21]. The PCE-based methods approximate uncertainty and its effect on model responses with a spectral expansion as a function of PCE coefficients. The PCE coefficients of uncertainty in models are given by a prior probability density function (PDF), but coefficients of model outputs are unknown. Thus, the key of PCE is to solve the PCE coefficients of model responses. Once the coefficients of model responses (e.g., performance function) are obtained, they can be used to approximate the PDF of performance function [19,20] and to evaluate failure probability in reliability analysis [18,22,23]. Since each model response can be represented explicitly with random variables of input uncertainties, it is useful to obtain the statistics of the response. Depending on how the PCE coefficients of model responses are calculated, it is categorized into intrusive and nonintrusive methods [20,24,25].

For the intrusive method, the coefficients of model outputs are obtained by projecting the governing equations into PCE, which requires calculating the inner product between a basis polynomial and the equations described with a truncated PCE in a stochastic Galerkin approach. The inner product generates a family of nested deterministic models to represent the original stochastic models. Calculating the coefficients with the inner product can be computationally prohibitive when models have complex functions (e.g., nonpolynomial functions) and many uncertainties. Nonintrusive PCE is another way, which has been used in the reliability analysis of bridge, truss, and frame structures [4,19,26,27]. Similar to MC, the convergence and accuracy of nonintrusive PCE depend on samples used for UQ. These samples are often referred to as collocation points. While practical, the convergence and accuracy of the nonintrusive PCE-based UQ can be affected by the total number of collocation points and by how the collocation points are generated [28,29]. The nonintrusive methods can be further categorized as a regression-based approach and projection-based approach, depending on how the PCE expressions of output responses are calculated. The former estimates the PCE coefficients by minimizing the mean square error of the approximation of model responses, while each PCE coefficient for the latter is defined with a multidimensional integral. It should be noted that as the number of uncertainties increase, both the size of dimension in the integral and the total number of PCE terms increase. Thus, the projection-based approach can be computationally demanding.

To address these issues, our previous works [29,30] have presented algorithms to combine the PCE with the generalized dimension reduction method (gDRM) or M-variate DRM [31]. This approach approximates high-dimensional integrals in the spectral projection (SP) [28] with a few low-dimensional ones. These low-dimensional integrals involve at most M integration variables [29] that are much smaller than the number of uncertainties. To further reduce the computational cost, the lower-dimensional integrals resulting from the gDRM were calculated with a sampling-based method (i.e., Gaussian quadrature rules) [30]. However, our previous works only deal with uncertainty that follows a standard distribution as done in the classic PCE theory.

Uncertainty may have nonstandard distributions. For example, lognormal distribution describes several natural phenomena, e.g., the bubble size distributions in the gas-liquid and gas-liquid-solid systems [32]. Because there is an explicit relationship between the lognormal and normal variables, we capitalize on the relationship following the work by Ghanem in [33] to efficiently deal with the lognormal uncertainty. Several tools were developed to approximate lognormal random variables [20,24,34–39], which include the isoprobabilistic transformation, such as the Rosenblatt or Nataf transformation [37,38], that transforms the original random variable into independent normal variables. Besides, the lognormal distribution was described in [39] by constructing orthogonal polynomials using the Stieltjes–Wigert polynomials. Compared to existing techniques, our algorithm approximates a lognormal distribution with a normal random variable based on the explicit relationship [33] and couples the PCE-based approximation with the gDRM to quantify the effect of many uncertainties on model responses in a computationally efficient way.

In summary, this work expands our previous work to couple the gDRM with the PCE to quantify the effect of nonstandard uncertainties on model outputs. For algorithm validation, lognormal and Weibull distributions are chosen as the testbed, which are approximated with standard random variables [20,33]. In addition, four examples in structural reliability analysis were chosen to show the efficiency of the algorithm and to discuss the UQ accuracy by comparing the results with recent works in the literature.

This paper is organized as follows. A brief description of the PCE-based UQ is given in Section 2. Section 3 shows procedures to approximate nonstandard uncertainty, followed by the gDRM-based PCE to deal with high-dimensional UQ problems. For algorithm verification, four numerical examples in structural reliability analysis are presented in Section 4, and a brief conclusion is given in Section 5.

2. Background of Polynomial Chaos Expansion (PCE)

Suppose that a stochastic process can be defined as $u = K(g)$, where $g = (g_1, \ldots, g_n) \in \mathbb{R}^n$ is a Gaussian vector involving n uncertain parameters ($n \geq 1$) and K is the function to describe the relationship between a model response u and g. It is assumed each parameter in g is independent, i.e., any correlation among uncertain parameters is not considered. In this work, uncertainty in model parameters, i.e., input random variable of the models, is hereafter referred to as parametric uncertainty.

To obtain a PCE expression of u, the first step is to rewrite each parametric uncertainty g_i as a function of a Gaussian variable x_i as in [20]:

$$g_i = g_i(x_i) = \sum_{k=0}^{p} \overline{g}_{i,k} H_k(x_i) \tag{1}$$

where $\left\{\overline{g}_{i,k}\right\}$ are the PCE coefficients to estimate the ith parametric uncertainty, which is defined by the ith standard normal distribution, i.e., $x_i \sim \mathcal{N}(0,1)$. These coefficients $\left\{\overline{g}_{i,k}\right\}$ are often assumed to be a given a priori or can be estimated with parameter estimation techniques [40]. In (1), H_k is the one-dimensional Hermite polynomial basis function, and the polynomial order p defines the number of terms to approximate g_i. Uncertainty in model parameters will introduce uncertainty in model responses. Thus, once the PCE coefficients of uncertainty in (1) are available, the PCE expansion of model response u can be subsequently defined with random variables as [20]:

$$u(x) = K(g) = \sum_{j=0}^{m-1} \overline{u}_j H_{j,n}(x) \tag{2}$$

where $x = (x_1, \ldots, x_i, \ldots, x_n)$ and $\left\{H_{j,n}(x)\right\}$ are the jth n-dimensional orthogonal polynomials defined by basis functions $\{H_k(x_i)\}$ for each uncertainty. Explicit expressions of $\left\{H_{j,n}(x)\right\}$ can be found in [15]. In (2), $\left\{\overline{u}_j\right\}$ are the deterministic PCE coefficients of model response u, which are unknown and have to be calculated. Further, m is the total number of terms to approximate $u(x)$, which can be calculated with the number of parametric uncertainty n and the polynomial order p as in [20]:

$$m = \frac{(n+p)!}{p!n!} \tag{3}$$

Unlike $\left\{ \overline{g}_{i,k} \right\}$, PCE coefficients of model response, $\{\overline{u}_j\}$, are unknown and have to be calculated by projecting (2) onto each polynomial basis function $H_{j,n}$ and by using the spectral projection (SP), which can be defined as in [20]:

$$\overline{u}_j = \frac{\langle u(x)H_{j,n}(x)\rangle}{\langle H_{j,n}^2(x)\rangle} \ (\forall j \in \{0, \ldots, m-1\}) \tag{4}$$

where the angle brackets $\langle \cdot \rangle$ represent the inner product operator for the n-dimensional random space $\mathbb{R}^n$ defined by x. For example, the inner product of a function $f(x)$ is defined as:

$$\langle f(x)\rangle = \int_{\mathbb{R}^n} f(x)\mathcal{W}(x)\,dx \tag{5}$$

where $\mathcal{W}(x)$ is the weighted function defined by the joint PDFs of random variables x. Note that the inner product of $f(x)$ in (5) can be further specified by the expectation operator $E[\cdot]$, i.e., $E[f(x)] = \langle f(x)\rangle$. Accordingly, the PCE coefficient $\overline{u}_j$ in (4) can be further written as in [20]:

$$\overline{u}_j = \frac{1}{\gamma_j} \int_{\mathbb{R}^n} u(x)H_{j,n}(x)\mathcal{W}(x)\,dx \tag{6}$$

where $\gamma_j = E\left[H_{j,n}^2\right]$ is the normalization factor. It is worth mentioning that the jth normalization factor of the Hermite polynomial basis can be easily calculated as $\gamma_j = j!$ [20]. In addition, the multidimensional integral in (6) is solved over the entire random domain $\mathbb{R}^n$ of x, and $\mathcal{W}(x)$ here is the Gaussian weighted function. By substituting the PCE coefficients in (6) into (2), uncertainty in model responses can be rapidly estimated. For example, the statistical central moments μ_u and σ_u^2, i.e., mean and variance of model response u, can be analytically calculated as in [20]:

$$\mu_u = E[u] = E\left[\sum_{j=0}^{m-1} \overline{u}_j H_{j,n}\right] = \overline{u}_0 E[H_{0,n}] + E\left[\sum_{j=1}^{m-1} \overline{u}_j H_{j,n}\right] = \overline{u}_0 \tag{7}$$

$$\begin{aligned}
\sigma_u^2 = E\left[\{u - E[u]\}^2\right] &= E\left[\left\{\sum_{j=0}^{m-1} \overline{u}_j H_{j,n} - \overline{u}_{j,0}\right\}^2\right] \\
&= E\left[\left\{\sum_{j=1}^{m-1} \overline{u}_j H_{j,n}\right\}^2\right] = \sum_{j=1}^{m-1} \{\overline{u}_j\}^2 E\left[H_{j,n}^2\right]
\end{aligned} \tag{8}$$

where $H_{0,n} \equiv 1$ is the constant polynomial. As seen in (7) and (8), the first PCE coefficient $\overline{u}_0$ can be used to estimate the mean, while the rest of the coefficients ($\overline{u}_{j\neq0}$) and the expectation of the squared basis functions ($E\left[H_{j\neq0,n}^2\right]$) can be used to estimate the uncertainty around the mean of u [20]. Similarly, other higher-order statistical central moments, such as skewness s_u and kurtosis κ_u, can be rapidly calculated with the high-order PCE coefficients as in [41].

$$s_u = E\left[\{u - E[u]\}^3\right]\frac{1}{\sigma_u^3} = E\left[\left\{\sum_{j=0}^{m-1} \overline{u}_j H_{j,n} - \overline{u}_{j,0}\right\}^3\right]\frac{1}{\sigma_u^3} = E\left[\left\{\sum_{j=1}^{m-1} \overline{u}_j H_{j,n}\right\}^3\right]\frac{1}{\sigma_u^3} \tag{9}$$

$$\kappa_u = E\left[\{u - E[u]\}^4\right]\frac{1}{\sigma_u^4} = E\left[\left\{\sum_{j=0}^{m-1} \overline{u}_j H_{j,n} - \overline{u}_{j,0}\right\}^4\right]\frac{1}{\sigma_u^4} = E\left[\left\{\sum_{j=1}^{m-1} \overline{u}_j H_{j,n}\right\}^4\right]\frac{1}{\sigma_u^4} \tag{10}$$

Details to calculate these statistical central moments can be found in [41–43]. These central moments are used to evaluate the reliability index and failure probability for structural reliability analysis in moment methods [18,22]. Thus, they are chosen in this work to discuss the performance of different UQ algorithms in terms of UQ accuracy and computational efficiency in Section 4.

3. Efficient UQ for Nonstandard Uncertainties

3.1. Approximation of Lognormal Uncertainties

Since lognormal uncertainty has an explicit relationship with Gaussian uncertainty, it will be transformed and estimated with the Gaussian variable as in [33]. For clarity, only a parametric uncertainty, which refers to an input random variable of the model, is considered, thus the subscript i in (1) will not be used.

Suppose a normally distributed uncertainty $g(x)$ is approximated with a random variable x; and a lognormal uncertainty $l(x)$ can be calculated by taking the exponential operator $exp[\cdot]$ of $g(x)$, which results in a relationship as in [33]:

$$g(x) = \ln l(x), \; l(x) = exp[g(x)] \tag{11}$$

In (11), $g(x)$ will be referred to as the normalized Gaussian distribution, and the mean and the variance of $g(x)$ are defined as μ_g and σ_g^2, respectively. As done for the Gaussian uncertainty in (1), let one assume that $l(x)$ can be approximated with a PCE expansion as in:

$$l(x) = \sum_{k=0}^{p} \bar{l}_k H_k(x) \tag{12}$$

where $\left\{ \bar{l}_k \right\}$ are the PCE coefficients, describing the original lognormal uncertainty, which will be derived based on normalized Gaussian variables. As defined in (1), H_k is the Hermite polynomial basis function. Since the first coefficient of PCE in (12) represents the mean value as described in (7), $\bar{l}_0$ can be easily given as $\mu_l = e^{\mu_g + (\sigma_g^2/2)}$ [33]. For other higher-order terms, we apply the SP to calculate PCE coefficients as a function of the normalized Gaussian random variables. Following the similar procedures to calculate the coefficients of model responses in (4), the resulting PCE expansion of the lognormal uncertainty in (12) can be represented as in [20,33]:

$$\begin{aligned}
l(x) &= \sum_{k=0}^{p} \bar{l}_k H_k(x) = \bar{l}_0 \sum_{k=0}^{p} \frac{\sigma_g^k}{k!} H_k(x) \\
&= \bar{l}_0 \left(1 + \sigma_g x + \tfrac{1}{2!}\sigma_g^2 (x^2 - 1) + \tfrac{1}{3!}\sigma_g^3 (x^3 - 3x) + \dots \right)
\end{aligned} \tag{13}$$

Once the PCE coefficients of a lognormal uncertainty are available, the resulting uncertainty in model response as in (2) can be approximated following the similar procedures as in Section 2.

Compared to the PDF of the Gaussian random variable that has a symmetrical shape, the required polynomial order p in (13) for approximating the original lognormal uncertainty needs to be carefully selected to capture its asymmetric property, such as the long tail. As an example, Figure 1 shows the simulation results of a lognormal uncertainty, where three different polynomial orders (i.e., $p = 2, 3$, and 4) were used. Note that the PDFs shown in Figure 1 are estimated with the kernel density estimation function in MATLAB. The accuracy to approximate a lognormal distribution with different polynomial orders is compared to MC with 10^7 samples.

As seen in Figure 1, the accuracy to approximate a lognormal uncertainty can be affected by p and the magnitude of uncertainty. When uncertainty is small, the approximation of a lognormal uncertainty with different p values exhibits almost identical results as in Figure 1a. Whereas, when uncertainty is large, there is a noticeable difference between the PCE-based approximation and MC. However, when p was increased (>2), as seen in Figure 1b, the difference between the PCE-based approximation and MC is insignificant. Thus, we focus on two different values of p (2 vs. 3) in this work to show the performance of the UQ method, which will be discussed in Section 4.

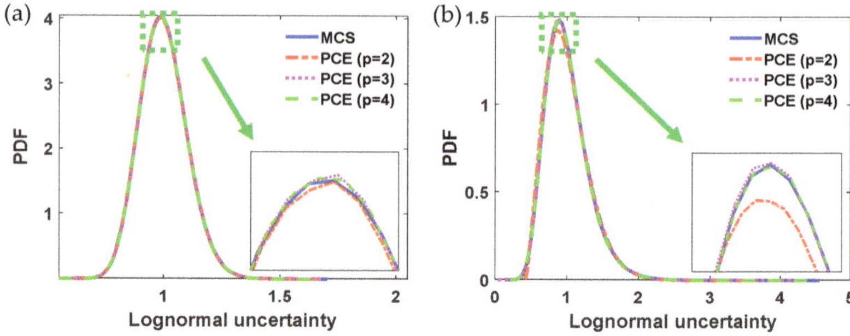

Figure 1. Illustration of the effect of polynomial order on the approximation accuracy of a lognormal uncertainty. The PCE-based approximation of a lognormal uncertainty is built with the normalized Gaussian variable with different polynomial orders and compared to the MC-based method. The mean value of *l* was set to 1 in both graphs, but two different standard deviations were used: 0.1 was used in (**a**) and 0.3 was used in (**b**).

3.2. Approximation of Other Nonstandard Uncertainties

Compared to approximating a lognormal uncertainty with a PCE expression that has an explicit correlation to a Gaussian random variable, the approximation of other nonstandard uncertainties can be made with its probability distribution function [20]. To estimate a nonstandard uncertainty in PCE, suppose Y is an uncertainty with nonstandard distributions (e.g., a Weibull distribution), which has a PCE expression as:

$$Y(x) = \sum_{k=0}^{p} \overline{Y}_k \Psi_k(x) \tag{14}$$

where $\{\overline{Y}_k\}$ are PCE coefficients, and $\{\Psi_k\}$ are polynomial basis functions. Considering Y follows a nonstandard distribution, the calculation of $\{\overline{Y}_k\}$ is difficult since the explicit correlation between Y and x is unknown [20]. In this work, a technique in [20] is used to estimate an uncertainty that follows a nonstandard distribution. The resulting PCE coefficients in (14) can be calculated as:

$$\overline{Y}_k = \frac{1}{\gamma_k} E\left[F_Y^{-1}(F_x(x)) \Psi_k(x) \right] = \frac{1}{\gamma_k} \int_{\mathbb{R}^n} F_Y^{-1}(F_x(x)) \Psi_k(x) W(x) dx \tag{15}$$

where F_x is the cumulative distribution function (CDF) defined as $F_x(x) = \int_{-\infty}^{x} W(t) dt$, and F_Y^{-1} is the inverse of a cumulative distribution function of Y, i.e., F_Y. For example, let one suppose that a Gaussian random variable x is used to approximate a Weibull-distributed uncertainty Y. Then, $W(x)$ is the probability distribution function of x; Ψ_k is the Hermite polynomial basis function; F_x is the CDF of the Gaussian random variable x; and F_Y^{-1} is the inverse of the CDF of Weibull-distributed uncertainty Y. Details about the derivation of (15) and its approximation are not given for brevity and can be found in [20].

3.3. Modified gDRM-Based PCE Using Quadrature Rules

When the PCE coefficients of uncertainty are available, such as (1) for the normal and (12) for lognormal uncertainty, SP is used to calculate the PCE coefficients of model response in (2), which requires calculating multivariate integrals as in (6). As in (3), the total number of terms (m) to accurately approximate uncertainty in model response in (2) is a function of the polynomial order (p) and the total number of uncertainties (n). When p is large (e.g., >2) and when the number of uncertainties increases, the number of terms in (2) can be greatly increased. This can increase the computational cost to calculate the PCE coefficients of model outputs. To reduce the computational cost, the gDRM will be used with quadrature rules to quickly calculate the PCE coefficients as in [30]. This approach is hereafter referred to as the modified gDRM-based PCE (or mgDRM-PCE).

Suppose that a continuous and differentiable function can be defined as $f(x) = u(x)H_{j,n}(x)$, which is a portion of the integrand in (6). Then, the multivariate integral in (6) is rewritten as:

$$E[f(x)] = \int_{\mathbb{R}^n} f(x)\mathcal{W}(x)\,dx \tag{16}$$

where $\mathcal{W}(x)$ represents the weighted function described by the joint PDFs of x. Note that the function $f(x)$ is defined to easily calculate a specific PCE coefficient $\overline{u}_j$ in (6), and this expression can also be reused for different $j = 0, 1, \ldots, m-1$.

Following the definition of gDRM [31] and the representation in (16), each of the unknown PCE coefficients of the model response in (6) can be calculated as in [29]:

$$\overline{u}_j(x) = \frac{1}{\gamma_j}E[f(x)] \cong \frac{1}{\gamma_j}\sum_{r=0}^{M}(-1)^r \binom{n-M+r-1}{r} E[f_{M-r}] \tag{17}$$

where f_{M-r} in the expectation operator $E[\cdot]$ is a function of $(M-r)$ random variables selected from x. In this way, several lower-dimensional functions, involving up to M random variables of x, can be used to approximate the original function f, which is written as in [31]:

$$f_{M-r} = \sum_{d_1<d_2<\ldots<d_{M-r}} f(0,\ldots,0,x_{d_1},0,\ldots,0,x_{d_2},0,\ldots,0,x_{d_{M-r}},0) \tag{18}$$

where $d_1, d_2, \ldots, d_{M-r} \in \{1, 2, \ldots, n\}$. Since the gDRM considers combinations of the random variables in x, the total number of terms in (18) can be calculated as $\binom{n}{M-r}$. Notably, the last term in (18), when $M = r$, can be calculated as $f_0 = f(0)$, which can be solved by setting the mean values of all random variables x to 0. For example, when a bi-variate dimension reduction method (BiDRM) is used, i.e., $M = 2$, the n-variate function $f(x)$ will be approximated with $\binom{n}{2}$ two-variate functions, $\binom{n}{1}$ one-variate functions, and a constant term f_0. Thus, the expectation of f_{M-r}, i.e., $E[f_{M-r}]$ in (17), results in a $(M-r)$-dimensional integral for each combination of $(M-r)$-random variables selected from x, which gives $\binom{n}{M-r}$ $(M-r)$-dimensional integrals in total.

As noted in [29,30], the accuracy to approximate a high-dimensional integral with several lower-dimensional ones increases as M increases. For example, compared to the BiDRM, a tri-variate DRM (TriDRM) approximates a high-dimensional integral in (16) with $\binom{n}{3}$ three-dimensional integrals, $\binom{n}{2}$ two-dimensional integrals, $\binom{n}{1}$ one-dimensional integrals, and a constant term f_0. As such, extra effort is required to calculate $\binom{n}{3}$ additional three-dimensional integrals. Since the total number of lower-dimensional integrals increases, the computational time to approximate (16) may increase. To address this issue, we will estimate the resulting lower-dimensional integrals with quadrature rules [44] to reduce the computational cost [30,31]. It should be noted that this approach reduces the computational difficulty to evaluate high-dimensional integrals using sampling methods, which appears to be methodologically similar to the anchored analysis-of-variance (ANOVA) expansion in [45]. However, the anchored ANOVA expansion is dependent on the choice of the anchor point, which impacts the accuracy of the expansion and the truncation dimension [45], while the mgDRM does not require any strategy for the decomposition procedure in (18). A detailed description on the ANOVA expansion can be found in [45,46].

To calculate the expectation of one-variate function $E[f_1]$—an integral only involves a single random variable, e.g., x_{d_1} of f_1 in (18), a one-dimensional quadrature rule can be defined as [30]:

$$\int_{-\infty}^{\infty} f(0,\ldots,0,x_{d_1},0,\ldots,0)\mathcal{W}(x_{d_1})dx_{d_1} \cong \sum_{q_1=1}^{\theta_1} f\left(0,\ldots,0,x_{d_1}^{q_1},0,\ldots,0\right)\cdot\alpha_{d_1}^{q_1} \tag{19}$$

where $\left\{ x^{q_1}_{d_1}, \, \alpha^{q_1}_{d_1} \right\}^{\theta_1}_{q_1=1}$ is a set of quadrature points $(x^{q_1}_{d_1})$ and their corresponding weights $(\alpha^{q_1}_{d_1})$ used to compute the one-dimensional integral resulting from the gDRM. In this work, Gauss–Hermite quadrature rules [44] are used to estimate low-dimensional integrals in (17) in the gDRM step.

To generate quadrature points to approximate multidimensional integrals in (17), a full tensor product grid can be constructed based on the one-dimensional quadrature rules. Note that these lower-dimensional integrals resulting from the gDRM only involve a few integration variables. Thus, the approximation with the full tensor product grid can be finished in real time.

To better illustrate the quadrature points-based approximation, suppose the PCE coefficients of model outputs in (17) are calculated with the TriDRM. This implies that these lower-dimensional integrals to approximate the multidimensional integrals in SP involve at most three random variables in x (or $M = 3$). That is, a multidimensional integral $(E[f(x)])$ is estimated with $\binom{n}{1}$ one-dimensional, $\binom{n}{2}$ two-dimensional, and $\binom{n}{3}$ three-dimensional integrals in total, which can be computed with the quadrature rules as in [30]:

$$E[f_1] \cong \sum_{d_1} \left\{ \sum_{q_1=1}^{\theta_1} f\left(0,\dots,0,x^{q_1}_{d_1},0,\dots,0\right) \cdot \alpha^{q_1}_{d_1} \right\} \tag{20}$$

$$E[f_2] \cong \sum_{d_1<d_2} \left\{ \sum_{q_1=1}^{\theta_1} \sum_{q_2=1}^{\theta_2} f\left(0,\dots,0,x^{q_1}_{d_1},0,\dots,0,x^{q_2}_{d_2},0,\dots,0\right) \cdot \left(\alpha^{q_1}_{d_1} \otimes \alpha^{q_2}_{d_2}\right) \right\} \tag{21}$$

$$E[f_3] \cong \sum_{d_1<d_2<d_3} \left\{ \sum_{q_1=1}^{\theta_1} \sum_{q_2=1}^{\theta_2} \sum_{q_3=1}^{\theta_3} f\left(0,\dots,0,x^{q_1}_{d_1},0,\dots,0,x^{q_2}_{d_2},0,\dots,0,x^{q_3}_{d_3},0\right) \right.$$
$$\left. \cdot \left(\alpha^{q_1}_{d_1} \otimes \alpha^{q_2}_{d_2} \otimes \alpha^{q_3}_{d_3}\right) \right\} \tag{22}$$

where θ_i $(i = 1, 2, 3)$ is the number of quadrature points, $x^{q_i}_{d_i}$, for each integration variable; $\alpha^{q_i}_{d_i}$ is the weight of each quadrature point; and $\otimes$ means the tensor product. In summary, the total number of quadrature points for low-dimensional integrals can be defined as $Q = \prod_{i=1}^{M-r} \theta_i = \theta_i^{M-r}$ [44]. It is important to note that as the dimension of a random space increases, the number of integrals resulting from gDRM increases. In this case, the total number of evaluations of these integrals with quadrature points is also increased, which may increase the computational cost. This issue will be discussed with benchmark examples in Section 4.

3.4. Summary of the UQ Algorithm

Our algorithm quantifies the effect of nonstandard uncertainty on model responses by coupling PCE with the gDRM. A flowchart summarizing the procedures of the modified gDRM-based PCE to deal with nonstandard random variables is shown in Figure 2.

As in Figure 2, nonstandard uncertainties are firstly approximated with standard random variables. Once the PCE-based surrogate model of parametric uncertainty, i.e., an input random variable of the models is available, model responses can be approximated with coupled PCE coefficients as in (2). The PCE coefficients of model responses are calculated with (4). This produces high-dimensional integrals, which are not trivial to solve, especially when the number of uncertainties is large and when the model involves nonpolynomial terms. To address this, the gDRM, such as the BiDRM and TriDRM noted in Section 3.3, is used to convert the high-dimensional integral into few low-dimensional ones. To further decrease the computational cost, quadrature points generated with Gaussian-quadrature rules are used to numerically solve these low-dimensional integrals. The approximation of lower-dimensional integrals is highlighted in green boxes in Figure 2. Once the PCE coefficients of model responses are available, the statistical moments of model outputs (e.g., mean, variance, skewness, and kurtosis) can be quickly calculated with (7)~(10).

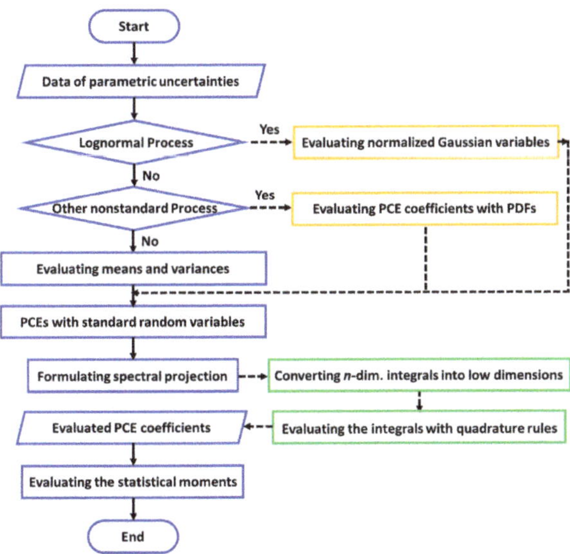

Figure 2. Schematic of the modified gDRM-based PCE for UQ under many uncertainties that follow nonstandard distributions.

4. Benchmark Examples in Structural Reliability Analysis

Four examples in structural reliability analysis are chosen to illustrate the performance of the UQ algorithm to deal with nonstandard uncertainties. The correlation among uncertainties is not considered, since our objective is to study the applicability of the algorithm to tackle many uncertainties. When uncertainties are correlated, the number of random variables to approximate uncertainties will be decreased, thus reducing the number of dimensions. To evaluate the UQ accuracy, statistical moments (mean, standard deviation, skewness, and kurtosis) are calculated and compared with other techniques, which include MC simulations and the least angle regression-based PCE.

4.1. Example 1: Linear Performance Function

As given in Figure 3, a linear performance function is used to study the plastic collapse mechanisms of a one-bay frame, which is mathematically described as in [47,48]:

$$Z = u(X) = X_1 + 2X_2 + 2X_3 + X_4 - 5X_5 - 5X_6 \tag{23}$$

Following [47,48], $X_1 \sim X_6$ in (23) are assumed to be independent and lognormally distributed; the mean and standard deviation of each variable are listed in Table 1. In this case study, all parameters are considered as parametric uncertainties, resulting in a six-dimensional random space. Since uncertainty follows a lognormal distribution, Hermite polynomials were used as the basis functions to build the PCE-based surrogate model of the response Z in (23).

The first step to build a surrogate model of Z in (23) is to formulate the PCE expressions of lognormally distributed uncertainties, $X_1 \sim X_6$. In this case study, (13) was used to build a PCE model for each uncertainty with normalized Gaussian variables that are related to the lognormal uncertainty as in Section 3.1. The second coefficients of the normalized Gaussian variables (i.e., σ_g for $X_1 \sim X_6$) were calculated as 0.0998 for $X_1 \sim X_4$ and 0.2936 for X_5 and X_6, respectively. These were also used to derive the explicit PCE expressions of lognormal uncertainties as in (13). Once the PCE expressions of uncertainties ($X_1 \sim X_6$) were available, the mgDRM-PCE in Section 3.3 was used to compute the PCE coefficients of Z in (23).

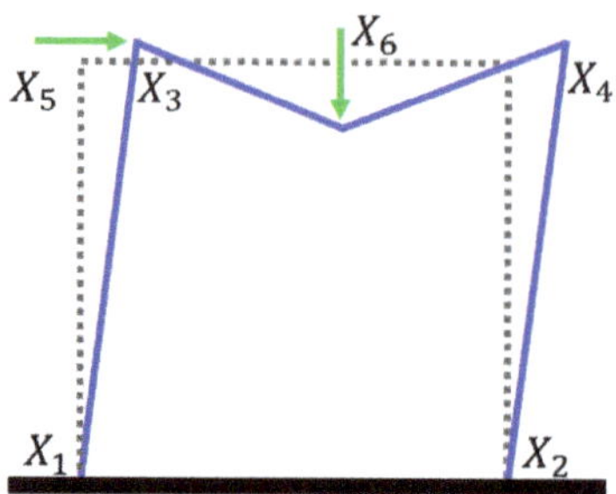

Figure 3. Illustration of the one-bay frame for a plastic collapse mechanism [48]. Reprinted from [48], Copyright (2019), with permission from Elsevier.

Table 1. Details of the uncertain variables in Example 1 [47,48]. Reprinted from [48], Copyright (2019), with permission from Elsevier.

Variable	Distribution Type	Mean	Standard Deviation
X_1	Lognormal	120	12
X_2	Lognormal	120	12
X_3	Lognormal	120	12
X_4	Lognormal	120	12
X_5	Lognormal	50	15
X_6	Lognormal	40	12

In this case study, different polynomial orders of $X_1 \sim X_6$ ($p = 2$ vs. 3) and different numbers of random variables in the gDRM ($M = 2$ vs. 3) were used. The UQ accuracy with different combinations of p and M was compared to the results of MC by calculating the relative error (ϵ_R), which is defined as in:

$$\epsilon_R = \left| \frac{e_{MC} - e}{e_{MC}} \right| \tag{24}$$

where e_{MC} is a reference of a specific statistical moment calculated with MC, and e represents the corresponding statistical moment calculated with other UQ methods. The simulation results are summarized in Table 2 and compared with existing algorithms: a recent work in [48] that integrates the BiDRM with a high-order unscented transformation [49] and the least angle regression-based PCE (LAR-based PCE) [50,51]. For MC, 10^7 samples were used for each uncertainty to ensure UQ accuracy.

In Table 2, mBiDRM-PCE means that the quadrature rules were used to estimate integrals from the BiDRM, which involve at most two integration variables, while mTriDRM-PCE means that the quadrature rules were applied to the TriDRM, which has at most three integration variables. The BiDRM converted a six-dimensional integral into 15 two- and 6 one-dimensional ones, while 20 three-, 15 two-, and 6 one-dimensional integrals were used in the TriDRM. In addition, five quadrature points of each dimension were generated following the Gauss–Hermite quadrature rules (i.e., $\theta_i = 5$), for which a full-tensor product grid was built to calculate the low-dimensional integrals from the BiDRM and the TriDRM.

In addition, for the LAR-based PCE, the LAR procedure was implemented to select a set of basis polynomials to build a sparse PCE and estimate the unknown PCE coefficients of the response Z in (23) with the least square regression. It is important to note the LAR-based PCE in this work does not involve the adaptive solver as in [52] because our objective is to assess the accuracy of the surrogate model derived by the gDRM-based PCE. To obtain model evaluations of Z, 1000 pairs of random samples of parametric uncertainties were used. Thus, the total number of model runs is 1000. Further, the LAR method was used to identify the sensitive basis polynomials of output Z until the maximum correlation between the residual and the basis polynomials is less than a preselected effective value, which was set to 1.0×10^{-6}. In this case study, two different polynomial orders ($p = 2$ and 3) were studied. Thus, the number of PCE terms of LAR-based PCE was 13 and 19, respectively. In contrast, the number of PCE terms for our algorithm was 28 and 84, respectively.

Table 2. Summary of statistical moments and their relative errors in Example 1.

UQ Methods	μ_Z [ϵ_R]	σ_Z [ϵ_R]	s_Z [ϵ_R]	κ_Z [ϵ_R]
MC	269.9226	103.2812	−0.5276	3.6135
mBiDRM-PCE ($p = 2$)	270.0000 [0.0287%]	103.2173 [0.0619%]	−0.4840 [8.2758%]	3.3813 [6.4250%]
mBiDRM-PCE ($p = 3$)	270.0000 [0.0287%]	103.2703 [0.0105%]	−0.5261 [0.2911%]	3.5897 [0.6577%]
mTriDRM-PCE ($p = 2$)	270.0000 [0.0287%]	103.2173 [0.0619%]	−0.4840 [8.2758%]	3.3813 [6.4250%]
mTriDRM-PCE ($p = 3$)	270.0000 [0.0287%]	103.2703 [0.0105%]	−0.5261 [0.2911%]	3.5897 [0.6577%]
LAR-based PCE ($p = 2$)	270.0000 [0.0287%]	103.2173 [0.0619%]	−0.4840 [8.2758%]	3.3813 [6.4250%]
LAR-based PCE ($p = 3$)	270.0000 [0.0287%]	103.2703 [0.0105%]	−0.5261 [0.2911%]	3.5897 [0.6577%]
Reference [48]	270.0010 [0.0291%]	103.2146 [0.0645%]	−0.5165 [2.1124%]	3.6980 [2.3383%]

As seen in Table 2, the difference in the mean value of Z (μ_Z) calculated with the mBiDRM-PCE and mTriDRM-PCE is insignificant, and their results are almost identical to MC. For the other higher-order statistical moments, it was found that the relative error is relatively larger when p was set to 2, as compared to the results when p was set to 3. This indicates that more polynomial terms to approximate a lognormal uncertainty can improve the UQ accuracy. Besides, compared to the LAR-based PCE, our algorithm shows identical results for the four statistical moments, which demonstrates the accuracy of the gDRM-based PCE. It was also found that, when p was set to 3, the mBiDRM-PCE outperforms the results in a recent work [48], for which only the BiDRM was used. This shows the advantage of combining the PCE with the gDRM, because PCE can not only provide an analytical expression for model responses as a function of random variables, but also explicitly quantify the impact of approximation on UQ accuracy.

For comparison, the simulation results of mBiDRM-PCE and mTriDRM-PCE with different polynomial orders and MC are shown in Figure 4, where the cumulative density function (CDF) of each UQ method was approximated with the built-in kernel density estimation function in MATLAB. As seen, the CDF of Z in (23) is almost identical to the results obtained with MC, when p was 3. This shows our algorithm can accurately quantify the effect of lognormal uncertainties on model responses.

4.2. Example 2: Roof Structure

The performance function of a roof truss structure as illustrated in Figure 5 was used to study the UQ accuracy of our algorithm to deal with different nonstandard uncertainties, when the system involves nonlinear polynomial functions. The performance function is described as in [19,53]:

$$Z = u(X) = 0.03 - 0.5ql^2 \left(\frac{3.81}{A_2 E_2} + \frac{1.13}{A_1 E_1} \right) \tag{25}$$

where q and l represent a vertical load and roof span that are used to calculate the nodal load $P = ql/4$, and A and E are the cross-sectional area and elastic modulus, respectively. In this case study, all parameters in the performance function in (25) are uncertain with predefined PDFs. This yields a six-dimensional random space ($n = 6$). Details about the statistical description of uncertainties are given in Table 3.

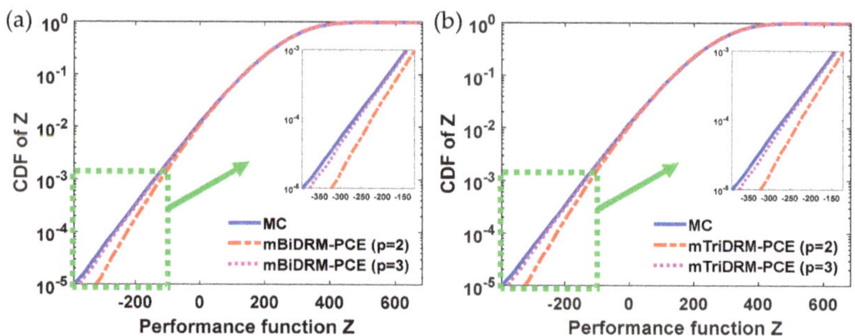

Figure 4. Comparison of the cumulative density functions of Z calculated with the algorithm in this work and MC in Example 1. (**a**) mBiDRM-PCE and (**b**) mTriDRM-PCE.

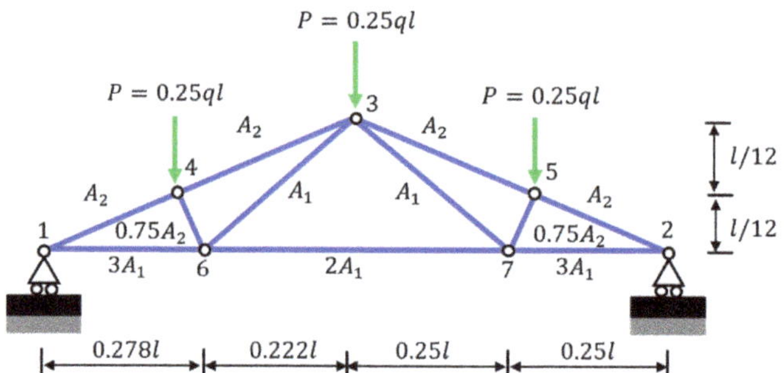

Figure 5. Illustration of a roof truss structure [19,53]. Reprinted from [19], Copyright (2018), with permission from Elsevier.

Table 3. Details of the uncertain variables in Example 2 [19,53]. Reprinted from [19], Copyright (2018), with permission from Elsevier.

Variable	Distribution Type	Mean	Standard Deviation	Unit
q	Lognormal	20,000	1400	N/m
l	Weibull	12	0.12	m
A_1	Lognormal	9.8×10^{-4}	5.89×10^{-5}	m^2
A_2	Lognormal	400×10^{-4}	48×10^{-4}	m^2
E_1	Lognormal	1×10^{11}	6×10^9	N/m^2
E_2	Lognormal	2×10^{10}	1.2×10^9	N/m^2

Like Example 1, Hermite polynomials were used as the basis functions, and the PCE expressions of lognormal uncertainties were constructed with normalized Gaussian variables. In addition, it is assumed that l follows a Weibull distribution, for which the PCE expression was approximated with standard random variables using its PDF as described in Section 3.2. Additionally, to compute the unknown PCE coefficients of the performance function Z in (25), the mBiDRM and mTriDRM were used to reduce the computational cost by converting the high-dimensional integrals into low-dimensional ones, which were further approximated with quadrature points constructed by the Gaussian–Hermite quadrature rules. Note that the number of quadrature points for each dimension θ_i was set to 5 as done in Example 1. To quantify the UQ accuracy, the relative errors ϵ_R defined in (24) were calculated by referring to the results of MC. For MC, 10^7 samples were used to ensure UQ accuracy.

The results of Example 2 are summarized in Table 4. It is worth mentioning that the results from previous works (e.g., [19,53]) were not given. This is because the uncertainty in [19,53] follows a normal distribution and we intentionally introduced lognormal and Weibull distributions to test the performance of our method to deal with different types of nonstandard uncertainties. Following Example 1 in the previous section, however, the LAR-based PCE was chosen to compare the UQ accuracy with two different polynomial orders ($p = 2$ and 3). The simulation results are given in Table 4. The total number of PCE terms for our algorithm is 28 and 84, when the polynomial order for each uncertain parameter was set to 2 and 3, respectively, while the number of PCE terms of the LAR-based PCE is 28 and 77 for each polynomial order, respectively.

Table 4. Summary of statistical moments and their relative errors in Example 2.

UQ Methods	μ_Z [ϵ_R]	σ_Z [ϵ_R]	s_Z [ϵ_R]	κ_Z [ϵ_R]
MC	0.0064	0.0024	−0.3004	3.1614
mBiDRM-PCE ($p = 2$)	0.0064 [0.0065%]	0.0024 [0.0134%]	−0.2955 [1.6162%]	3.1189 [1.3427%]
mBiDRM-PCE ($p = 3$)	0.0064 [0.0060%]	0.0024 [0.0030%]	−0.3000 [0.1340%]	3.1485 [0.4067%]
mTriDRM-PCE ($p = 2$)	0.0064 [0.0065%]	0.0024 [0.0156%]	−0.2952 [1.7095%]	3.1186 [1.3512%]
mTriDRM-PCE ($p = 3$)	0.0064 [0.0060%]	0.0024 [0.0011%]	−0.3000 [0.1374%]	3.1620 [0.0202%]
LAR-based PCE ($p = 2$)	0.0064 [0.0106%]	0.0024 [0.0231%]	−0.2933 [2.3358%]	3.1172 [1.3961%]
LAR-based PCE ($p = 3$)	0.0064 [0.0058%]	0.0024 [0.0033%]	−0.2995 [0.2836%]	3.1610 [0.0120%]

Similar to Example 1, it was found that the UQ accuracy can be affected by the polynomial order (p) and the number of random variables in the gDRM (M). It is important to note that the mean values (μ_Z) of different PCE-based gDRM methods in Table 4 reached the same result (0.0064), since only two significant figures were shown. Specifically, as the number of p and M increases, the UQ accuracy can be improved. For example, when M was set to 3 for the mTriDRM and when p was set to 3, the relative error ϵ_R of κ_Z is one order of magnitude smaller than other methods (~0.0202%). It was also found that when the polynomial order p was set to 2, the UQ accuracy of LAR-based PCE is close to our method (mBiDRM and mTriDRM) for all four statistical moments. When p was set to 3, the UQ accuracy of mTriDRM-PCE has the same order of magnitude as the LAR-based PCE. This clearly shows the algorithm in this work can deal with nonlinear problems with nonstandard uncertainties and provide accurate results as one of the most popular techniques in the literature.

For comparison, Figure 6 shows the CDFs of the performance function Z in (25), with different combinations of p and M. As seen, the UQ results of our algorithm converge to MC, as the number of p and M increases. It is important to note that the accuracy of MC is related to the total number of samples used for UQ. When the model is highly nonlinear and when the number of uncertainties is large, a large number of samples is required, thereby increasing the computational cost for MC. This is further discussed with two more complicated examples below.

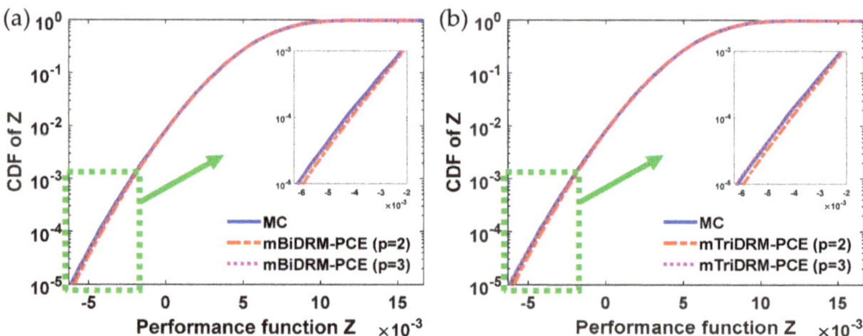

Figure 6. Comparison of the cumulative density functions of Z calculated with the algorithm in this work and with MC in Example 2. (**a**) mBiDRM-PCE and (**b**) mTriDRM-PCE.

4.3. Example 3: Truss Structure with 13 Members

A 13-bar truss structure [48] was chosen to study the performance of the UQ algorithm for dealing with a combination of different types of uncertainties. A schematic of the truss structure is shown in Figure 7, which involves 8 nodes and 13 members.

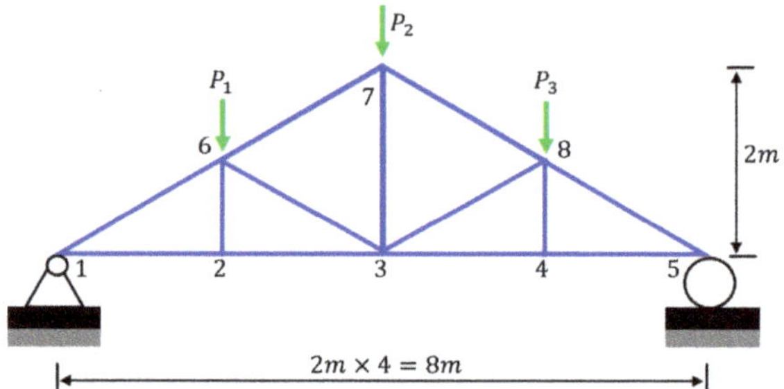

Figure 7. Illustration of a 13-bar truss structure [48]. Reprinted from [48], Copyright (2019), with permission from Elsevier.

In this example, it is assumed that there are three external loads, i.e., P_1, P_2, and P_3, imposed on nodes 6, 7, and 8, respectively, which follow a normal distribution. Additionally, the elastance E and the sectional area A for each member are assumed to be independent and lognormally distributed. The statistical description of each parameter is listed in Table 5. Finite element analysis was used to quantify uncertainty in the performance function Z of the 13-bar truss structure, which is mathematically defined as in [48]:

$$Z = u(X) = h_{max} - h \tag{26}$$

where h_{max} is the threshold describing the maximum allowable deflection, i.e., displacement, and h is the vertical displacement on node 3. In reliability analysis, when the performance function Z exceeds a limit, e.g., zero in (26), it would be considered as a failure event. Additionally, the probability of failure events is referred to as the failure probability, which is often used for structural reliability analysis [54]. Thus, it is essential to calculate the statistics of the prediction h in a precise and computationally efficient way. Here we focus on evaluating the displacement h with different UQ methods. The proposed methods in this work were integrated with finite element analysis to assess the precise prediction h, and the performance was validated with other UQ techniques, such as MC, LAR-based PCE, and the method in [48].

Table 5. Details of the uncertain variables used in Example 3 [48]. Reprinted from [48], Copyright (2019), with permission from Elsevier.

Variable	Distribution Type	Mean	Standard Deviation	Unit
E	Lognormal	206	20.6	Gpa
A	Lognormal	500	50	mm^2
P_1	Normal	20	3	kN
P_2	Normal	20	3	kN
P_3	Normal	20	3	kN

As seen in Table 5, there are two lognormal uncertainties and three normal uncertainties. For the lognormal uncertainty, the normalized Gaussian variables and Hermite basis functions were used to build the PCE surrogate models in (13), while for normal uncertainty, the formulation of PCE models is defined in (1). Once these PCE models of uncertainties are available, the resulting surrogate model of h in (26) is described with unknown PCE coefficients that can be solved with the mgDRM involving five quadrature points in each dimension (i.e., $\theta_i = 5$). Two different polynomial orders (2 vs. 3) and two different values of M (2 vs. 3) were considered. Table 6 summarizes the simulation results. It is important to note that the total number of samples for each parametric uncertainty in MC was set to 10^7 to ensure UQ accuracy.

Table 6. Summary of statistical moments and their relative errors in Example 3.

UQ Methods	μ_Z [ϵ_R]	σ_Z [ϵ_R]	s_Z [ϵ_R]	κ_Z [ϵ_R]
MC	10.6866	1.7808	0.4661	3.3907
mBiDRM-PCE ($p = 2$)	10.6864 [0.0021%]	1.7802 [0.0306%]	0.4507 [3.3024%]	3.2762 [3.3761%]
mBiDRM-PCE ($p = 3$)	10.6864 [0.0019%]	1.7807 [0.0030%]	0.4590 [1.5131%]	3.3243 [1.9594%]
mTriDRM-PCE ($p = 2$)	10.6864 [0.0021%]	1.7803 [0.0273%]	0.4518 [3.0748%]	3.2774 [3.3414%]
mTriDRM-PCE ($p = 3$)	10.6864 [0.0019%]	1.7808 [0.0000%]	0.4655 [0.1377%]	3.3852 [0.1628%]
LAR-based PCE ($p = 2$)	10.6863 [0.0032%]	1.7802 [0.0302%]	0.4545 [2.4969%]	3.2810 [3.2371%]
LAR-based PCE ($p = 3$)	10.6865 [0.0016%]	1.7808 [0.0004%]	0.4652 [0.1860%]	3.3841 [0.1954%]
Reference [48]	10.6865 [0.0013%]	1.7809 [0.0083%]	0.4685 [0.5153%]	3.4052 [0.4273%]

For the implementation of LAR-based PCE, 1000 model runs were used to evaluate the truss structure model as in Figure 7. Further, the sparse PCE terms were determined with the LAR algorithm to find the sensitive basis polynomials with a predefined value (i.e., 1.0×10^{-6}) as done in previous examples. Note that the value was used to terminate the LAR algorithm when the maximum correlation between the residual and the basis polynomials was below the value. Details of the LAR procedures can be found in [50,51]. In addition, the resulting number of PCE terms for the LAR-based PCE was 21 and 40, when the polynomial order p was 2 and 3, respectively. In contrast, the total numbers of PCE coefficients of our method were 21 and 56, which were determined using (3).

As seen in Table 6, it was found that UQ accuracy can be affected by the polynomial order (p) and the total number of integration variables (M) used in the gDRM. For example, when the mTriDRM-PCE was used ($p = 3$ and $M = 3$), the relative error ϵ_R of κ_h is ~0.1628% in Table 6. This is smaller than the results (~0.4273%) in [48]. As compared to the LAR-based PCE, our algorithm provides comparable results. This shows the potential of the gDRM-based PCE to deal with more complicated problems, especially when estimating the higher-order statistical moments since it was previously considered challenging [55].

To graphically study the UQ accuracy with different combinations of p and M, Figure 8 shows the estimated CDFs of the displacement h. As seen, the mTriDRM-PCE, when p was 3, can provide almost identical results, as compared to MC.

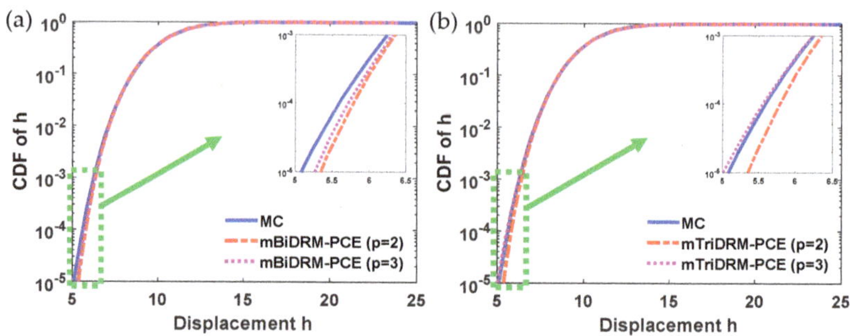

Figure 8. Comparison of the cumulative density function of the displacement h calculated with the algorithm in this work and with MC in Example 3. (**a**) mBiDRM-PCE and (**b**) mTriDRM-PCE.

As compared to the first two examples, UQ in this case study requires integrating the finite element analysis with the PCE, which may increase the computational burden for UQ. We further studied the computational time to calculate the displacement h with an office desktop (Core i5-8400 central processing unit (CPU) at 2.80 GHz). Using MC with 10^7 samples for each uncertainty, it was found that ~31.03 min were required to simulate all nodes in Figure 7. For the mBiDRM-PCE, it took ~21.36 and 48.06 s to simulate all nodes, when the polynomial order (p) was set to 2 and 3, respectively. In addition, for the mTriDRM-PCE, it was found that ~26.28 and 67.31 s were required, when p was 2 and 3, respectively. The mTriDRM-PCE took longer than mBiDRM-PCE due to the larger number of integrals that need to be solved. Further, depending on the polynomial order p, the requisite number of terms in PCE to approximate the displacement h differs as shown in (3), which leads to the difference in computational times of each method (e.g., mBiDRM- and mTriDRM-PCE).

4.4. Example 4: Planar Truss Structure with 23 Members

A planar truss structure with 23 members as in Figure 9 is considered in this case study. It is assumed that all parameters cannot be known with certainty, resulting in a nine-dimensional random space. This allows us to validate the UQ accuracy of the proposed method for dealing with many uncertainties. Details of model parameters are given in Table 7.

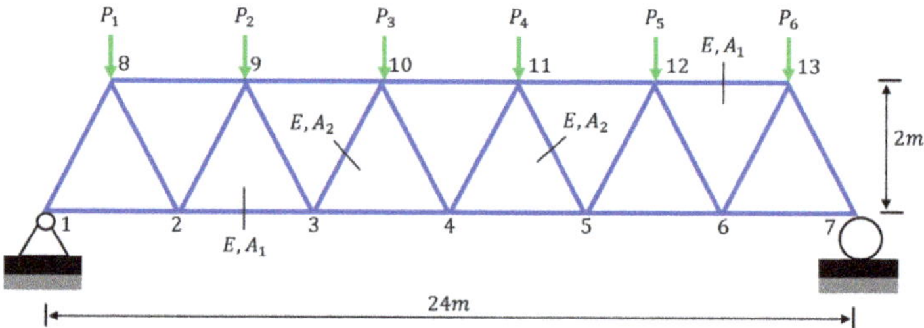

Figure 9. Illustration of a planar truss structure with 23 members [19,56]. Reprinted from [56], Copyright (2006), with permission from Elsevier.

Table 7. Details of the uncertain variables in Example 4 [19,56]. Reprinted from [56], Copyright (2006), with permission from Elsevier.

Variable	Distribution Type	Mean	Standard Deviation	Unit
E	Lognormal	2.10×10^{11}	2.10×10^{10}	Pa
A_1	Lognormal	2.0×10^{-3}	2.0×10^{-4}	m^2
A_2	Lognormal	1.0×10^{-3}	1.0×10^{-4}	m^2
$P_1 - P_6$	Weibull	5.0×10^4	7.5×10^3	N

In this case study, E is the elastic modulus; A_1 and A_2 are the cross-sectional area corresponding to the specific member; and P_1 to P_6 represent external loads imposed on nodes from 8 to 13. As done in Example 3, the finite-element analysis was used to calculate the performance function Z as in [19,56]:

$$Z = u(X) = v_{max} - v \tag{27}$$

where v_{max} represents a specific threshold, i.e., the maximum deflection, and v is the vertical displacement on node 4 that will be approximated with different UQ methods. Our objective in this example is to accurately approximate uncertainty in v such that the failure probability can be quantified for structural reliability analysis.

As seen in Table 7, three parameters (E, A_1, and A_2) follow a lognormal distribution, while the rest of the parameters follow a Weibull distribution. Thus, the resulting random space is nine-dimensional, i.e., $n = 9$. To derive a PCE expression of v and to compute the statistical moments, the finite element analysis was coupled with the proposed methods, i.e., mBiDRM- and mTriDRM-PCE, and two different polynomial orders, i.e., $p = 2$ and 3, were investigated. Note that in these methods, the BiDRM approximates a nine-dimensional integral with 36 two- and 9 one-dimensional integrals, while the TriDRM requires the evaluations of 84 three-, 36 two-, and 9 one-dimensional ones. These integrals were calculated with the Gauss–Hermite quadrature rule [44], where the number of quadrature points for each random variable was set to 5. In addition, the number of samples for MC was set to 10^7 in order to compute the relative errors ϵ_R in (24) for comparison purposes. The UQ results and relative errors (ϵ_R) are shown in Table 8. Since the distribution of uncertainties is different as compared to [19,56], the results in these works were not given. However, LAR-based PCE was simulated for two different polynomial orders ($p = 2$ and 3) for comparison. The simulation results are summarized in Table 8. For the LAR-based PCE, the total number of PCE terms was 55 and 212, when p was set to 2 and 3, respectively. In constrast, 55 and 220 PCE coefficients were used in our algorithm for the output response, when p was set to 2 and 3, respectively.

Table 8. Summary of statistical moments and their relative errors in Example 4.

UQ Methods	μ_Z [ϵ_R]	σ_Z [ϵ_R]	s_Z [ϵ_R]	κ_Z [ϵ_R]
MC	0.0794	0.0118	0.4067	3.3010
mBiDRM-PCE ($p = 2$)	0.0794 [0.0002%]	0.0118 [0.0224%]	0.3976 [2.2522%]	3.2234 [2.3512%]
mBiDRM-PCE ($p = 3$)	0.0794 [0.0000%]	0.0118 [0.0181%]	0.4131 [1.5571%]	3.3429 [1.2683%]
mTriDRM-PCE ($p = 2$)	0.0794 [0.0002%]	0.0118 [0.0282%]	0.3951 [2.8698%]	3.2205 [2.4406%]
mTriDRM-PCE ($p = 3$)	0.0794 [0.0000%]	0.0118 [0.0027%]	0.4086 [0.4477%]	3.3039 [0.0863%]
LAR-based PCE ($p = 2$)	0.0794 [0.0006%]	0.0118 [0.0162%]	0.3961 [2.6238%]	3.2210 [2.4256%]
LAR-based PCE ($p = 3$)	0.0794 [0.0008%]	0.0118 [0.0080%]	0.4084 [0.4159%]	3.3032 [0.0667%]

Similar to previous examples, our method provides accurate results, as the number of polynomial order (p) and the integration variables (M) in gDRM increase. For example, the mTriDRM outperforms others when p was 3, since the relative errors are smaller as in Table 8. In addition, it was found that the mTriDRM-PCE can provide comparable results to the LAR-based PCE for all statistical moments, thus confirming the accuracy of the mTriDRM-PCE-based algorithm in this work. For comparison, the CDF of the displacement v on node 4 was approximated with the proposed method and is shown in Figure 10. As compared to MC, it was found that mTriDRM-PCE provided the most accurate result, when p was set to 3. This shows the proposed algorithm can deal with complicated problems involving many uncertainties.

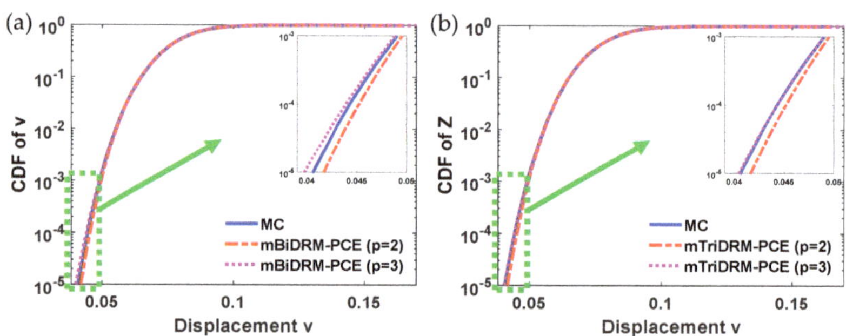

Figure 10. Comparison of the cumulative density functions of the displacement v calculated with the algorithm in this work and with MC in Example 4. (**a**) mBiDRM-PCE and (**b**) mTriDRM-PCE.

We further studied the computational time of the proposed method and compared the results to MC. Similar to the previous case study, it was found that the computational time for MC is larger, as compared to the gDRM-based method. For example, ~55.54 min were required in this case study, when 10^7 samples were used for each parametric uncertainty for MC. However, the computational time for the mTriDRM-PCE was found to be ~11.52 min, when p was set to 3, which is about 80% lower than MC. It is also important to note that the gDRM-PCE based UQ method, as compared to sampling-based MC, can provide an analytical expression of the displacement v as a function of uncertainties. This provides mathematical explanations to gain a deep understanding of the problem for improved reliability analysis. The analytical expression, for instance, can be combined with sensitivity analysis techniques to find the most sensitive random variable that can significantly affect the system's performance.

5. Conclusions

Using the polynomial chaos expansion (PCE), an uncertainty quantification (UQ) algorithm is presented to deal with many uncertainties that follow nonstandard distributions. To build PCE-based surrogate models, standard random variables are used to identify the relationship between nonstandard and standard distributions. To reduce the computational cost for calculating the PCE coefficients of model outputs, a generalized dimension reduction method is used to transform a high-dimensional integral resulting from the spectral projection (SP) into a few lower-dimensional integrals, which can be rapidly solved with quadrature rules in real-time. To show the UQ accuracy of our algorithms, four examples of structural reliability analysis were used. As compared to Monte Carlo simulations and other works in the literature, our results show the superior performance of the algorithms in terms of UQ accuracy and computational time. This shows the potential of the algorithm to tackle UQ in more complicated engineering problems that require consideration of many uncertainties. However, when uncertainty has a nonstandard distribution with a large variance, more PCE coefficients of uncertainty might be needed in the proposed approach to ensure UQ accuracy. In this case, it can lead to intensive computational burden in the presence of many uncertainties. Future work will improve the proposed algorithm to address this issue by identifying orthogonal polynomial functions only for a given uncertainty.

Author Contributions: Conceptualization, Y.D.; methodology, Y.D. and J.S.; validation, J.S.; investigation, J.S.; writing—original draft preparation, J.S.; writing—review and editing, Y.D.; supervision, Y.D. All authors have read and agreed to the published version of the manuscript.

Funding: This work was partially supported by the National Science Foundation, Division of Civil, Mechanical and Manufacturing Innovation (CMMI), under the award No. 1727487.

Institutional Review Board Statement: Not applicable.

Informed Consent Statement: Not applicable.

Data Availability Statement: Not applicable.

Conflicts of Interest: The authors declare no conflict of interest.

Abbreviations

ANOVA	Analysis of variance
BiDRM	Bivariate dimension reduction method
CDF	Cumulative distribution function
gDRM	Generalized dimension reduction method
gDRM-PCE	gDRM-based PCE
LAR	Least angle regression
MC	Mote Carlo
mgDRM-PCE	Modified gDRM-based PCE
PCE	Polynomial chaos expansion
PDF	Probability density function
SP	Spectral projection
TriDRM	Trivariate dimension reduction method
UQ	Uncertainty quantification

References

1. Coit, D.W.; Jin, T.; Wattanapongsakorn, N. System optimization with component reliability estimation uncertainty: A multi-criteria approach. *IEEE Trans. Reliab.* **2004**, *53*, 369–380. [CrossRef]
2. Zaman, K.; Mahadevan, S. Reliability-based design optimization of multidisciplinary system under aleatory and epistemic uncertainty. *Struct. Multidiscip. Optim.* **2017**, *55*, 681–699. [CrossRef]
3. Xu, J.; Li, J. Stochastic dynamic response and reliability assessment of controlled structures with fractional derivative model of viscoelastic dampers. *Mech. Syst. Signal Process.* **2016**, *72–73*, 865–896. [CrossRef]
4. Ni, P.; Xia, Y.; Li, J.; Hao, H. Using polynomial chaos expansion for uncertainty and sensitivity analysis of bridge structures. *Mech. Syst. Signal Process.* **2019**, *119*, 293–311. [CrossRef]
5. Kwon, K.; Ryu, N.; Seo, M.; Kim, S.; Lee, T.H.; Min, S. Efficient uncertainty quantification for integrated performance of complex vehicle system. *Mech. Syst. Signal Process.* **2020**, *139*, 106601. [CrossRef]
6. Zhang, Z.; Deng, W.; Jiang, C. Sequential approximate reliability-based design optimization for structures with multimodal random variables. *Struct. Multidiscip. Optim.* **2020**, *62*, 511–528. [CrossRef]
7. Dubourg, V.; Sudret, B.; Bourinet, J.-M. Reliability-based design optimization using kriging surrogates and subset simulation. *Struct. Multidiscip. Optim.* **2011**, *44*, 673–690. [CrossRef]
8. Echard, B.; Gayton, N.; Lemaire, M. AK-MCS: An active learning reliability method combining kriging and monte Carlo simulation. *Struct. Saf.* **2011**, *33*, 145–154. [CrossRef]
9. Wang, C.; Qiu, Z.; Xu, M.; Li, Y. Novel reliability-based optimization method for thermal structure with hybrid random, interval and fuzzy parameters. *Appl. Math. Model.* **2017**, *47*, 573–586. [CrossRef]
10. Fishman, G.S. *Monte Carlo: Concepts, Algorithms, and Applications*; Springer: New York, NY, USA, 1996.
11. Schueller, G.I. Efficient Monte Carlo simulation procedures in structural uncertainty and reliability analysis—Recent advances. *Struct. Eng. Mech.* **2009**, *32*, 1–20. [CrossRef]
12. Zhang, Z.; Jiang, C.; Wang, G.G.; Han, X. First and second order approximate reliability analysis methods using evidence theory. *Reliab. Eng. Syst. Saf.* **2015**, *137*, 40–49. [CrossRef]
13. Smith, C.L. Uncertainty propagation using taylor series expansion and a spreadsheet. *J. Ida. Acad. Sci.* **1994**, *30*, 93–105.
14. Wiener, N. The homogeneous chaos. *Am. J. Math.* **1938**, *60*, 897–936. [CrossRef]
15. Ghanem, R.G.; Spanos, P.D. *Stochastic Finite Elements: A Spectral Approach*; Springer Science and Business Media LLC: New York, NY, USA, 1991.

16. Hong, J.; Shaked, S.; Rosenbaum, R.K.; Jolliet, O. Analytical uncertainty propagation in life cycle inventory and impact assessment: Application to an automobile front panel. *Int. J. Life Cycle Assess.* **2010**, *15*, 499–510. [CrossRef]

17. MacLeod, M.; Fraser, A.J.; Mackay, D. Evaluating and expressing the propagation of uncertainty in chemical fate and bioaccumulation models. *Environ. Toxicol. Chem.* **2002**, *21*, 700–709. [CrossRef] [PubMed]

18. Zhao, Y.-G.; Ono, T. Moment methods for structural reliability. *Struct. Saf.* **2001**, *23*, 47–75. [CrossRef]

19. Xu, J.; Kong, F. A cubature collocation based sparse polynomial chaos expansion for efficient structural reliability analysis. *Struct. Saf.* **2018**, *74*, 24–31. [CrossRef]

20. Xiu, D. *Numerical Methods for Stochastic Computations: A Spectral Method Approach*; Princeton University Press: Princeton, NJ, USA, 2010.

21. Xiu, D.; Karniadakis, G.E. The Wiener–Askey polynomial chaos for stochastic differential equations. *SIAM J. Sci. Comput.* **2002**, *24*, 619–644. [CrossRef]

22. Wang, C.; Zhang, H.; Li, Q. Moment-based evaluation of structural reliability. *Reliab. Eng. Syst. Saf.* **2019**, *181*, 38–45. [CrossRef]

23. Pulch, R. Polynomial Chaos for the Computation of Failure Probabilities in Periodic Problems. In *Scientific Computing in Electrical Engineering SCEE 2008*; Roos, J., Costa, L.R.J., Eds.; Springer: Berlin/Heidelberg, Germany, 2010; pp. 191–198.

24. Le Maître, O.P.; Knio, O.M. *Spectral Methods for Uncertainty Quantification: With Applications to Computational Fluid Dynamics*; Springer Science & Business Media: Berlin/Heidelberg, Germany, 2010.

25. Najm, H.N. Uncertainty quantification and polynomial chaos techniques in computational fluid dynamics. *Annu. Rev. Fluid Mech.* **2009**, *41*, 35–52. [CrossRef]

26. Wan, H.-P.; Ren, W.-X.; Todd, M.D. Arbitrary polynomial chaos expansion method for uncertainty quantification and global sensitivity analysis in structural dynamics. *Mech. Syst. Signal Process.* **2020**, *142*, 106732. [CrossRef]

27. Wang, C.; Matthies, H.G.; Xu, M.; Li, Y. Dual interval-and-fuzzy analysis method for temperature prediction with hybrid epistemic uncertainties via polynomial chaos expansion. *Comput. Methods Appl. Mech. Eng.* **2018**, *336*, 171–186. [CrossRef]

28. Eldred, M.; Burkardt, J. Comparison of Non-Intrusive Polynomial Chaos and Stochastic Collocation Methods for Uncertainty Quantification. In Proceedings of the 47th AIAA Aerospace Sciences Meeting Including the New Horizons Forum and Aerospace Exposition, Orlando, FL, USA, 5–8 January 2009.

29. Son, J.; Du, Y. Comparison of intrusive and non-intrusive polynomial chaos expansion-based approaches for high dimensional parametric uncertainty quantification and propagation. *Comput. Chem. Eng.* **2020**, *134*, 106685. [CrossRef]

30. Son, J.; Du, D.; Du, Y. Modified polynomial chaos expansion for efficient uncertainty quantification in biological systems. *Appl. Mech.* **2020**, *1*, 153–173. [CrossRef]

31. Xu, H.; Rahman, S. A generalized dimension-reduction method for multidimensional integration in stochastic mechanics. *Int. J. Numer. Methods Eng.* **2004**, *61*, 1992–2019. [CrossRef]

32. Zhang, L.; Li, T.; Ying, W.; Fang, D. Rising and descending bubble size distributions in gas–liquid and gas–liquid–solid slurry bubble column reactor. *Chem. Eng. Res. Des.* **2008**, *86*, 1143–1154. [CrossRef]

33. Ghanem, R. The nonlinear gaussian spectrum of log-normal stochastic processes and variables. *J. Appl. Mech.* **1999**, *66*, 964–973. [CrossRef]

34. Kundu, A.; Adhikari, S.; Friswell, M.I. Stochastic finite elements of discretely parameterized random systems on domains with boundary uncertainty. *Int. J. Numer. Methods Eng.* **2014**, *100*, 183–221. [CrossRef]

35. Mohan, P.S.; Nair, P.B.; Keane, A.J. Stochastic projection schemes for deterministic linear elliptic partial differential equations on random domains. *Int. J. Numer. Methods Eng.* **2011**, *85*, 874–895. [CrossRef]

36. Witteveen, J.A.; Bijl, H. Modeling Arbitrary Uncertainties Using Gram-Schmidt Polynomial Chaos. In Proceedings of the 44th AIAA Aerospace Sciences Meeting and Exhibit, Reno, NV, USA, 9–12 January 2006.

37. Rosenblatt, M. Remarks on a multivariate transformation. *Ann. Math. Stat.* **1952**, *23*, 470–472. [CrossRef]

38. Nataf, A. Détermination des distributions de probabilité dont les marges sont données. *Comptes Rendus Acad. Sci.* **1962**, *225*, 42–43.

39. Ernst, O.G.; Mugler, A.; Starkloff, H.-J.; Ullmann, E. On the convergence of generalized polynomial chaos expansions. *ESAIM Math. Model. Numer. Anal.* **2012**, *46*, 317–339. [CrossRef]

40. Du, Y.; Budman, H.; Duever, T. Parameter estimation for an inverse nonlinear stochastic problem: Reactivity ratio studies in copolymerization. *Macromol. Theory Simul.* **2017**, *26*, 1600095. [CrossRef]

41. Lasota, R.; Stocki, R.; Tauzowski, P.; Szolc, T. Polynomial chaos expansion method in estimating probability distribution of rotor-shaft dynamic responses. *Bull. Pol. Acad. Sci. Tech. Sci.* **2015**, *63*, 413–422. [CrossRef]

42. Krishnamoorthy, K. *Handbook of Statistical Distributions with Applications*; CRC Press: Boca Raton, FL, USA, 2016.

43. Lomax, R.G.; Hahs-Vaughn, D.L. *An Introduction to Statistical Concepts*, 3rd ed.; Routledge: New York, NY, USA, 2013.

44. McClarren, R.G. Gauss Quadrature and Multi-dimensional Integrals. In *Computational Nuclear Engineering and Radiological Science Using Python*; Academic Press: Cambridge, MA, USA, 2018; pp. 287–299.

45. Gao, Z.; Hesthaven, J.S. On ANOVA expansions and strategies for choosing the anchor point. *Appl. Math. Comput.* **2010**, *217*, 3274–3285. [CrossRef]

46. Cao, Y.; Chen, Z.; Gunzburger, M. ANOVA expansions and efficient sampling methods for parameter dependent nonlinear PDEs. *Int. J. Numer. Anal. Model.* **2009**, *6*, 256–273.

47. Der Kiureghian, A.; Lin, H.-Z.; Hwang, S.-J. Second-order reliability approximations. *J. Eng. Mech.* **1987**, *113*, 1208–1225. [CrossRef]
48. Xu, J.; Dang, C. A new bivariate dimension reduction method for efficient structural reliability analysis. *Mech. Syst. Signal Process.* **2019**, *115*, 281–300. [CrossRef]
49. Zhang, Y.-G.; Huang, Y.-L.; Wu, Z.-M.; Li, N. A high order unscented Kalman filtering method. *Acta Autom. Sin.* **2014**, *40*, 838–848.
50. Chen, X. Welcome to Xiaohui Chen's Webpage, LARS: Least Angle Regression (LARS). Available online: https://publish.illinois.edu/xiaohuichen/code/lars/ (accessed on 28 March 2020).
51. Efron, B.; Hastie, T.; Johnstone, I.; Tibshirani, R. Least angle regression. *Ann Stat.* **2004**, *32*, 407–499. [CrossRef]
52. Blatman, G.; Sudret, B. Adaptive sparse polynomial chaos expansion based on least angle regression. *J. Comput. Phys.* **2011**, *230*, 2345–2367. [CrossRef]
53. Song, S.; Lu, Z.; Qiao, H. Subset simulation for structural reliability sensitivity analysis. *Reliab. Eng. Syst. Saf.* **2009**, *94*, 658–665. [CrossRef]
54. Ayyub, B.M.; McCuen, R.H. *Probability, Statistics, and Reliability for Engineers and Scientists*; CRC Press: Boca Raton, FL, USA, 2016.
55. Grigoriu, M. *Stochastic Calculus: Applications in Science and Engineering*; Springer Science & Business Media: Berlin/Heidelberg, Germany, 2002.
56. Lee, S.H.; Kwak, B.M. Response surface augmented moment method for efficient reliability analysis. *Struct. Saf.* **2006**, *28*, 261–272. [CrossRef]

Article

Sliding Mode Observer-Based Fault Detection and Isolation Approach for a Wind Turbine Benchmark

Vicente Borja-Jaimes [1],*, Manuel Adam-Medina [1], Betty Yolanda López-Zapata [2], Luis Gerardo Vela Valdés [1], Luisana Claudio Pachecano [1] and Eduardo Mael Sánchez Coronado [3]

1 Centro Nacional de Investigación y Desarrollo Tecnológico, Interior Internado Palmira s/n Col Palmira, Cuernavaca 62490, Morelos, Mexico; manuel.am@cenidet.tecnm.mx (M.A.-M.); luis.vv@cenidet.tecnm.mx (L.G.V.V.); luisana.cp@cenidet.tecnm.mx (L.C.P.)
2 Departamento de Mecatrónica, Universidad Politécnica de Chiapas, Tuxtla Gutiérrez 29082, Chiapas, Mexico; blopez@upchiapas.edu.mx
3 Departamento de Ingeniería Mecatrónica, Universidad Tecnológioca del Centro de Veracruz, Cuitláhuac 94910, Veracruz, Mexico; eduardo.sanchez@utcv.edu.mx
* Correspondence: vicentebj@cenidet.edu.mx

Abstract: A fault detection and isolation (FDI) approach based on nonlinear sliding mode observers for a wind turbine model is presented. Problems surrounding pitch and drive train system FDI are addressed. This topic has generated great interest because the early detection of faults in these components allows avoiding irreparable damage in wind turbines. A fault diagnosis strategy using nonlinear sliding mode observer banks is proposed due to its ability to handle model uncertainties and external disturbances. Unlike the reported solutions, the solution approach does not need a priori knowledge of the faults and considers system uncertainty. The robustness to disturbances, uncertainties, and measurement noise is shown in the dynamic of the generated residuals, which is sensible to only one kind of fault. To show the effectiveness of the proposed FDI approach, numerical examples based on a wind turbine benchmark model, considering closed loop applications, are presented.

Keywords: fault detection and isolation (FDI); sliding mode observer and wind turbine; nonlinear systems

Citation: Borja-Jaimes, V.; Adam-Medina, M.; López-Zapata, B.Y.; Vela Valdés, L.G.; Claudio Pachecano, L.; Sánchez Coronado, E.M. Sliding Mode Observer-Based Fault Detection and Isolation Approach for a Wind Turbine Benchmark. *Processes* **2022**, *10*, 54. https://doi.org/10.3390/pr10010054

Academic Editors: Francisco Ronay López-Estrada and Guillermo Valencia-Palomo

Received: 15 November 2021
Accepted: 8 December 2021
Published: 28 December 2021

Publisher's Note: MDPI stays neutral with regard to jurisdictional claims in published maps and institutional affiliations.

1. Introduction

Wind energy has been one of the fastest growing renewable energy sources in recent years [1]. Recently, alternative research has emerged, known as hybrid wind and photovoltaic (PV)-based systems, which provide information on the improvements of these two activities together, as described in [2]. Therefore, development of new wind turbines and wind energy conversion systems, to increase the efficiency and reliability of the energy conversion process, has become one of the main activities (and challenges) of the power industry, as well as the design of new approaches for fault diagnosis. From the reliability analysis carried out by leading countries in the production of wind energy, it is known that most of the faults that occur in wind turbines are located in the pitch system, drive train, and the control system [3]. Moreover, other faults that commonly occur are located on input and output sensors of the different subsystems that make up the wind energy conversion system [4].

In addition to their susceptibility to faults, wind turbines present significant control challenges due to the fact that they include several subsystems and due to the non-linear nature of their input [5]. Therefore, there are different approaches that describe the wind turbine model, such as in [6], where a method is proposed to analyze the parameters in the propeller design, such as drag force, lift force, thrust coefficient, power coefficient, and lift coefficient, among others. These mathematical formulations used in the QBlade software were based on the blade element moment (BEM) theory. In the work [7], specific airfoil simulations were performed for wind turbines of less than 1 kW, considering that the nominal wind speed was 8.4 m/s. In the presented simulations, lift and drag coefficients for different angles of attack are compared. Furthermore, a blade design and performance analysis were published in [8], where the main objective was to optimize the number of blades and tip speed ratio. A frequency regulation model for a wind farm and a thermal power plant was established in [9], where the main variable to be controlled was the load frequency of the produced power. To obtain an equivalent model of a wind farm, different wind turbines were compared, computing the gap metrics of 11 models under different operating conditions. One way to avoid wind turbine failures is through early fault detection methods. An FDI-based approach to dynamic systems, where faults can be detected and located, is described in [10]. This is based on data-driven, intelligent, and analytical modeling methods. In the literature, data-based diagnostic systems use multi-sensors that are a set of signals with fusion technology, which allows estimating, combining, and correlating the data obtained from sensors or information sources [11]. In the sense of the use of multi-sensors, but applied to fault diagnosis in engines, there is a proposal using infrared thermal imaging data, and vibration measurements for automatic fault detection [12]. In the multisensory integration to perform FDI in rotating machines, there is work using data fusion and artificial neural networks [13], FDI was also done using learning algorithms limited to rotor faults [14], the latter later proposed another work based on the same fault detection strategy, but on the gearbox [15]. Multi-sensor techniques represent an interesting and novel alternative to perform FDI in wind turbines.

On the other hand, model-based methods for fault diagnosis in wind turbines present good results, such as in [16], where the diagnosis is carried out using Kalman filters for residual generation. The model-based approach is more frequent in the literature than the data-driven approach, since it is possible to use existing mathematical methods based on differential equations to design control approaches based on fault diagnosis for wind turbine systems [17]. The methods proposed in the literature generally use thresholds for the residual values generated by the difference between the outputs of observer banks and the output signals of the system, as shown in [18]. Fault detection in wind turbine systems is reported in [19] and fault diagnosis based on state estimation is presented in [20]; these works describe the detection and isolation of faults in sensors and actuators of reference wind turbines. An approach to detect faults in actuators and sensors of wind turbines using unknown input observers (UIOs), is presented in [21]. Fault detection is carried out using residual signals and fault localization is achieved by estimating states, output signals, and controlled input signals without considering step faults. In [22], an adaptive diagnosis based on an observer is designed to perform FDI in wind turbines. In [23], three different FDD approaches are addressed, in the power train subsystem, in the generator, and in the wind turbine converter. A robust fault estimation scheme is presented in [24], where actuator and sensor faults in the pitch and drive train system are considered. Robustness is achieved by decoupling the faulty dynamics from the system using a series of coordinate transformations. For the non-faulty dynamics, an observed reduced order input is used for state estimation. While for dynamic faults, an accommodation approach based on sliding mode observers is presented. In [25], a wind speed estimator is developed to remedy the inaccuracies in the single point measurement in the WECS gondola by using the discrete Kalman filter to perform the aerodynamic torque estimation and subsequently calculate the effective wind speed. The design procedure, calculation, and estimation procedure are repeated by implementing an extended Kalman filter (EKF). The results of both algorithms are compared by simulation.

Sliding mode designs for observer and/or control are well known for their robustness properties [26]. Therefore, in [27], an approach based on adaptive sliding mode observers in wind turbines is proposed to perform fault diagnosis as part of an adaptive fault tolerant control (AFTC). In this approach, the adaptive sliding mode observer is used to simultaneously detect both the actuator and sensor faults. In the FTC design, model uncertainty and wind speed variations are considered as unknown disturbances. This allows accurate state estimation and reconstruction of the disturbances. However, the proposed scheme does not consider the faults in the pitch and drive train.

From the literature review, it can be observed that, to solve the problem of detecting faults in the pitch system and handling dynamics introduced by faults, most of the authors have presented methods of compensating for the dynamics introduced by faults. However, the main drawback of the methods presented is that a priori knowledge of the magnitude of the faults is required. On the other hand, the problem of the lack of precision in the measurement of the rotor torque has been addressed through estimation approaches, which do not consider the uncertainty present in the estimation and the system itself.

In this work, a benchmark model is used, which consists mainly of three parts: the blade and pitch system, the generator and converter, as well as the drive train, for which a three-bladed horizontal wind turbine coupled to an electric generator with a capacity of 4.8 MW is considered. The model allows proposing faults in the actuators, sensors, and systems in the pitch actuators, the drive train, as well as in the converter system. In addition, it offers the possibility to analyze and study the operation with different types of structures for the detection and localization of faults in the wind turbine model in a more realistic context [28]. A fault diagnosis approach is proposed on this model. The main contribution of this work is the design of a fault diagnosis for the pitch system and the drive train of a wind turbine model-based. The novelty of the proposed approach is that the problem of the dynamics introduced by the faults, and the uncertainty associated with the system, in a simple way, is addressed, and eliminates the need to know a priori the behavior of the faults. Fault detection and isolation is done using the concept of residual signals and fault signatures. For the residual generation, a bank of sliding mode observers is proposed, which are designed to be sensitive only to one type of fault and to be robust to both, the dynamics of the faults and system uncertainty. This is accomplished through proper handling of the observer's error injection term and a transformation of the model in which the error introduced by the faults and the uncertainty of the system act on the input signal. The results obtained in simulation show the effectiveness of the proposed scheme.

The document is organized as follows: in Section 2 we present the wind turbine model, sliding mode observer design, and FDI scheme based on sliding mode observers, Section 3 describes the results obtained from numeric examples; Section 4 is a discussion about the results analyzed and finally the conclusion is presented.

2. Materials and Methods

2.1. Wind Turbine Model

The nonlinear wind turbine model considered in this paper is the benchmark described in [28]. This benchmark model considers the wind turbine at the system level as shown in Figure 1. This figure shows the interconnection of the subsystems that make up the wind turbine. The input to the system is the wind speed v_ω, which causes the turbine blades to rotate. This motion propagates through the rotor at a speed ω_r, and a torque τ_r. The drive train couples this motion with the shaft of the generator, which rotates at a speed ω_g, and has a torque τ_g. The generator is in charge of transforming the mechanical energy into electrical energy. The electrical power generated by the system is controlled by modifying the turbine aerodynamics through pitch control or by controlling the rotational speed of the generator (torque control). In either case, the objective of the control system is to track the reference electric power P_{ref}, or, if not possible at least, to minimize the tracking error. The pitch control block in Figure 1 receives as inputs the measurements of the pitch angle β_i, the generator speed ω_g, and the generated electrical power P_g. The output of the block is the pitch position reference signal β_{ref}, which will allow the capture of the kinetic energy needed to produce the desired electrical power output. On the other hand, the torque control block has as input the generator speed measurement and the generated electrical power. The controller generates as output the reference torque τ_{ref}, which will allow us to reach the desired electrical power output.

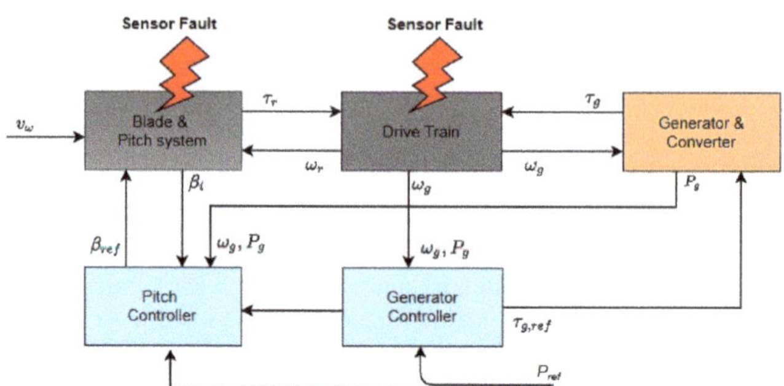

Figure 1. Interconnection of subsystems of the wind turbine in benchmark.

The mathematical model of each component of the wind turbine shown in Figure 1 is presented in the following lines.

2.1.1. Blade and Pitch Model

This model is a combination of the aerodynamic model and the wind and pitch model. The wind turbine nonlinear aerodynamic model is modeled as a torque acting on the blades. This aerodynamic torque is defined as

$$\tau_r(t) = \frac{1}{2}\,\pi\rho R^3 v_w^2\,(t) S\,(\lambda,\beta) \tag{1}$$

where $S(\lambda,\beta)$ is the power coefficient, R is the rotor ratio, ρ is the density of the air, and $v_w(t)$ is the wind speed.

The hydraulic pitch system is modeled as a second order transfer function

$$\frac{\beta_i(s)}{\beta_{ref}(s)} = \frac{\psi^2}{s^2 + 2\zeta\psi\,s + \psi^2} \tag{2}$$

where β_i is the ith pitch system whose position is measured with two sensors $\beta_{i,m1}$ and $\beta_{i,m2}$; $\beta_{ref}(t)$ is the reference value, ζ is the damping factor, and ψ is the natural frequency.

2.1.2. Drive Train Model

The drive train system is modeled by a simple two-mass model. Therefore, a drive train model can be represented by

$$J_r\dot{\omega}_r(t) = \tau_r(t) - K_{dt}\theta - (H_{dt} - F_r)\omega_r(t) + \frac{H_{dt}}{N_g}\omega_g(t)$$
$$J_g\dot{\omega}_g(t) = \frac{(\eta_{dt}K_{dt})}{N_g}\theta + \frac{(\eta_{dt}H_{dt})}{N_g}\omega_g(t) - \left(\frac{(\eta_{dt}H_{dt})}{N_g^2} + F_g\right)\omega_g(t) - \tau_g(t) \tag{3}$$
$$\dot{\theta}\,(t) = \omega_r(t) - \frac{1}{N_g}\omega_g(t)$$

where ω_r and ω_g are the rotor and generator speed, both measured by two sensors $\omega_{r,mi}$ and $\omega_{g,mi}$ (with $i = 1, 2$). Moreover, θ denotes the torsion angle of the drive train. τ_g and τ_r are the converter and aerodynamic torques. J_r y J_g are the moment of inertia of the low- and high-speed shafts, respectively. F_r and F_g are the viscous frictions of the low- and high-speed shafts, respectively. K_{dt} is the torsion stiffness of the drive train, H_{dt} is the torsion damping coefficient of the drive train, N_g is the gear ratio and η_{dt} is the efficiency of the drive train.

2.1.3. Generator and Converter Model

The generator and converter dynamics can be modeled by a first–order transfer function:

$$\frac{\tau_g(s)}{\tau_{g,ref}(s)} = \frac{\gamma}{s+\gamma} \tag{4}$$

where $\tau_{g,ref}$ is the reference value and γ is the cutoff frequency. The power produced by the generator is given by

$$P_g(t) = \eta_g \omega_g(t) \tau_g(t) \tag{5}$$

The efficiency of the generator is determined by η_g and P_g is the power produced by the generator.

2.2. Sliding Mode Observer

The objective of an observer is to estimate the states of a system, which are generally not measurable. In essence, an observer is a mathematical model of the system, which receives the inputs of the system and the estimation error obtained from the difference between the output of the system and the output of the observer himself.

The simplest form of an observer is the Luenberger observer, in which the estimation error is fed back linearly through a gain. However, the Luenberger observer is very sensitive to measurement noise and the success of the estimation is highly dependent on the accuracy of the system model. Therefore, if the model presents some uncertainty, or is affected by external disturbances, the estimation results will be very poor. To overcome this drawback, the design of a type of non-linear observer is proposed, known as a sliding mode observer. This observer shows to be insensitive to parametric variations and presents robustness to disturbances, which represents a great advantage over other types of observers [29]. In particular, in this work, these properties are used to obtain accurate estimates of the measurements made by the sensors, even in the presence of system faults and uncertainty.

To appreciate the properties of the sliding mode observers, namely, their robustness to modeling uncertainties and disturbances, the design considers a system with uncertainty with representation in the state space given by:

$$\begin{aligned}\dot{x}(t) &= Ax(t) + Bu(t) + D\xi(t,x,u)\\ y(t) &= Cx(t)\end{aligned} \tag{6}$$

where $A \in R^{n\times n}$, $B \in R^{n\times m}$, $C \in R^{p\times n}$, and $D \in R^{n\times q}$ with $p \geq q$. Assume that the matrices B, C are the full rank and the function $\xi : R_+ \times R^n \times R^m \to R^q$ is unknown, but bounded so that $\| \xi(t,x,u) \| \leq r_1 \| u \| + \alpha(t,y)$ and $\alpha : R_+ \times R^p \to R_+$.

Suppose that exits a liner change of coordinates T so that the system can be written as

$$\begin{aligned}\dot{x}_1(t) &= A_{11}x_1(t) + A_{12}x_2(t) + B_1u(t)\\ \dot{x}_2(t) &= A_{21}x_1(t) + A_{22}x_2(t) + B_2u(t) + D_2\,\xi\\ y(t) &= Cx(t)\end{aligned} \tag{7}$$

where $x_1 \in R^{(n-p)}$, $x_2 \in R^p$, and the matrix A_{11} has stable eigenvalues. The observer structure that will be considered can be written in form

$$\begin{aligned}\dot{\hat{x}}_1(t) &= A_{11}\hat{x}_1(t) + A_{12}\,\hat{x}_2(t) + B_1u(t) - A_{12}e_y(t)\\ \dot{\hat{x}}_2(t) &= A_{21}\hat{x}_1(t) + A_{22}\hat{x}_2(t) + B_2u(t) - (A_{22} - A_{22}^s)e_y(t) + w\\ \hat{y}(t) &= \hat{x}_2(t)\end{aligned} \tag{8}$$

where A_{22}^s is a stable design matrix, $e_y(t) = \hat{y}(t) - y(t)$ and the discontinuous vector w is defined as

$$w = \begin{cases} -\sigma \| D_2 \| \frac{P_2 e_y(t)}{\|P_2 e_y(t)\|} & \text{if } e_y \neq 0\\ 0 & \text{otherwise}\end{cases} \tag{9}$$

where $P_2 \in R^{p \times p}$ is a symmetric positive defined Lyapunov matrix for A_{22}^s.

The scalar σ is chosen so that $\|\xi\| < \sigma$. If the state estimation error is defined as $e_1(t) = \hat{x}_1(t) - x_1(t)$, then it is straightforward to show

$$\begin{aligned}
\dot{e}_1(t) &= A_{11}\, e_1(t) \\
\dot{e}_y(t) &= A_{21}\, e_1(t) + A_{22}^s e_1(t) + w - D_2\, \xi
\end{aligned} \tag{10}$$

It is shown in [30] that the error system Equation (10) is quadratically stable and a sliding motion takes place, forcing $e_y(t) = 0$ in finite time.

If $\hat{x}(t)$ represents the state estimate for $x(t)$ and $e(t) = \hat{x}(t) - x(t)$, it follows that if

$$G_l = T^{-1}\begin{bmatrix} A_{12} \\ A_{22} - A_{22}^s \end{bmatrix} \text{ and } G_{nl} = [D_2]T^{-1}\begin{bmatrix} 0 \\ I_p \end{bmatrix} \tag{11}$$

then the robust observer given in Equation (8) can be written in terms of the original coordinates in the form

$$\dot{\hat{x}}(t) = Ax(t) + Bu(t) - G_l Ce(t) + G_{nl}w \tag{12}$$

where the nonlinear function is given by

$$w = \begin{cases} -\sigma \dfrac{P_2 Ce(t)}{\|P_2 Ce(t)\|} & \text{if } Ce \neq 0 \\ 0 & \text{otherwise} \end{cases} \tag{13}$$

It is important to note that, even when the model considered for the design of the sliding mode observer has an uncertainty term, the dynamics of the observer does not depend on the uncertainty; therefore, the convergence of the observer is achieved even in the presence of uncertainty.

The methodology for the design of the sliding mode observer described in this section is applied in the following sections for the development of fault diagnosis schemes in the pitch and drive train system.

2.3. FDI Scheme Based on Sliding Mode Observers

The principal purpose of the FDI scheme is to generate an alarm when a fault develops in the system and to identify the source and location. To detect and isolate faults on the wind turbine, a model based FDI is proposed. To achieve fault detection and isolation in the model-based approach with the observer-based method, some authors have proposed particular configurations of observer banks. In this work, the proposed scheme is based on the configuration presented in [31], and is shown in Figure 2. In this scheme, a set of observers dedicated to isolate p sensor faults are designed, the proposed configuration allows observer p to use all inputs and only output p. In this way, observer p is sensitive to sensor p faults. The residue generator block generates a set of structured residues from the difference of the measured variables and those estimated by the observer bank. In the decision block, the decision is made to detect the fault if the residual exceeds a predefined threshold.

Since the wind turbine system is subject to modeling uncertainty and unknown disturbance, to achieve the effective fault detection, a robust fault detection scheme is needed. Since the sliding mode techniques are well known for their robustness properties, the bank of the observer is designed using these techniques.

The novelty of the proposed scheme is that it does not require a priori knowledge of the failures and considers the uncertainty of the system. This is achieved by treating the dynamics of the faults in the pitch system and the uncertainty in the estimation of the rotor torque, as coupled disturbances acting on the input signal. Under this condition, it is possible to observe the characteristics of the sliding mode observer. The faults considered in this paper cover the pitch systems and drive train, and these are independent. Therefore, an FDI scheme was designed separately for each one.

The configuration proposed for the development of the fault diagnosis scheme in the pitch system is presented below, considering the general structure described in Figure 2.

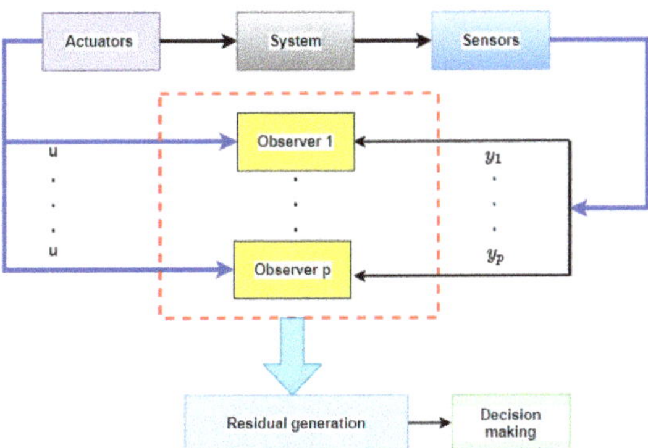

Figure 2. Architecture of observer-based FDI.

2.3.1. FDI Architecture for Pitch System

In this section, the FDI scheme for each pitch position system is presented. It considers only sensor faults, which are characterized by either a fixed value or a change gain factor on the sensor measurements. Since all three pitch positions are measured with two sensors, the aim of the FDI scheme is to detect and isolate faults in one of the two sensors. In order to perform the detection and isolation of faults in the pitch system, the architecture shown in Figure 3 is proposed. The figure shows the particular configuration of the observer bank used to estimate the pitch angle measurements, which is measured with two sensors for reasons due to hardware redundancy. Since the pitch system is integrated by three identical subsystems, there are a total of six sensors. The configuration of the observer bank allows the observer in charge of estimating the output 1 of pitch i, i.e., observer $i, 1$, to use only the measurement of sensor 1 of pitch i. In this way, the observer is sensitive only to faults in that sensor. The residual generator block generates a set of structured residues from the difference between the variables measured and those estimated by the observers. In the decision block, an alarm is generated if any of the residuals exceeds a pre-established threshold, which is established from the behavior of the residuals before and after the fault.

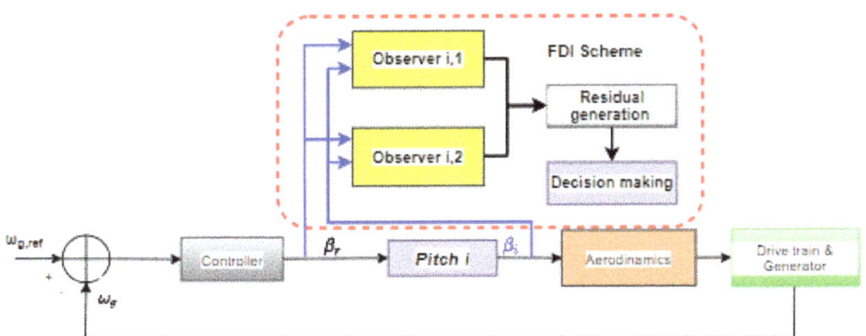

Figure 3. Architecture for FDI in the pitch system.

The FDI design only considers that one fault occurs at a time and, therefore, two fault scenarios for each pitch position are considered. The first assumes that sensor 1 is faulty and sensor 2 is healthy. The second assumes sensor 2 is faulty and sensor 2 is healthy. Under this consideration, two residual signals are needed for each pitch position system. Because residual signals are generated from the difference between the estimates of the measurement and real measurement, a set of two sliding mode observers for each pitch position system is needed.

Since sensor faults directly influence the pitch system through the feedback, in order to consider this influence, the function ξ^i_{pb} is used to model the position error introduced by the sensor fault. Then, the continuous-time state space model of ith pitch position is as follows:

$$\dot{x}^i_{pb}(t) = A^i_{pb}x^i_{pb}(t) + B^i_{pb}\beta^i_{ref}(t) + D\xi^i_{pb}$$
$$y^i_{pb}(t) = C^i_{pb}x^i_p(t)$$

$$(14)$$

where $x^i_{pb}(t) = \left[\dot{\beta}_i(t),\ \beta_i\right]^T$. A_{pb}, B_{pb} and C_{pb} are the system matrices and ξ^i_{pb} represents the position error introduced by the sensor fault given by $\xi^i_{pb} = \beta_i - 0.5i(\beta_{i,m1} + \beta_{i,m2})$ with $i = 1,\ 2,\ 3$.

For system (14), we used the sliding mode observer described in Section 2.2. From Equation (12), two sliding mode observers for the ith pitch position can be designed as:

$$\dot{\hat{x}}^i_{pb1}(t) = A^i_{pb}\hat{x}^i_{pb1}(t) + B^i_{pb}\beta^i_{ref}(t) - G^i_l e^i_{y1} + G^i_{nl} w^i_1$$
$$\hat{y}^i_{pb1}(t) = \hat{\beta}_{i,m1}(t)$$

$$(15)$$

$$\dot{\hat{x}}^i_{pb2}(t) = A^i_{pb}\hat{x}^i_{pb2}(t) + B^i_{pb}\beta^i_{ref}(t) - G^i_l e^i_{y2} + G^i_{nl} w^i_2$$
$$\hat{y}^i_{pb2}(t) = \hat{\beta}_{i,m2}(t)$$

$$(16)$$

where $\hat{x}^i_{pb}(t)$ and $\hat{y}^i_{pb}$ denote, respectively, the estimate of state and the estimate of output. G^i_l and G^i_{nl} are appropriate gain matrices to guarantee the stability and convergence of the dynamics of error. The estimation error is defined as $e^i_{y1} = \hat{\beta}_{i,m1}(t) - \beta_{i,m1}(t)$ and $e^i_{y2} = \hat{\beta}_{i,m2}(t) - \beta_{i,m2}(t)$. The w represents a discontinuous switched component to induce a sliding motion.

The observer proposed has the capacity to handle the influences of a fault if $\|\xi\| < \sigma$, where σ represents the gain in the discontinues injection term w and is obtained from Equation (12).

Under this consideration, the sliding mode observer proposed can estimate the state accurately, even in the presence of sensor faults; therefore, it is clear that the estimated i^{th} pitch position must be close to the real value before the fault occurrence and after the sensor fault. To isolate sensor faults, a structured set of residual signals is constructed to be sensitive to only one fault and is given in Table 1.

Table 1. Pitch system FDI logic.

Residuals	Definition	Fault 1	Fault 2	Fault3
$r_{1,1}$	$\| \hat{\beta}_{1,m1} - \beta_{1,m1} \|$	1	0	0
$r_{1,2}$	$\| \hat{\beta}_{1,m2} - \beta_{1,m2} \|$	0	0	0
$r_{2,1}$	$\| \hat{\beta}_{2,m1} - \beta_{2,m1} \|$	0	0	0
$r_{2,2}$	$\| \hat{\beta}_{2,m2} - \beta_{2,m2} \|$	0	1	0
$r_{3,1}$	$\| \hat{\beta}_{3,m1} - \beta_{3,m1} \|$	0	0	0
$r_{3,2}$	$\| \hat{\beta}_{3,m2} - \beta_{3,m2} \|$	0	0	1

The residual signal $r_{i,j}$ is used to indicate a fault in the sensor j of the ith-pitch position system, with $i = 1,\ 2,\ 3$ and $j = 1, 2$. Each residual is sensitive to only one fault. The details of the residual combinations, which are used to isolate different faults, are given by the fault signature.

The following subsection describes the proposed architecture for the transmission train fault diagnosis scheme. In a similar way to the scheme proposed for the pitch system, for the generation of residual signals, an observer bank configuration is proposed.

2.3.2. FDI Configurations for Drive Train System

To achieve fault detection and isolation in the drive train, the architecture shown in Figure 4 is proposed, where each observer has all drive train inputs, i.e., rotor torque and generator torque, and only the corresponding sensor measurement. The drive train has hardware redundancy so that the rotor and generator speed measurement is performed with two sensors respectively. Therefore, a bank of four observers is required so that each observer is sensitive only to faults present in the sensor feeding the output signal corresponding to the observer. Since the rotor torque cannot be measured, an estimator based on an algebraic map that depends on the measurement of wind speed, pitch angle, and generator speed is used in this scheme. This algebraic map is described in more detail in [19]. In the residual generator block, the estimation errors of each of the outputs are used as residuals. Then in the decision-making block, an alarm is generated if any of the residuals exceed a preset threshold.

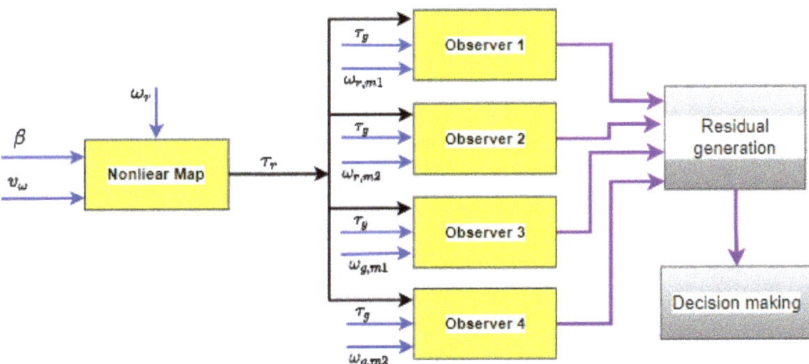

Figure 4. FDI architecture for drive train.

To ensure physical redundancy, both the generator and rotor speed are measured with two sensors. Therefore, the aim of the FDI scheme is to detect and isolate faults in one of the four sensors. Since both the rotor and generator speed are measured using encoders, the fault signals are characterized by either a fixed value or a change gain factor on the sensor measurements. The FDI design only considers that one fault occurs at a time and, therefore, two fault scenarios for both rotor and generator speed sensors are considered. The first assumes that sensor 2 is faulty and sensor 1 is healthy. The second assumes that sensor 1 is faulty and that sensor 2 is healthy. Under this consideration, four residual signals are needed to isolate a fault in the drive train system. Therefore, a set of four sliding mode observers to estimate each sensor measurement was designed.

On the other hand, since the aerodynamic torque is not possible to measure, it depends on wind speed, which can only be roughly measured, and imposes a challenge to the design of the FDI scheme. To overcome this issue, the nonlinear aerodynamics torque is obtained using an aerodynamic torque estimator based on the measured rotor speed ω_r, pitch position β_i, wind speed v_ω, and a mapping of the toque coefficient S (λ, β). It is important to note that due to the difficulty to accurately estimate the aerodynamics torque, an unknown bounded disturbance, ξ_{dt}, is added in the model. The disturbance is modeled as a bounded matching uncertainty entering the system through the input channel.

For the design of the sliding mode observers, consider the model in state space of the drive train system written as

$$\dot{x}_{dt}(t) = A_{dt}x_{dt}(t) + B_{dt}u_{dt}(t) + D\xi_{dt}$$
$$y_{dt}(t) = C_{dt}x_{dt}(t)$$

(17)

where $x_{dt}(t) = \left[\omega_r(t),\ \omega_g(t),\ \theta_\Delta(t)\right]^T, u = \left[\ \tau_r(t)\quad \tau_g(t)\ \right]^T$ and A_{dt}, B_{dt}, and C_{dt} are the system matrices, and ζ_{dt}, represents represent the system uncertainty, which are unknown but bounded.

Based on Equation (12), two sliding mode observers of FDI on the drive train are designed, in which $\omega_r(t)$ and $\omega_g(t)$ measurements are used separately.

The proposed sliding mode observers to estimate the rotor speed are given by:

$$\dot{\hat{x}}_{dt}^j(t) = A_{dt}(t)\hat{x}_{dt}^j(t) + B_{dt}u_{dt}(t) - G_l^j e_{yr}^j + G_{nl}^j w_r^j$$
$$\hat{y}_{dt}^{rj}(t) = \hat{\omega}_{r,mj} \tag{18}$$

and the proposed sliding mode observers to estimate the generator speed are given by:

$$\dot{\hat{x}}_{dt}^j(t) = A_{dt}(t)\hat{x}_{dt}^j(t) + B_{dt}u_{dt}(t) - G_l^j e_{yg}^j + G_n^j v_g^j$$
$$\hat{y}_{dt}^{gj}(t) = \hat{\omega}_{g,mj} \tag{19}$$

with $e_{yr}^j(t) = \hat{\omega}_{r,mj} - \omega_{r,mj}$, $e_{yg}^j(t) = \hat{\omega}_{g,mj} - \omega_{g,mj}$, and $j = 1, 2$.
where $\hat{\omega}_{r,mj}$ and $\hat{\omega}_{g,mj}$ are the rotor speed and generator speed estimations and are computed by observer 1 and 2, respectively. G_l and G_{nl} are the observer gains matrices.

The residuals of the FDI scheme are presented in Table 2. From Table 2, each residual is sensitive to only one fault. The residuals $r_{4,1}$ and $r_{4,2}$ indicate the rotor speed sensor faults, while the residuals $r_{5,1}$ and $r_{5,2}$ indicate the generator speed sensor faults.

Table 2. Drive train system FDI logic.

Residuals	Definition	Fault 4	Fault 5
$r_{4,1}$	$\| \hat{y}_{dt}^{r1} - \omega_{r,m1} \|$	1	0
$r_{4,2}$	$\| \hat{y}_{dt}^{r2} - \omega_{r,m2} \|$	0	0
$r_{5,1}$	$\| \hat{y}_{dt}^{g1} - \omega_{g,m1} \|$	0	0
$r_{5,2}$	$\| \hat{y}_{dt}^{g2} - \omega_{g,m2} \|$	0	1

In the next section, the effectiveness of the proposed configuration to diagnose faults in both the pitch system and the transmission train is put to the test.

3. Results

In this section, the proposed FDI scheme based on sliding mode observers is tested on the benchmark model of the wind turbine.

The test only considers sensor faults in the pitch system and the drive train. The type of fault is either electrical or mechanical and may result in a fixed value or a change gain factor on the measurements. In general, the severity of these faults is low, but if these are not handled correctly, can result in a breakdown. In order to prevent this situation, early fault detection and isolation are needed.

The wind turbine is a nonlinear system, driven by a disturbance, namely the wind. The residual signals can be nonzero, even when there are no faults. Therefore, the design of the FDI scheme is expected to be robust, in order to achieve a low false alarm rate and a high fault detection rate.

To solve this challenge, the proposed FDI scheme is based on a set of sliding mode observers, robust against disturbances and model uncertainties. The effectiveness of the proposed scheme is demonstrated by a test that includes five sensor faults, three for the pitch and two for the drive train system. These faults are defined in [25] and set as:

- First fault: this fault is present at the first pitch system in sensor 1 and outcome in a fixed value: $\beta_{1,m1} = 5°$ at time 2000–2100 s.
- Second fault: the sensor 2 in second pitch system is faulty and outcomes in a gain factor on the measurements: $\beta_{2,m2} = 1.2\beta_{2,m2}$ at time 2300–2400 s.

- Third fault: the sensor 1 in third pitch system is faulty and outcomes in a fixed value: $\beta_{3,m1} = 10°$ at time 2600–2700 s.
- Fourth fault: the rotor speed sensor signal is faulty and outcomes in a fixed value: $\omega_{r,m1} = 1.4$ rad/s at time 1500–1600 s.
- Fifth fault: the generator speed sensor signal is faulty and outcomes in a gain factor on the measurements: $\omega_{g,m2} = 0.9\,\omega_{g,m2}$ at time 1000–1100 s.

The results of the FDI scheme for the pitch 1 system when the fault occurs are shown in Figure 5. Fault 1 occurs in sensor 1 of pitch 1, in the time interval from 2000 to 2100 s. As can be seen in the residuals plot in Figure 5, the behavior of the residuals allows identifying the presence of fault 1. It can be observed that the residual $r_{1,1}$ presents a remarkable abrupt change during the time interval of the occurrence of fault 1, while the residual $r_{1,2}$ remains very close to zero. Therefore, it can be concluded that a fault occurred in the pitch 1 system, specifically in sensor 1. Ideally, it would be expected that the residuals remain at zero, while there is no fault in the sensors; however, due to the presence of noise in the sensors, the residuals present small deviations around zero, even when the fault has not occurred.

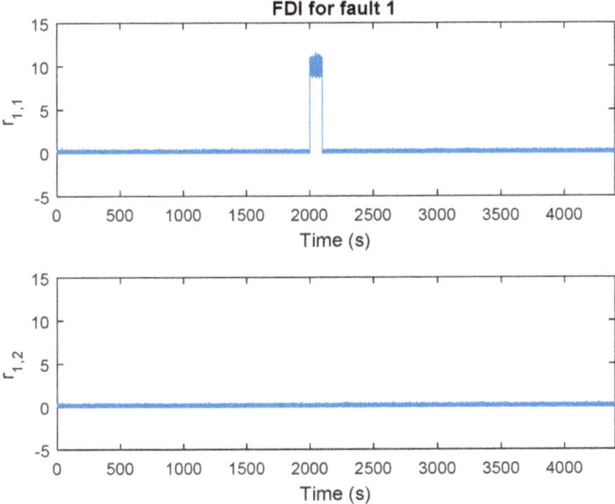

Figure 5. Faults in the first pitch system.

For the pitch 2 system, a scaling fault is considered during a time interval of 2300–2400 s, acting on sensor 2. Figure 6 shows the behavior of the residuals generated from the observer bank in sliding modes. In the graph, it can be noticed that the residual $r_{2,2}$ presents an erratic behavior in the fault incidence interval, while the residual $r_{2,1}$ does not present any atypical behavior and remains at a value close to zero. Therefore, from the fault signature of the two residuals, it is shown that the fault occurred in sensor 2 of pitch 2. We should note that the magnitude of the change of the fault-sensitive residual 2 is much smaller with respect to the behavior of the fault-sensitive residual 1. This is because a scaling error fault does not have as noticeable an effect on the sensor signal as a fixed value fault does.

For the pitch 3 system, the fault occurs in sensor 1. It is a fixed value fault acting in the time interval 2600–2700 s. The results of the fault detection scheme in the pitch 3 system are presented in Figure 7. It can be observed that fault 3 can be detected by means of the residual $r_{3,1}$, which presents a significant deviation in the time interval of the fault. While the residual $r_{3,2}$ does not present significant changes, it remains very close to zero. From the behavior of the residuals of the pitch system, it can be observed that these were shown to be sensitive to the presence of faults, thus solving the problem of fault detection in the pitch system. Moreover, since each of the presented residuals was shown to be sensitive to only one fault, fault isolation is achieved with the help of the fault signature table.

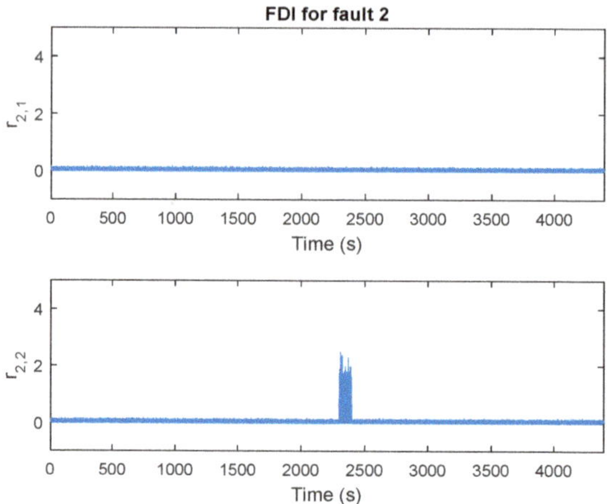

Figure 6. Fault bias in the second pitch during the time 2300 to 2400 s.

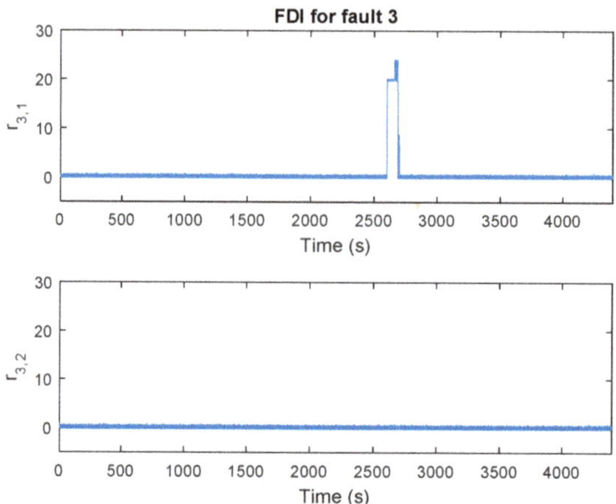

Figure 7. Fault bias in the third pitch during the time 2600 to 2700 s.

The proposed scheme for the detection and isolation of faults in the drive train employs a bank of four observers whose function is to estimate the turbine rotor speed and the generator speed. From the difference between the estimates made by the observers and the measurements obtained by the sensors, a set of four residuals is generated. Figure 8 shows the behavior of the residuals designed to be sensitive to fault 4. Fault 4 corresponds to a scaling error in the signal of sensor 1 in charge of measuring the rotor speed. Since sensor 2 is fault free, it can be observed that the residual $r_{4,2}$ remains practically invariant during the whole simulation interval. While the residual $r_{4,1}$ shows an atypical behavior during the 1500–1600 s interval. The magnitude of the deviation observed in the residual in the mentioned interval is high enough to infer the presence of a fault in the sensor.

Finally, the behavior of the residuals sensitive to fault 5 is shown in Figure 9. Fault 5 is a fixed value in sensor 2 responsible for measuring the generator speed, during the time interval 1000–1100 s. It is important to note that, from the point of view of fault dynamics, in a fixed value fault a more significant deviation of the residual can be observed, this compared to the case of a scaling error type fault. This is evident in the magnitude of the deviation of the residual $r_{5,2}$. From which it can be concluded that a fault has occurred in sensor 2 responsible for the measurement generator speed. While the behavior of the residual $r_{5,1}$ confirms that sensor 1 is free of fault.

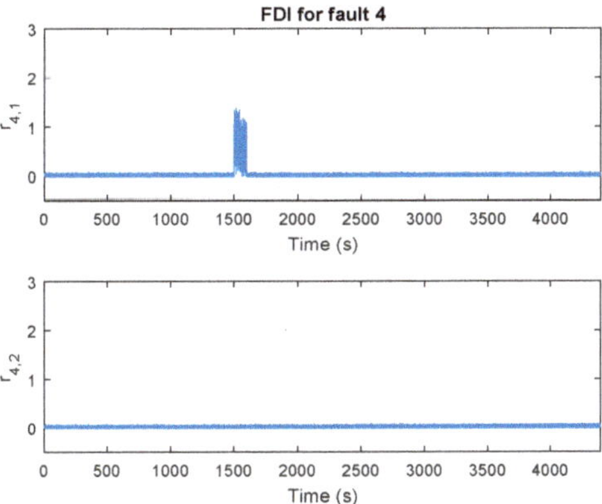

Figure 8. The abrupt fault of rotor speed sensor 1 in the time period of 1500 to 1600 s.

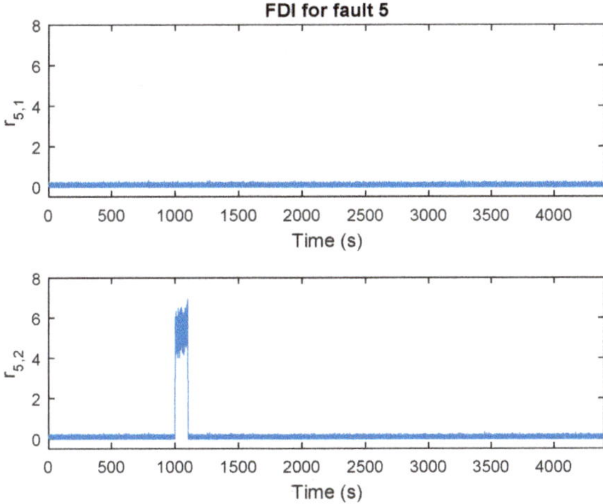

Figure 9. The abrupt fault in the speed of the generator in the period time of 1000 to 1100 s.

4. Discussion

The faults considered in this paper only cover the pitch system and drive train, and since these are modeled independent, an FDI scheme was designed separately for each one. The FDI scheme proposed for the pitch system considers the error introduced by sensor faults as an uncertainty acting on the input channel. Under this condition, a sliding mode observer bank was capable of estimating the system outputs, even in the presence of faults is proposed. The differences between the estimates made by the observers and the sensor measurements were used to regenerate a structured set of residuals for each pitch system. The structure of the FDI scheme for the pitch system is shown in Figure 3. It considers a set of six observers. To detect sensor faults, each sliding mode observer is dedicated to detect one sensor fault and generate one residual. The residual signals are generated from the measurements and estimates made by the bank of observers. After successful fault detection, the next task is to isolate the faults, which is done using a signature Table 1.

Secondly, to overcome the problem of coupled dynamics and the absence of aerodynamic torque measurement, the proposed FDI approach for the drive train employed an aerodynamic torque estimator based on an algebraic map. The uncertainty associated with the estimation was modeled as a disturbance in the input channel of the system model. The aerodynamic torque estimate was fed to a bank of observers in sliding modes each of which employed different sensor measurements in order to be sensitive to only one fault. Each observer was dedicated to one sensor fault; consequently, the generated residual was sensitive to only one sensor fault. For the task of isolating the faults, a signature Table 2 was used.

The effectiveness of the FDI schemes is demonstrated from the evaluation of five sensor fault scenarios as described in the results section. Due to the measurement noises and the uncertainties in the system, the residual signals are not zero, even for the fault-free case. To overcome this issue, residual evaluation and threshold are employed. The thresholds are designed based on the expressions of the residuals and uncertainties and measurement noise in the system. From the results presented in the last section, it can be easily seen that the residual remains around zero before and after the occurrence of fault, but during the fault interval, the residual rises significantly. To achieve a low false alarm, the thresholds are set at the maximum absolute value of the residuals. Therefore, a fault is presented in the system if it causes significant deviations from zero on some of the residuals.

5. Conclusions

In this article, a fault detection and isolation approach is proposed for the sensors of the pitch system and drive train in a benchmark wind turbine model. This approach employs a bank of sliding mode observers, designed to receive all of the inputs of the system and only one measurement of the set of available sensors. This allows each observer to be sensitive only to faults present in the sensor input signal. The dynamics introduced by the fault in a sensor of the pitch system, as well as the uncertainty associated with the estimation of the rotor torque in the transmission train, are modeled as disturbances acting on the input channel of their respective subsystem. Under this condition, the convergence of the observers turns out to be independent of the fault dynamics and the uncertainty present in the system. The difference between observer bank outputs and the measurements of the sensors is used to generate a structured set of residues for the pitch system and the drive train. It is shown that the residuals are capable of detecting a sensor fault occurrence, and, on the other hand, they remain very close to zero when the system is fault-free. This allows decision-making on fault detection using the concept of a fixed threshold. The value of the thresholds was selected based on the behavior of the residues before and after the appearance of the fault. To achieve fault isolation, fault signature matrix constructed from the residuals was used.

Author Contributions: Conceptualization, V.B.-J. and M.A.-M.; methodology, V.B.-J. and B.Y.L.-Z.; software, V.B.-J. and E.M.S.C.; validation, V.B.-J. and M.A.-M.; formal analysis, M.A.-M. and V.B.-J.; investigation, V.B.-J., M.A.-M. and B.Y.L.-Z.; resources, M.A.-M., B.Y.L.-Z., E.M.S.C., L.C.P.; writing—original draft preparation, V.B.-J., B.Y.L.-Z., L.G.V.V.; writing—review and editing, M.A.-M., E.M.S.C., V.B.-J., L.G.V.V., L.C.P.; supervision, M.A.-M. All authors have read and agreed to the published version of the manuscript.

Funding: This research received no external funding.

Institutional Review Board Statement: Not applicable.

Informed Consent Statement: Not applicable.

Data Availability Statement: Not applicable.

Acknowledgments: The development of this project is the product of the support provided by the National Council of Science and Technology (CONACYT), the National Technological Institute of Mexico (TecNM)/CENIDET.

Conflicts of Interest: The authors declare no conflict of interest.

Nomenclature

A	System matrix
B	Input matrix
C	Output matrix
D	Distribution matrix
e	Estimation error
F	Viscous friction
G	Observer gain
H	Torsion damping coefficient
J	Moment of inertia
K	Torsion stiffness
N	Gear ratio
P	Power
R	Rotor ratio
r	Residual signal
S	Power coefficient
T	Change of Coordinates Matrix
v	Wind speed
w	Discontinue function
x	System states
$\hat{x}$	Estimate of x
y	Output system
$\hat{y}$	Estimate of y
α	a known function
β	Pitch angle
γ	Cutoff frequency
ζ	Damping coefficient
η	Efficiency
θ	Torsion angle
λ	Tip speed ratio
ξ	Bounded uncertainty
ρ	Wind density
σ	Scalar
τ	Torque
ψ	Natural frequency
ω	Speed

Subscripts

dt	Drive train
g	Generator
l	Linear
nl	Nonlinear
pb	Pitch blade
r	Rotor
ref	Reference
ω	Wind

References

1. Sawant, M.; Thakare, S.; Rao, A.; Feijóo-Lorenzo, A.; Bokde, N. A Review on State-of-the-Art Reviews in Wind-Turbine- and Wind-Farm-Related Topics. *Energies* **2021**, *14*, 2041. [CrossRef]
2. Mazzeo, D.; Matera, N.; De Luca, P.; Baglivo, C.; Congedo, P.M.; Oliveti, G. A literature review and statistical analysis of photovoltaic-wind hybrid renewable system research by considering the most relevant 550 articles: An upgradable matrix literature database. *J. Clean. Prod.* **2021**, *295*, 126070. [CrossRef]
3. Pfaffel, S.; Faulstich, S.; Rohrig, K. Performance and Reliability of Wind Turbines: A Review. *Energies* **2017**, *10*, 1904. [CrossRef]
4. Márquez, F.P.G.; Tobias, A.M.; Pérez, J.M.P.; Papaelias, M. Condition monitoring of wind turbines: Techniques and methods. *Renew. Energy* **2012**, *46*, 169–178. [CrossRef]
5. Bianchi, F.D.; De Battista, H.; Mantz, R.J. *Wind Turbine Control Systems: Principles, Modelling and Gain Scheduling Design*; Springer: London, UK, 2007; Volume 19.

6. Alaskari, M.; Abdullah, O.; Majeed, M.H. Analysis of Wind Turbine Using QBlade Software. *IOP Conf. Series Mater. Sci. Eng.* **2019**, *518*, 032020. [CrossRef]

7. Raut, S. Simulation of Micro Wind Turbine Blade in Q-Blade. *Int. J. Res. Appl. Sci. Eng. Technol.* **2017**, *5*, 256–262. [CrossRef]

8. Chaudhary, M.; Roy, A. Design & optimization of a small wind turbine blade for operation at low wind speed. *World J. Eng.* **2015**, *12*, 83–94. [CrossRef]

9. Liu, J.; Yao, Q.; Hu, Y. Model predictive control for load frequency of hybrid power system with wind power and thermal power. *Energy* **2019**, *172*, 555–565. [CrossRef]

10. Chen, J.; Patton, R.J. *Robust Model-Based Fault Diagnosis for Dynamic Systems*; Springer Science & Business Media: Berlin/Heidelberg, Germany, 1999; Volume 11.

11. Luo, G.; Hurwitz, J.; Habetler, T.G. A survey of multi-sensor systems for online fault detection of electric machines. In Proceedings of the 2019 IEEE 12th International Symposium on Diagnostics for Electrical Machines, Power Electronics and Drives (SDEMPED), Toulouse, France, 27–30 August 2019; pp. 338–343.

12. Janssens, O.; Loccufier, M.; Van Hoecke, S. Thermal Imaging and Vibration-Based Multisensor Fault Detection for Rotating Machinery. *IEEE Trans. Ind. Inform.* **2019**, *15*, 434–444. [CrossRef]

13. Luwei, K.C.; Yunusa-Kaltungo, A.; Sha'Aban, Y.A. Integrated fault detection framework for classifying rotating machine faults using frequency domain data fusion and Artificial Neural Networks. *Machines* **2018**, *6*, 59. [CrossRef]

14. Yunusa-Kaltungo, A.; Cao, R. Towards Developing an Automated Faults Characterisation Framework for Rotating Machines. Part 1: Rotor-Related Faults. *Energies* **2020**, *13*, 1394. [CrossRef]

15. Cao, R.; Yunusa-Kaltungo, A. An Automated Data Fusion-Based Gear Faults Classification Framework in Rotating Machines. *Sensors* **2021**, *21*, 2957. [CrossRef] [PubMed]

16. Chen, W.; Ding, S.; Haghani, A.; Naik, A.; Khan, A.; Yin, S. Observer-based FDI Schemes for Wind Turbine Benchmark. *IFAC Proc. Vol.* **2011**, *44*, 7073–7078. [CrossRef]

17. Hwas, A.; Katebi, R. Model-based fault detection and isolation for wind turbine. In Proceedings of the 2012 UKACC International Conference on Control, Cardiff, UK, 3–5 September 2012; pp. 876–881.

18. XiaHou, K.; Li, M.S.; Liu, Y.; Wu, Q.H. Sensor Fault Tolerance Enhancement of DFIG-WTs via Perturbation Observer-Based DPC and Two-Stage Kalman Filters. *IEEE Trans. Energy Convers.* **2017**, *33*, 483–495. [CrossRef]

19. Sharan, B.; Jain, T. Actuator and Sensor Fault Diagnosis for Wind Energy Conversion Systems. In Proceedings of the 2018 15th International Conference on Control, Automation, Robotics and Vision (ICARCV), Singapore, 18–21 November 2018; pp. 955–959.

20. Zhang, X.; Zhang, Q.; Zhao, S.; Ferrari, R.; Polycarpou, M.; Parisini, T. Fault detection and isolation of the wind turbine benchmark: An estimation-based approach. *IFAC Proc. Vol.* **2011**, *44*, 8295–8300. [CrossRef]

21. Odgaard, P.F.; Stoustrup, J. Unknown input observer based detection of sensor faults in a wind turbine. In Proceedings of the 2010 IEEE International Conference on Control Applications, Yokohama, Japan, 8–10 September 2010; pp. 310–315.

22. Mokhtari, A.; Belkheiri, M. An adapative observer based FDI for wind turbine benchmark model. In Proceedings of the 2016 8th International Conference on Modelling, Identification and Control (ICMIC), Algiers, Algeria, 15–17 November 2016; pp. 742–746.

23. Dey, S.; Pisu, P.; Ayalew, B. A Comparative Study of Three Fault Diagnosis Schemes for Wind Turbines. *IEEE Trans. Control Syst. Technol.* **2015**, *23*, 1853–1868. [CrossRef]

24. Ziyabari, S.H.S.; Shoorehdeli, M.A.; Karimirad, M. Robust fault estimation of a blade pitch and drivetrain system in wind turbine model. *J. Vib. Control* **2021**, *27*, 277–294. [CrossRef]

25. Song, D.; Yang, J.; Dong, M.; Joo, Y.H. Kalman filter-based wind speed estimation for wind turbine control. *Int. J. Control Autom. Syst.* **2017**, *15*, 1089–1096. [CrossRef]

26. Edwards, C.; Spurgeon, S. *Sliding Mode Control: Theory and Applications*; CRC Press: Boca Raton, FL, USA, 1998.

27. Habibi, H.; Howard, I.; Simani, S.; Fekih, A. Decoupling Adaptive Sliding Mode Observer Design for Wind Turbines Subject to Simultaneous Faults in Sensors and Actuators. *IEEE/CAA J. Autom. Sin.* **2021**, *8*, 837–847. [CrossRef]

28. Odgaard, P.F.; Stoustrup, J.; Kinnaert, M. Fault-tolerant control of wind turbines: A benchmark model. *IEEE Trans. Control Syst. Technol.* **2013**, *21*, 1168–1182. [CrossRef]

29. Edwards, C.; Spurgeon, S. On the development of discontinuous observers. *Int. J. Control* **1994**, *59*, 1211–1229. [CrossRef]

30. Edwards, C.; Spurgeon, S.; Patton, R.J. Sliding mode observers for fault detection and isolation. *Automatic* **2000**, *36*, 541–553. [CrossRef]

31. Simani, S.; Fantuzzi, C.; Patton, R.J. Model-Based Fault Diagnosis Techniques. In *Advances in Industrial Control*; Springer: London, UK, 2003; pp. 19–60.

 *processes*

Article

Predictive Control in Water Distribution Systems for Leak Reduction and Pressure Management via a Pressure Reducing Valve

Jose-Roberto Bermúdez [1], **Francisco-Ronay López-Estrada** [1], **Gildas Besançon** [2], **Guillermo Valencia-Palomo** [3,*] **and Ildeberto Santos-Ruiz** [1]

[1] Tecnológico Nacional de Mexico, I.T. Tuxtla Gutiérrez, TURIX Diagnosis and Control Group, Carretera Panamericana km 1080, SN, Tuxtla Gutierrez 29050, Mexico; bermudez_r10@hotmail.com (J.-R.B.); frlopez@tuxtla.tecnm.mx (F.-R.L.-E.); ildeberto.dr@tuxtla.tecnm.mx (I.S.-R.)

[2] GIPSA-Lab, CNRS, Grenoble INP, Université Grenoble Alpes, 38000 Grenoble, France; gildas.besancon@gipsa-lab.grenoble-inp.fr

[3] Tecnológico Nacional de México, IT Hermosillo, Av. Tecnológico y Periférico Poniente S/N, Hermosillo 83170, Mexico

* Correspondence: gvalencia@hermosillo.tecnm.mx

Abstract: This work proposes a model predictive control (MPC) strategy for pressure management and leakage reduction in a water distribution system (WDS). Unlike most of the reported models that mainly consider EPANET-based models, the proposed method considers its dynamic representation given by ordinary differential equations. The proposed MPC uses a pressure-reducing valve (PRV) as a control element to regulate the pressure in the WDS to track the demand. The control scheme proposes a strategy to manage the high nonlinearity of the PRV and takes into account the demand profile throughout the day as well as the leaks that occur in the pipeline. The estimates of magnitude and location of the leak are provided by an Extended Kalman Filter from previous work and with the aid of a rule-based set point manager reduces the fluid loss in the event of a leak. Different scenarios are studied to illustrate the effectiveness of the proposed control system, achieving an approximate reduction of up to 5% of water losses, demonstrating robustness in the case of uncertainty in the leak location estimate.

Keywords: pipelines; control valves; leak reduction; water distribution system; pressure management

Citation: Bermúdez, J.-R.; López-Estrada, F.-R.; Besanşon, G.; Valencia-Palomo, G.; Santos-Ruiz, I. Predictive Control in Water Distribution Systems for Leak Reduction and Pressure Management via a Pressure Reducing Valve. *Processes* **2022**, *10*, 1355. https://doi.org/10.3390/pr10071355

Academic Editor: Alfredo Iranzo

Received: 1 July 2022
Accepted: 8 July 2022
Published: 12 July 2022

Publisher's Note: MDPI stays neutral with regard to jurisdictional claims in published maps and institutional affiliations.

1. Introduction

Water distribution systems (WDS) are the most sustainable and efficient means of transporting fluids such as drinking water, natural gas, and oil [1]. These systems are composed of pipelines, pipe joints, connection nodes, and other components such as valves and pumps, which are prone to damage due to aging or unwanted events, such as earthquakes, floods, and lack of management, among others [2]. According to a study by the Organisation for Economic Co-operation and Development [3], these abnormal events cause water losses by leaks that reach almost 21%. The study examined water usage in 48 major cities across 17 countries, finding that in some cities of Mexico, the percentage was more than 40%. This highlights the need to propose technological developments to mitigate such losses.

From the control theory point of view, it is essential to investigate techniques to reduce the leaks without affecting the demand. In the literature, it has been found that the most effective approach to find a trade-off between maintaining the desired demand and reducing the water losses is by considering pressure water controllers on critical nodes of the water distribution network [4]. However, many challenges are associated with increasing demand and managing the pressure levels. In particular, pressure-reducing valves (PRV) are the recommended actuators to minimize these undesirable effects and operate the WDS effectively by following a pattern of demands [5].

Modeling the WDS is vital for designing any control algorithm [6,7]. These models are analyzed taking into account the location and number of PRVs for a correct evaluation of the pressure [8]. For instance, Mazumder et al. [9] developed an optimization method based on genetic algorithms for pressure management in a WDS by adjusting a PRV of a hydraulic network designed in EPANET. Parra et al. [10] proposed a pressure management system in EPANET composed of a PRV and a pump as a turbine where the hydraulic model played a decisive role identifying critical nodes and predicting hydraulics properties in the network. García-Ávila et al. [11] designed a water leakage minimization system by optimizing pressure using a PRV. The hydraulic model of the WDN was developed using EPANET and WaterNetGen software. Dini and Asadi [12] designed a methodology based on particle swarm optimization (PSO) to identify the PRV that requires adjustment and obtain pressure management in the system; the network model was designed in EPANET. Hernández et al. [13] proposed a detrended fluctuation analysis (or DFA) to highlight some of the traits such as the head loss of high-viscosity gas–liquid flows. Navarro et al. [14] designed a leak diagnosis system for a pipeline and residue analysis using genetic algorithms (GA) to minimize location error. Jara-Arriagada and Stoianov [15] designed a sensitivity analysis system to evaluate the potential impact of pressure control in a WDS to reduce pipe breaks by applying a logistic regression technique. Mosetlhe et al. [16] proposed a review of techniques for locating PRVs and controlling pressure in a WDS, minimizing problems such as excessive pressure in the system. Mathye et al. [17] designed a pressure management system to reduce leaks through PRVs, taking into account consumption and leak flow. All those works developed algorithms for pressure management using hydraulic models based on EPANET and waterNetGen software, whose main limitation is that the models only work in steady-state behavior, while the effects of leaks and pressure changes are dynamic.

The present work proposes a dynamical approach to pressure management in a WDS. The transient effects due to pressure changes or leaks are modeled on the basis of water-hammer equations. Then, a constrained model predictive control (MPC) system is proposed to track the desired pressure profile driven by a set point manager that considers the water loss due to leaks and the demanded pressure profile. Moreover, a strategy to handle the high nonlinearity of the PRV control input is proposed. Finally, some simulations are proposed by considering the mathematical model of a real distribution system that can be configured as a single pipeline or a branched system. The results illustrate the effectiveness of the proposed method in the presence of physical constraints, noise, and transient behaviors due to leaks, saving up to 5% of water losses in the event of a leak and demonstrating robustness in the case of uncertainty in the leak location estimate. The rest of the document is organized as follows: Section 2 presents the considered case studies; Section 3 describes the problem formulation; Section 4 is devoted to the control strategy; Section 5 presents the simulation results; finally, Section 6 draws the conclusions.

2. Case Studies

Three different cases will be considered in this paper, corresponding to different hydraulic system structures with the pressure to be controlled at a specific position (called controlled node) and some leaks affecting the system at some other position.

2.1. Case 1: Pipeline with a Leak before the Controlled Node

In this case, the system is a pipeline under a leak like the one shown in Figure 1. It is composed of a reservoir that provides the fluid to the pipe divided into three sections for convenience. The first section is related to the distance between the inlet node and the leak (z_ℓ); the second one is related to the distance from the leak to the controlled node ($z_2 = z_1 - z_\ell$); and finally, the third section is related to the distance between the controlled node and the outlet node (z_3).

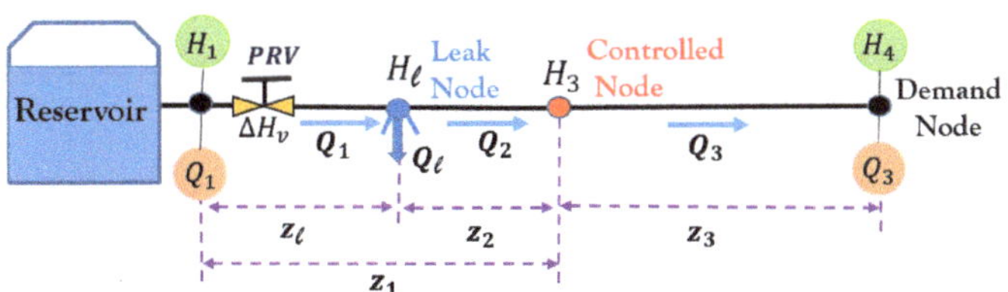

Figure 1. Case 1: Pipeline with a leak before the controlled node.

The mathematical model can be determined by considering water-hammer equations [18], which can be approximated on the basis of the three considered sections as follows (e.g., [19,20]):

$$\dot{Q}_1 = \frac{\alpha_1}{z_\ell}(H_1 - H_\ell) + \mu_1(Q_1)Q_1|Q_1| - \frac{\alpha_1}{z_1}\Delta H_v, \tag{1}$$

$$\dot{H}_\ell = \frac{\alpha_2}{z_\ell}(Q_1 - Q_2 - Q_\ell), \tag{2}$$

$$\dot{Q}_2 = \frac{\alpha_1}{z_2}(H_\ell - H_3) + \mu_2(Q_2)Q_2|Q_2|, \tag{3}$$

$$\dot{H}_3 = \frac{\alpha_2}{z_2}(Q_2 - Q_3), \tag{4}$$

$$\dot{Q}_3 = \frac{\alpha_1}{z_3}(H_3 - H_4) + \mu_3(Q_3)Q_3|Q_3|, \tag{5}$$

with

$$\alpha_1 = gA_r, \quad \alpha_2 = \frac{c^2}{gA_r}, \quad Q_\ell = \lambda\sqrt{|H_\ell|},$$

$$\mu_i(Q_i) = \frac{-f(Q_i)}{2DA_r}, \quad f(Q_i) = \frac{1.325}{\left[\ln\left(\frac{\epsilon}{3.7d} + \frac{5.74}{(\frac{4Q_i}{\pi d v})^{0.9}}\right)\right]^2},$$

where H_1, H_ℓ, H_3, and H_4 are the piezometric heads (m) at the inlet, leak node, controlled node, and outlet, respectively; Q_1, Q_2, Q_3 are the volumetric flow rates in each section (m^3/s); g is the gravitational constant (m/s^2); A_r is the cross-sectional area of the pipe (m^2); c is the wave speed (m/s); d is the pipeline diameter (m); v represents the kinematic viscosity; the friction term is $f(Q_i)$; ϵ is the roughness of the pipe; and λ is the leak coefficient. Finally, ΔH_v describes the PRV effect.

A PRV is an actuator used to reduce downstream pressure. Internally, a PRV is made of a fixed orifice, a pilot valve, and a needle valve [21]. The shutter is the outer mechanism for adjusting the outlet pressure, either increasing or decreasing the pressure in a range from 0 to 100%. Figure 2 shows the schematic view of a PRV, where H_{v1} and H_{v2} are the piezometric head at valve ends, $r \in (0,1]$ is the valve adjustment (control input), and Q_1 is the flow through the valve.

The differential pressure in a PRV is described as [22]:

$$\Delta H_v = H_{v1} - H_{v2} = \frac{Q_1|Q_1|}{(rE)^2}, \tag{6}$$

where $E = C_v A_v(2g)^{1/2}$ is the Torricelli expression, C_v is the discharge coefficient of the valve, and A_v is the cross-sectional area of the valve (m^2).

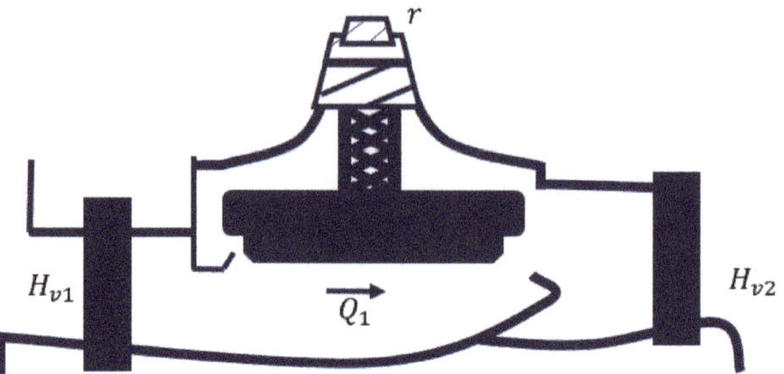

Figure 2. Schematic view of a PRV.

2.2. Case 2: Pipeline with a Leak after the Controlled Node

When the leak appears after the controlled node, as shown in Figure 3, then $z_2 = z_3 - z_\ell$, and the model becomes

$$\dot{Q}_1 = \frac{\alpha_1}{z_1}(H_1 - H_2) + \mu_1(Q_1)Q_1|Q_1| - \frac{\alpha_1}{z_1}\Delta H_v, \tag{7}$$

$$\dot{H}_2 = \frac{\alpha_2}{z_1}(Q_1 - Q_2), \tag{8}$$

$$\dot{Q}_2 = \frac{\alpha_1}{z_\ell}(H_2 - H_3) + \mu_2(Q_2)Q_2|Q_2|, \tag{9}$$

$$\dot{H}_\ell = \frac{\alpha_2}{z_\ell}(Q_2 - Q_3 - \lambda\sqrt{H_\ell}), \tag{10}$$

$$\dot{Q}_3 = \frac{\alpha_1}{z_2}(H_3 - H_4) + \mu_3(Q_3)Q_3|Q_3|. \tag{11}$$

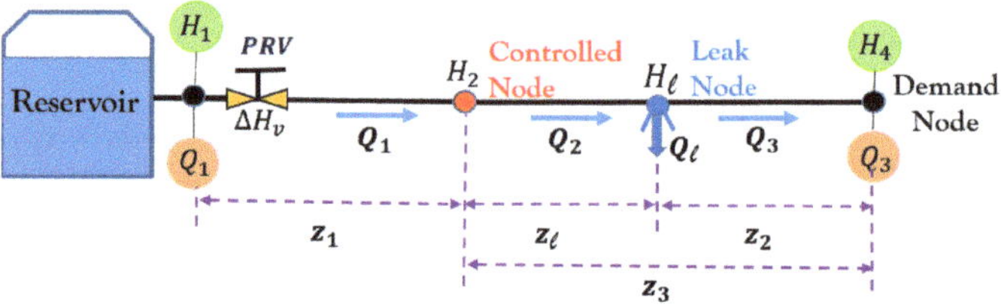

Figure 3. Case 2: Pipeline with a leak after the controlled node.

2.3. Case 3: Branched Water Distribution Network

In this third case, a branched water distribution network is considered with the topology shown in Figure 4. With the same notations as before, the mathematical model describing branch flows Q_i and node pressures H_i can be given by:

$$\dot{Q}_1(t) = \frac{\alpha_1}{z_1}(H_{B1} - H_2) + \mu_1(Q_1)Q_1|Q_1| - \Delta H_v, \tag{12}$$

$$\dot{H}_2(t) = \frac{\alpha_2}{z_1}(Q_1 - Q_2 - Q_4), \tag{13}$$

$$\dot{Q}_2(t) = \frac{\alpha_1}{z_2}(H_2 - H_\ell) + \mu_2(Q_2)Q_2|Q_2|, \tag{14}$$

$$\dot{H}_\ell(t) = \frac{\alpha_2}{z_2}(Q_2 - Q_5 - Q_3 - Q_\ell), \tag{15}$$

$$\dot{Q}_3(t) = \frac{\alpha_1}{z_3}(H_3 - H_{B2}) + \mu_3(Q_3)Q_3|Q_3|, \tag{16}$$

$$\dot{Q}_4(t) = \frac{\alpha_1}{z_4}(H_2 - H_{B3}) + \mu_4(Q_4)Q_4|Q_4|, \tag{17}$$

$$\dot{Q}_5(t) = \frac{\alpha_1}{z_5}(H_3 - H_{B4}) + \mu_5(Q_5)Q_5|Q_5|. \tag{18}$$

The mathematical models developed here are valid for any pipeline system with the configurations described in Figures 1–4. For validation tests, the physical parameters were taken from a real system located at the Hydroinformatics Laboratory of the Technological Institute of Tuxtla Gutiérrez, whose mathematical model was presented in [23].

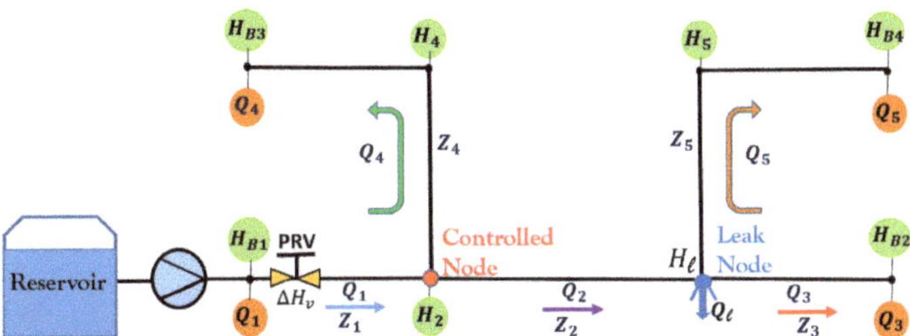

Figure 4. Case 3: Branched water distribution network.

Remark 1. *Notice that λ and z_ℓ are unknown parameters that must be estimated by considering a leak detection and localization method. For Cases 1 and 2, we consider our previous result published in [24] where a leak location and estimation method was designed using an extended Kalman filter (EKF). For Case 3, an EKF as the one in [25] can be used. Therefore, for the sake of simplicity, λ and z_ℓ are assumed to be known.*

3. Problem Formulation

WDS are designed to meet the desired demands at their ends, even when affected by leaks. It is important to note that these demands have different profiles depending on the time and day. Typically, the demand is higher during the day and lower during the early morning [26]. In case of a leak event, the fluid loss rate (leak magnitude) can be reduced by reducing the pressure on the controlled node. However, this also reduces the flow at the node of demand, compromising the main objective of the WDS. In this regard, the control strategy for reducing leaks must consider a time-varying profile, with the primary objective to maintain a trade-off between reducing the leak magnitude and maintaining the desired demand.

To address this problem, the control scheme shown in Figure 5 is proposed. This scheme is made of three components: an extended Kalman filter (EKF), a pressure controller, and a set point manager. The EKF is used to estimate the flows and pressures along the WDS and to detect and estimate the leak position and its magnitude by using only pressure head and flow rate measurements at the pipeline ends. The EKF considered for Cases 1 and 2 is the one reported in our previous work in [24] and, for Case 3, the one in [27] is considered; however, any other leak location and estimation method can be used for the proposed scheme. The pressure controller is an MPC that takes into account physical constraints and the leak dynamics. For its operation, the MPC uses the estimated values obtained from the EKF ($\hat{z}_\ell$, $\hat{\lambda}$, $\hat{H}_\ell$) and measured flows and pressures from the WDS. It is important to mention that the MPC also considers the PRV model whose behavior is highly nonlinear due to the inverse quadratic term of its control variable (r), which represents a challenge for the control system. Finally, the set point manager block provides the reference pressure (s_k) to the MPC; it handles the trade-off between fluid loss and fulfilment of a demanded pressure profile (DP) in the event of a leak.

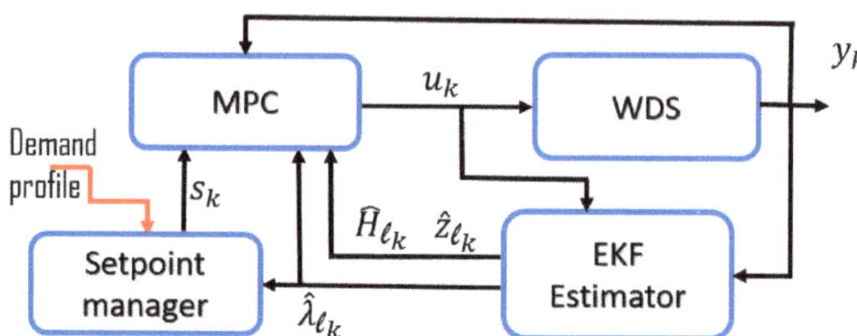

Figure 5. Block diagram representation of MPC used in WDS.

4. Pressure Control

The control strategy adopted to regulate the pressure in the WDS is an MPC. Figure 6 shows the implemented scheme which will be detailed in this section. The basic idea in MPC is to calculate the control action at each sampling instant through the solution of an optimization problem, which is written in terms of a prediction model.

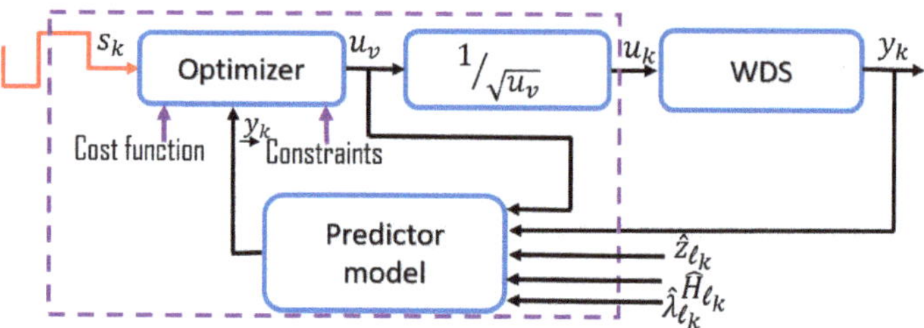

Figure 6. MPC scheme for pressure control.

4.1. PRV Handling

The PRV has the function of regulating the incoming water for a safer constant predetermined downstream level. The control signal r establishes the downstream pressure. One of the main challenges in using a PRV is its high nonlinearity as an inverse quadratic term (r) represents its control variable. This term affects the control input considerably because the more the PRV is closed, the greater the effects on the pressure. Then, to use linear models in the MPC formulation, the PRV control variable r is substituted in (6) by a virtual input u_v to artificially hide the nonlinear behavior of the PRV for the prediction model. This is accomplished with:

$$\Delta H_v = \frac{Q_1|Q_1|u_v}{E^2}, \quad \text{with} \quad u_v = \frac{1}{r^2}, \tag{19}$$

where $u_v \in [1, \infty)$, but for practical reasons u_v ranges from 1 (for the PRV fully open, $r = 1$) to 100 (for the PRV 90% closed, $r = 0.1$). This new term is substituted in the three models discussed in Section 2. In this way, the MPC calculates u_v instead of r. Next, u_v is transformed back to r before it is applied to the WDS with $u = r = \frac{1}{\sqrt{u_v}}$, as shown in Figure 6.

4.2. Prediction Models

The standard MPC strategy considers linear prediction models [28,29], so the models of the case studies listed before need to be linearized.

For a pipeline with a leak before the controlled node, the state and input vectors are defined as $\bar{x} = [x_1,\ x_2,\ x_3,\ x_4,\ x_5] = [Q_1,\ H_\ell,\ Q_2,\ H_3,\ Q_3]$, and $\bar{u} = [H_1,\ H_4,\ \lambda,\ u_v]^T$. The system (1)–(5) is linearized at the operating point $(\bar{x}^*, \bar{u}^*)$ taking the form $\dot{x}(t) = A_c\bar{x}(t) + B_c\bar{u}(t) + \delta_c$, where δ_c is the offset caused by linearization. The Jacobian matrices are given by:

$$
A_c =
\begin{bmatrix}
a_{11} & -\frac{\alpha_1}{z_\ell} & 0 & 0 & 0 \\
\frac{\alpha_2}{z_\ell} & 0 & -\frac{\alpha_2}{z_\ell} & 0 & 0 \\
0 & \frac{\alpha_1}{z_2} & a_{33} & -\frac{\alpha_1}{z_2} & 0 \\
0 & 0 & \frac{\alpha_2}{z_2} & 0 & -\frac{\alpha_2}{z_2} \\
0 & 0 & 0 & \frac{\alpha_1}{z_3} & a_{55}
\end{bmatrix},
$$

$$
B_c =
\begin{bmatrix}
\frac{\alpha_1}{z_\ell} & 0 & 0 & \frac{-g A_r x_1^* |x_1^*|}{E^2 z_\ell} \\
0 & 0 & -\frac{\alpha_2 \sqrt{H_2^*}}{z_2} & 0 \\
0 & 0 & 0 & 0 \\
0 & 0 & 0 & 0 \\
0 & \frac{-\alpha_1}{z_3} & 0 & 0
\end{bmatrix},
$$

with

$$
a_{11} = 2|x_1^*|\mu(x_1^*) + \frac{2(1.325)}{0.9\frac{5.74}{(\frac{4}{\pi dv})^{0.9}}} \frac{(x_1^*)^{1.9}}{\ln\left((\frac{\epsilon}{3.7d}) + \frac{5.74}{(\frac{4x_1^*}{\pi dv})^{0.9}} \right)^3}
$$

$$
- \frac{2 A_r g |x_1^*| u_v^*}{z_1 E^2},
$$

$$
a_{33} = 2|x_3^*|\mu(x_3^*) + \frac{2(1.325)}{0.9\frac{5.74}{(\frac{4}{\pi dv})^{0.9}}} \frac{(x_3^*)^{1.9}}{\ln\left((\frac{\epsilon}{3.7d}) + \frac{5.74}{(\frac{4x_3^*}{\pi dv})^{0.9}} \right)^3},
$$

$$
a_{55} = 2|x_5^*|\mu(x_5^*) + \frac{2(1.135)}{0.9\frac{5.74}{(\frac{4}{\pi dv})^{0.9}}} \frac{(x_5^*)^{1.9}}{\ln\left((\frac{\epsilon}{3.7d}) + \frac{5.74}{(\frac{4x_5^*}{\pi dv})^{0.9}} \right)^3},
$$

$$
\delta_c = f_1(\bar{x}^*, \bar{u}^*) - A_c\bar{x}^* - B_c\bar{u}^* + \Delta_1,
$$

where $f_1(\bar{x}^*, \bar{u}^*)$ is the nonlinear model of Case 1 evaluated at the linearization point, and Δ_1 gathers terms of order larger than 1.

For a pipeline with a leak after the controlled node, the state and input vectors are defined as $\bar{x} = [x_1,\ x_2,\ x_3,\ x_4,\ x_5] = [Q_1,\ H_2,\ Q_2\ H_\ell,\ Q_3]$, $\bar{u} = [H_1,\ H_4,\ \lambda, u_v]^T$. Using a_{11}, a_{33}, a_{55} from the previous case, the Jacobian matrices for system (7)–(11) are:

$$
A_c =
\begin{bmatrix}
a_{11} & -\frac{\alpha_1}{z_1} & 0 & 0 & 0 \\
\frac{\alpha_2}{z_1} & 0 & -\frac{\alpha_2}{z_1} & 0 & 0 \\
0 & \frac{\alpha_1}{z_\ell} & a_{33} & -\frac{\alpha_1}{z_\ell} & 0 \\
0 & 0 & \frac{\alpha_2}{z_\ell} & 0 & \frac{\alpha_2}{z_\ell} \\
0 & 0 & 0 & \frac{\alpha_1}{z_2} & a_{55}
\end{bmatrix},
$$

$$
B_c = \begin{bmatrix}
\frac{a_1}{z_1} & 0 & 0 & \frac{-gA_rx_1^*|x_1^*|}{E^2z_1} \\
0 & 0 & 0 & 0 \\
0 & 0 & 0 & 0 \\
0 & 0 & -\frac{a_2\sqrt{H_4^*}}{z_\ell} & 0 \\
0 & \frac{-a_1}{z_2} & 0 & 0
\end{bmatrix},
$$

$$
\delta_c = f_2(\bar{x}^*, \bar{u}^*) - A_c\bar{x}^* - B_c\bar{u}^* + \Delta_2,
$$

where $f_2(\bar{x}^*, \bar{u}^*)$ and Δ_2 refer to the nonlinear model of Case 2 as before.

For the branched water distribution network, the states and the inputs are defined as $\bar{x} = [x_1, x_2, x_3, x_4, x_5, x_6, x_7] = [Q_1, H_2, Q_2, H_\ell, Q_3, Q_4, Q_5]$, $\bar{u} = [H_{B1}, H_{B2}, H_{B3}, H_{B4}, \lambda, u_v]^T$. The Jacobian matrices for system (12)–(18) are:

$$
A_c = \begin{bmatrix}
a_{11} & -\frac{\alpha_1}{z_1} & 0 & 0 & 0 & 0 & 0 \\
\frac{\alpha_2}{z_1} & 0 & -\frac{\alpha_2}{z_2} & 0 & 0 & -\frac{\alpha_2}{z_1} & 0 \\
0 & \frac{\alpha_1}{z_2} & a_{33} & -\frac{\alpha_1}{z_2} & 0 & 0 & 0 \\
0 & 0 & \frac{\alpha_2}{z_2} & 0 & -\frac{\alpha_2}{z_2} & 0 & -\frac{\alpha_2}{z_2} \\
0 & 0 & 0 & \frac{\alpha_1}{z_3} & a_{55} & 0 & 0 \\
0 & \frac{\alpha_1}{z_4} & 0 & 0 & 0 & a_{66} & 0 \\
0 & 0 & 0 & \frac{\alpha_1}{z_5} & 0 & 0 & a_{77}
\end{bmatrix},
$$

$$
B_c = \begin{bmatrix}
\frac{\alpha_1}{z_1} & 0 & 0 & 0 & 0 & \frac{-gA_rx_1^*|x_1^*|}{E^2z_1} \\
0 & 0 & 0 & 0 & 0 & 0 \\
0 & 0 & 0 & 0 & 0 & 0 \\
0 & 0 & 0 & 0 & \frac{-\alpha_2\sqrt{H_3^*}}{z_1} & 0 \\
0 & \frac{-\alpha_1}{z_3} & 0 & 0 & 0 & 0 \\
0 & 0 & \frac{-\alpha_1}{z_4} & 0 & 0 & 0 \\
0 & 0 & 0 & \frac{-\alpha_1}{z_5} & 0 & 0
\end{bmatrix},
$$

with

$$
a_{11} = 2|x_1^*|\mu(x_1^*) + \frac{2(1.135)}{0.9\frac{5.74}{(\frac{4}{\pi dv})^{0.9}}} \frac{(x_1^*)^{1.9}}{\ln\left((\frac{\epsilon}{3.7d}) + \frac{5.74}{(\frac{4x_1^*}{\pi dv})^{0.9}}\right)^3}
$$

$$
- \frac{2A_r g|x_1^*|u_v^*}{z_1 E^2},
$$

$$
a_{33} = 2|x_3^*|\mu(x_3^*) + \frac{2(1.135)}{0.9\frac{5.74}{(\frac{4}{\pi dv})^{0.9}}} \frac{(x_3^*)^{1.9}}{\ln\left((\frac{\epsilon}{3.7d}) + \frac{5.74}{(\frac{4x_3^*}{\pi dv})^{0.9}}\right)^3},
$$

$$
a_{55} = 2|x_5^*|\mu(x_5^*) + \frac{2(1.135)}{0.9\frac{5.74}{(\frac{4}{\pi dv})^{0.9}}} \frac{(x_5^*)^{1.9}}{\ln\left((\frac{\epsilon}{3.7d}) + \frac{5.74}{(\frac{4x_5^*}{\pi dv})^{0.9}}\right)^3},
$$

$$
a_{66} = 2|x_6^*|\mu(x_6^*) + \frac{2(1.135)}{0.9\frac{5.74}{(\frac{4}{\pi dv})^{0.9}}} \frac{(x_6^*)^{1.9}}{\ln\left((\frac{\epsilon}{3.7d}) + \frac{5.74}{(\frac{4x_6^*}{\pi dv})^{0.9}}\right)^3}
$$

$$a_{77} = 2|x_3^*|\mu(x_7^*) + \frac{2(1.135)}{0.9\frac{5.74}{(\frac{4}{\pi d v})^{0.9}}} \frac{(x_7^*)^{1.9}}{\ln\left((\frac{\epsilon}{3.7d}) + \frac{5.74}{(\frac{4x_7^*}{\pi d v})^{0.9}}\right)^3},$$

$$\delta_c = f_3(\bar{x}^*, \bar{u}^*) - A_c\bar{x}^* - B_c\bar{u}^* + \Delta_3,$$

where $f_3(\bar{x}^*, \bar{u}^*)$ and Δ_3 refer to the nonlinear model of Case 3.

For all cases, the controlled output is $y(t) = C_c\bar{x}(t) = x_2(t)$. It is also noted that the only manipulated input is u_v. The rest of the elements in $\bar{u}$ are non-manipulated in the three cases.

Now for a discrete-time implementation, the previous linear models are discretized, with a sample time T_s, leading to systems of the form

$$x_{k+1} = A_d x_k + B_d u_k + \delta_d; \quad y_k = C_d x_k. \tag{20}$$

The MPC calculates control input increments; then each model is augmented to express it in those terms:

$$\zeta_{k+1} = \underbrace{\begin{bmatrix} A_d & B_d \\ 0 & I \end{bmatrix}}_{A} \zeta_k + \underbrace{\begin{bmatrix} B_d \\ I \end{bmatrix}}_{B} \Delta u_k + \underbrace{\begin{bmatrix} \delta_d \\ 0 \end{bmatrix}}_{\delta}, \tag{21}$$

$$y_k = \underbrace{\begin{bmatrix} C_d & 0 \end{bmatrix}}_{C} \zeta_k, \tag{22}$$

where $\zeta = [x_k^T \ u_{k-1}^T]^T$ and $\Delta u_k = u_k - u_{k-1}$.

4.3. Prediction Equations

The predictions equations of the augmented-state, output and input over a prediction horizon n_y, and control horizon n_u are:

$$\overrightarrow{\zeta}_k = P_{\zeta\zeta}\zeta_k + P_{\zeta\Delta u}\overrightarrow{\Delta u}_{k-1} + P_{\zeta\delta}\delta, \tag{23}$$

$$\overrightarrow{y}_k = P_{y\zeta}\zeta_k + P_{y\Delta u}\overrightarrow{\Delta u}_{k-1} + P_{y\delta}\delta, \tag{24}$$

$$\overrightarrow{u}_{k-1} = P_{u\Delta u}\overrightarrow{\Delta u}_{k-1} + P_{u\zeta}\zeta_k, \tag{25}$$

where the arrow notation denotes prediction and is defined as $\overrightarrow{x}_k = [x_{k+1}^T \ x_{k+2}^T \ \ldots]^T$.
Augmented state prediction matrices are

$$P_{\zeta\zeta} = \begin{bmatrix} A \\ A^2 \\ \vdots \\ A^{n_y} \end{bmatrix}, \ P_{\zeta\Delta u} = \begin{bmatrix} B & \cdots & 0 \\ AB & \cdots & 0 \\ \vdots & \ddots & \vdots \\ A^{n_y-1}B & \cdots & A^{n_y-n_u}B \end{bmatrix}, \ P_{\zeta\delta} = \begin{bmatrix} I \\ I+A \\ \vdots \\ I + \sum_{i=1}^{n_y-1} A^i \end{bmatrix}.$$

Output prediction matrices are $P_{y\zeta} = \mathrm{diag}(C)P_{\zeta\zeta}$, $P_{y\Delta u} = \mathrm{diag}(C)P_{\zeta\Delta u}$, $P_{y\delta} = \mathrm{diag}(C)P_{\zeta\delta}$. Input predictions matrices are given by:

$$\overrightarrow{u}_{k-1} = \underbrace{\begin{bmatrix} I & 0 & \cdots & 0 \\ I & I & \cdots & 0 \\ \vdots & \vdots & \ddots & \vdots \\ I & I & \cdots & I \end{bmatrix}}_{P_{u\Delta u}} \overrightarrow{\Delta u}_{k-1} + \underbrace{\mathrm{col}(\begin{bmatrix} 0 & I \end{bmatrix})}_{P_{u\zeta}}\zeta_k, \tag{26}$$

and for the states $\underset{\rightarrow}{x}_k = P_{x\zeta}\,\underset{\rightarrow}{\zeta}_k = \mathrm{diag}\!\left(\begin{bmatrix} I & 0 \end{bmatrix}\right)\underset{\rightarrow}{\zeta}_k$.

To construct the prediction vectors, full state availability is required. Although $H_\ell(t)$ is not a measured state and λ, z_ℓ are unknown parameters, all are considered known as they could be provided by the EKF estimator. The rest of the states are measured flows and pressures.

4.4. Cost Function and Constraints

The cost function to be optimized penalizes the future output error with respect to the desired output value s_k and control input along the prediction and control horizons:

$$J = \sum_{i=1}^{n_y}(y_{k+i}-s_k)^T \mathcal{Q}(y_{k+i}-s_k) + \sum_{i=0}^{n_u-1}\Delta u_{k+i}^T \mathcal{R}\Delta u_{k+i} \tag{27}$$

with $\mathcal{Q}>0$ and $\mathcal{R}>0$.

Constraints of the form:

$$\mathrm{diag}(A_x)\,\underset{\rightarrow}{x}_k \le \mathrm{col}(b_x); \quad \mathrm{diag}(A_y)\,\underset{\rightarrow}{y}_k \le \mathrm{col}(b_y);$$

$$\mathrm{diag}(A_u)\,\underset{\rightarrow}{u}_{k-1} \le \mathrm{col}(b_u); \quad \mathrm{diag}(A_{\Delta u})\underset{\rightarrow}{\Delta u}_{k-1} \le \mathrm{col}(b_{\Delta u}) \tag{28}$$

are considered, where $A_x, A_y, A_u, A_{\Delta u} = [I, -I]^T$ and $b_x, b_y, b_u, b_{\Delta u}$ are vectors that contain the maximum and minimum values allowed in the form $[max \ min]^T$.

4.5. Control Law

Cost function (27) and constraints (28) can be expressed in terms of the decision variable Δu [30], and the following optimization problem is obtained:

$$\underset{\rightarrow}{\Delta u}_{k-1}^* = \underset{\underset{\rightarrow}{\Delta u}_{k-1}}{\arg\min}\left\{\frac{1}{2}\underset{\rightarrow}{\Delta u}_{k-1}^T \mathcal{H}\underset{\rightarrow}{\Delta u}_{k-1} + \mathcal{F}^T\underset{\rightarrow}{\Delta u}_{k-1}\right\}$$

$$\text{s.t. } M_c\zeta_k + N_c\underset{\rightarrow}{\Delta u}_{k-1} \le f_c,$$

with

$$\mathcal{H} = P_{y\Delta u}^T\,\mathrm{diag}(\mathcal{Q})P_{y\Delta u} + \mathrm{diag}(\mathcal{R});$$

$$\mathcal{F} = P_{y\Delta u}\,\mathrm{diag}(\mathcal{Q})(P_{yx}\zeta_k + P_{y\delta}\delta - \underset{\rightarrow}{s}_k);$$

$$M_c = \begin{bmatrix} \mathrm{diag}(A_x)P_{x\zeta}P_{\zeta\zeta} \\ \mathrm{diag}(A_y)P_{y\zeta} \\ \mathrm{diag}(A_u)P_{u\zeta} \\ 0 \end{bmatrix};$$

$$N_c = \begin{bmatrix} \mathrm{diag}(A_x)P_{x\zeta}P_{\zeta\Delta u} \\ \mathrm{diag}(A_y)P_{y\Delta u} \\ \mathrm{diag}(A_u)P_{u\Delta u} \\ \mathrm{diag}(A_{\Delta u}) \end{bmatrix};$$

$$f_c = \begin{bmatrix} \mathrm{col}(b_x) \\ \mathrm{col}(b_y) \\ \mathrm{col}(b_u) \\ \mathrm{col}(b_{\Delta u}) \end{bmatrix} - \begin{bmatrix} \mathrm{diag}(A_x)P_{x\zeta}P_{\zeta\delta} \\ \mathrm{diag}(A_y)P_{y\delta} \\ 0 \\ 0 \end{bmatrix}\delta.$$

Finally, the control law is given by

$$u_k = u_{k-1} + \Delta u_k^*, \tag{29}$$

where Δu_k^* is the first element of $\underset{\rightarrow}{\Delta u}_{k-1}^*$.

4.6. Offset-Free Tracking

The MPC presented earlier does not contain an explicit mechanism to deal with disturbances and modeling errors that arise from set point changes that move the system away from the original operation point where it was linearized. Therefore, an integral action must be embedded into the control law to deal with these situations and guarantee zero steady-state error. This is performed including in the state space model an integrating state

$$\begin{bmatrix} \dot{\bar{x}}(t) \\ \dot{\xi}(t) \end{bmatrix} = \begin{bmatrix} A_c & 0 \\ -C_c & 0 \end{bmatrix} \begin{bmatrix} \bar{x}(t) \\ \xi(t) \end{bmatrix} + \begin{bmatrix} B_c \\ 0 \end{bmatrix} \bar{u}(t) + \begin{bmatrix} \delta_c \\ s(t) \end{bmatrix},$$

$$y(t) = \begin{bmatrix} C_c & 0 \end{bmatrix} \begin{bmatrix} \bar{x}(t) \\ \xi(t) \end{bmatrix}.$$

A stabilizing feedback gain can be computed to guarantee that in steady-state $y(t) = x_2(t) = s(t)$. Another approach to obtain integral action could be to estimate the disturbance with the use of a disturbance observer [31].

4.7. Set Point Manager

The set point manager (SPM) is a ruled-based algorithm that takes into account the periods of expected maximum and minimum demand through time conditions, that is, it identifies the hours (early morning) in which it is possible to reduce the pressure to reduce the leak magnitude without affecting the demand. Therefore, when the system is leak-free (i.e., $\lambda = 0$), the SPM computes the pressure reference (s_k) by solving the hydraulic model to ensure that the demand is satisfied. Then, if there is no fluid loss, the priority is to supply to the WDS the pressure needed to fulfil the users' demand in both maximum and minimum demand hours.

On the other hand, the SPM receives the leak magnitude estimate through the leak coefficient λ_k when a leak occurs. The SPM identifies if it is possible to reduce the pressure reference according to the day's time, prioritizing meeting the DP. If the DP is maximum, the pressure reference is set to the control pressure (H_{ctrl}) calculated by simulation. Otherwise, the pressure reference is reduced to meet the minimum DP to minimize fluid loss without neglecting the required demand. This procedure is formally described in Algorithm 1.

Algorithm 1: Set point manager for pressure management

Input: One-day demand profile (DP), time-varying leak coefficient (λ_k), leak detection threshold (λ_{th}).
Output: Time-varying set point (s_k). Initialization:
    for each index i in DP **do**
        Compute the control pressure (H_{ctrl}) to ensure
        scheduled demand in DP assuming leak-free
        conditions:
            $H_{\text{ctrl}}[i] \leftarrow \text{solveHydraulics}(\text{DP}[i])$
    end for
Online operation:
    for each time-step k **do**
      if $\lambda_k > \lambda_{\text{th}}$
        $S_k \leftarrow \min(H_{\text{ctrl}})$
      else
        $t_k \leftarrow \text{getCurrentTime}(\,)$
        $S_k \leftarrow \text{interp}(H_{\text{ctrl}}, t_k)$
      end if
    end for

The solveHydraulics() subroutine in Algorithm 1 solves the hydraulic model to compute the required pressure in the control node, ensuring that the demand is satisfied at each hour of the day. The interp() subroutine computes by interpolation the control pressure corresponding to each time step t_k.

5. Simulation Results

This section presents simulations of the WDS built in the Hydroinformatics Laboratory at the Technological Institute of Tuxtla Gutierrez, whose mathematical model was validated in [23]. The system parameters are given in Table 1. The system can be configured as a single horizontal pipeline and a branched network as in Figures 1, 3, and 4. An EKF-based method [24] has been considered to estimate the leak position z_ℓ, its magnitude $\hat{\lambda}$, and the leak pressure H_ℓ for Cases 1 and 2. A leak occurring in an accessory (pipeline joint) located in an unknown position is assumed for Case 3; nevertheless, it can be estimated by an EKF, e.g., [25]. However, it is important to note that the proposed method is not attached to any leak estimation method, and it could be generalized to any WDS within the topology presented here.

The initial conditions for the flows $Q_i(0)$ and pressures $H_i(0)$ for all cases are presented in Table 2. The sampling period is $T_s = 0.01$ s. Gaussian noise with a variance of 2.53×10^{-10} m^6/s^2 for the flow rate and 3.72×10^{-4} m^2 for the pressure were added to the signals. These noise levels were characterized according to the response of the Yokogawa sensors installed in the physical system as described in [32]. Cases 1 and 2 take into account an uncertainty for z_ℓ, λ, and H_ℓ of $\pm15\%$, that is, an error in the estimation of the leakage variables. For Case 3, the same value for λ as the previous cases is considered. For the MPC, the prediction horizon is $n_y = 15$, the control horizon is $n_u = 3$, and the weights are $\mathcal{Q} = 1$ and $\mathcal{R} = 0.5$. These values are based on well-known tuning methods [33]. The constraints were proposed according to the physical behavior and expected DP. For all cases, $0.1 \leq r \leq 1$. For Case 1, the constraints on the states are

$$1 \times 10^{-3} \, \text{m}^3/\text{s} \leq Q_{1,2,3} \leq 1.3 \times 10^{-3} \, \text{m}^3/\text{s},$$

$$2 \, \text{m} \leq H_3 \leq 2.87 \, \text{m}.$$

Table 1. System parameters.

Parameter	Value
Relative roughness (ϵ)	0.1×10^{-4}
Fluid (water) density (ρ)	995.736 kg/m^3
Kinematic viscosity (ν)	0.803×10^{-6} m^2/s
Leak coefficient (λ)	0.1×10^{-4} m$^{5/2}$/s
Gravity acceleration (g)	9.81 m/s^2
Valve coefficient (C_v)	1.156
Pipeline diameter (d)	0.048 m
Case 1 lengths ($z_{1,2,3}$)	$11.278, 27.662, 75.7$ m
Case 2 lengths ($z_{1,2,3}$)	$38.94, 14.79, 60.91$ m
Case 3 lengths ($z_{1,\ldots,5}$)	$34.456, 31.056, 34.456, 35.456, 35.456$ m

Table 2. Initial conditions for the three simulation cases.

Variable	Case 1	Case 2	Case 3	Units
$H_1(0)$	3.5480	7.04	–	m
$H_2(0)$	–	5.2908	1.99	m
$H_3(0)$	3.2	–	–	m
$H_4(0)$	1.5884	2.21	–	m
$H_{B1}(0)$	–	–	5.6542	m
$H_{B2,B3,B4}(0)$	–	–	1	m
$Q_1(0)$	0.0012	0.0019	0.0030	m^3/s
$Q_2(0)$	0.0012	0.0019	0.0014	m^3/s
$Q_3(0)$	0.0012	0.0019	0.0007	m^3/s
$Q_4(0)$	–	–	0.0015	m^3/s
$Q_5(0)$	–	–	0.0007	m^3/s
$H_\ell(0)$	3.2	4.6889	1.2255	m

For Case 2,

$$1.5 \times 10^{-3}\,m^3/s \leq Q_{1,2,3} \leq 2 \times 10^{-3}\,m^3/s,$$

$$4.3\,m \leq H_2 \leq 5.4\,m.$$

Finally, for Case 3,

$$0.17 \times 10^{-3}\,m^3/s \leq Q_1 \leq 3.2 \times 10^{-3}\,m^3/s,$$

$$0.7 \times 10^{-3}\,m^3/s \leq Q_2 \leq 1.6 \times 10^{-3}\,m^3/s,$$

$$0.4 \times 10^{-3}\,m^3/s \leq Q_3 \leq 0.7 \times 10^{-3}\,m^3/s,$$

$$0.95 \times 10^{-3}\,m^3/s \leq Q_4 \leq 1.5 \times 10^{-3}\,m^3/s,$$

$$0.4 \times 10^{-3}\,m^3/s \leq Q_5 \leq 0.7 \times 10^{-3}\,m^3/s,$$

$$1.2\,m \leq H_2 \leq 2\,m.$$

These constraints were chosen to satisfy a minimum demand, even in the case of a leak. The simulation covers a period of 24 h whose maximum and minimal magnitudes are concerning a typical demand profile. For Case 1, the results are displayed in Figure 7. The top plot of Figure 7 shows the pressure in the demand node. The dotted line is the DP, the dashed is the reference driven by the SPM that adapts S_k during the leak period to reduce the water losses, and the solid line represents the controlled pressure (H_3) in the node. The leak occurs at $t = 3$ h with the conditions given in Table 1. The SPM identifies a time of minimum demand and reduces the pressure. This pressure reduction remains until $t = 7$ h where the maximum demand period starts. During the period of maximum demand, the SPM does not adjust the reference because the priority is to satisfy the demand. After $t = 17$ h, the SPM again reduces S_k to reduce the leak magnitude. The middle plot of Figure 7 shows the effect of leak reduction due to the SPM. The solid line is the leak magnitude without the SPM and the dashed-line with the SPM. The bottom plot of Figure 7 shows the PRV opening to track the desired reference. It is important to note that all the constraints are satisfied for the MPC. We can conclude that the MPC tracks the set point with good performance even during the leak period, which demonstrates the effectiveness of the proposed method.

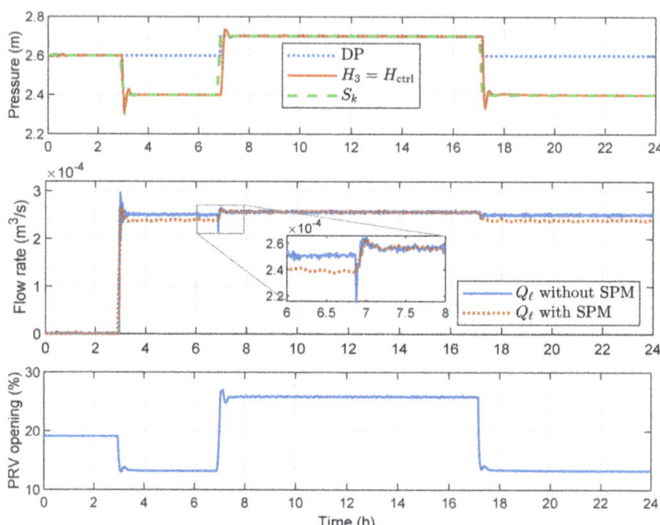

Figure 7. Case 1: Pressure control and management in a pipeline with a leak before the controlled node.

For Case 2, the results are displayed in Figure 8. As can be seen, the MPC tracks the pressure reference provided by the SPM even in a leakage scenario. Similar to Case 1, the SPM automatically adjusts the controlled node pressure at the maximum and minimum demand periods to reduce water losses, and all the constraints are satisfied for the MPC. To test the robustness of the controller, $\pm15\%$ of uncertainty in the location of the leak is added, demonstrating the performance of the controller even when there is a leak location error.

To contrast the results with the traditional method of control and management of pressure in a pipe, Figure 9 presents a comparison between a traditional PID controller and the proposed predictive control scheme. The PID gains were computed with the Matlab® PIDtuner®, obtaining the optimal values for proportional $K_P = 1$, integral $K_I = 0.5$, and derivative $K_D = 0.5$ gains. It can be seen that the PID tracks the demand profile with good performance. However, in the event of a leak, the set point remains the same, and the leak is seen by the controller as a disturbance. So the PID adjusts (increases) the pressure to track the original demand profile. This way of operating guarantees the pressure in the demand node but at the cost of fluid loss. In contrast, the MPC is aware of the leak event and with the aid of the set point manager adjusts the pressure operating point to a lower value reducing the fluid loss and delivering the minimum allowed pressure to the demand node.

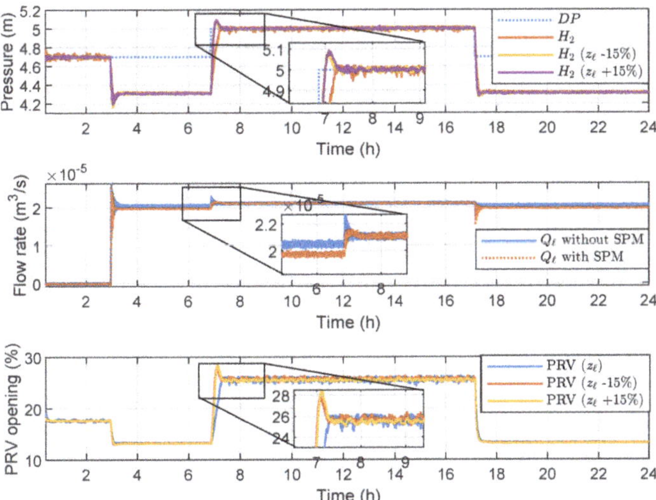

Figure 8. Case 2: Pressure control and management in a pipeline with a leak after the controlled node. Including ±15% of uncertainty in the leak location.

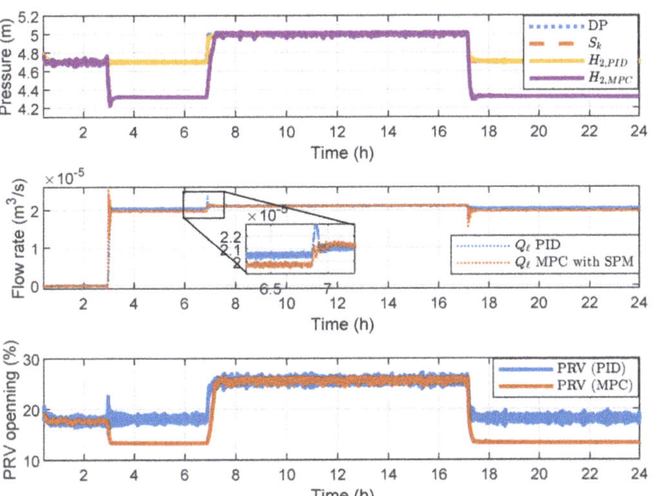

Figure 9. Case 2: Comparison between a traditional PID controller and the proposed control scheme.

Results for Case 3 are shown in Figure 10, with a leak being simulated at a connection. The MPC still achieves a good tracking of S_k that the SPM sets, which proposes a pressure reduction at the time of the leak. This pressure reduction remains constant in the first period until the maximum demand. The middle plot of Figure 10 shows the leak magnitude with and without the SPM. Note that Case 3 considers three demand nodes (due to its branches) controlled by adjusting the pressure on the controlled node. Therefore, it is essential to analyze these flows to evaluate the MPC effectiveness. The magnitudes with and without the SPM are represented by a dashed and a solid line, respectively. As can be seen, the priority is to satisfy each node demand during the maximum demand periods. However, during the minimum demand periods, the pressure is reduced, which reduces water losses.

Finally, Figure 11 displays the flow rates at the demand nodes for Case 3: Q_3, Q_4, and Q_5. The solid lines are the flows without the effect of the SPM, and the dotted lines represent the reduction due to the SPM. The reduction of flows is within the minimum permissible limits given by constraints on the states taken into account by the MPC and only during the hours of minimum demand.

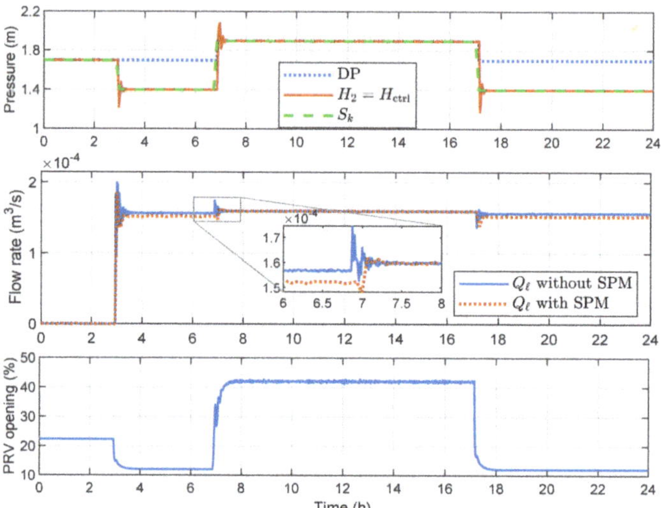

Figure 10. Case 3: Pressure control and management in a branched water distribution network.

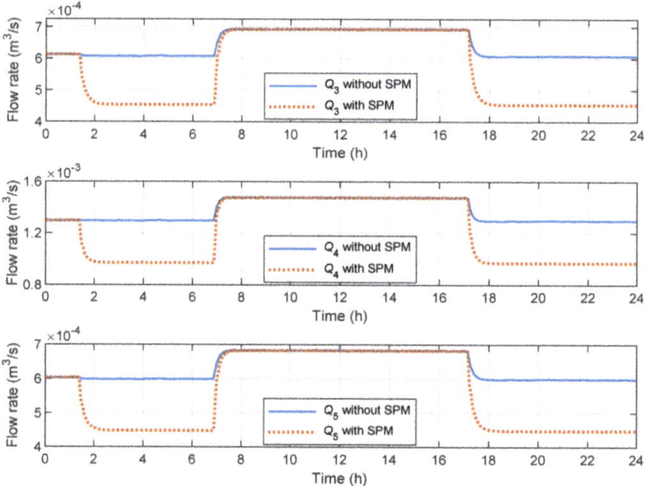

Figure 11. Case 3: Flows in the main water intakes for the WDS.

For all the cases, the trade-off between maintaining a minimum demand and reducing the leak magnitude was achieved due to the good performance of the MPC and the SPM algorithm. As a result, a reduction of water losses of ≈5% was accomplished, which is important due to the fact that water distribution systems must operate without interruptions throughout the year.

6. Conclusions

This work proposed a model-based predictive controller for managing and controlling water distribution systems. The proposed method seeks a trade-off between maintaining water supply and reducing water losses due to leaks. To achieve this goal, the MPC is based on the water-hammer equations of the hydraulic system together with physical constraints and an adaptive demand profile managed by a set point manager algorithm. The control input was calculated by minimizing the output errors with respect to the demand profile driven by the set point manager. The control objective was formulated with state, input, and output constraints on the cost function. Moreover, the control scheme includes a strategy to handle the strong nonlinear behavior of the PRV. The controller was tested with numerical simulations in a model characterized by a real pipeline and pipe network located at the Hydroinformatics Laboratory of the Technological Institute of Tuxtla Gutiérrez. Therefore, this work presents a mathematical model based on a real system and realistic operating conditions. The simulation results illustrate the performance and robustness of the MPC for pressure management in the system and the reduction of leaks due to water losses, with an average of 5% in the presence of noise, disturbances, and uncertainty in the leak location estimate. Future work will extend this work to an integrated methodology of a multi-leak tolerant control algorithm.

Author Contributions: Conceptualization, J.-R.B., F.-R.L.-E. and G.B.; methodology, J.-R.B., F.-R.L.-E. and G.B.; software, J.-R.B. and G.V.-P.; validation, J.-R.B., G.V.-P. and I.S.-R.; formal analysis, J.-R.B., F.-R.L.-E. and G.B.; data curation, G.V.-P. and I.S.-R.; writing—original draft preparation, J.-R.B., F.-R.L.-E. and G.B.; writing—review and editing, G.V.-P. and I.S.-R.; visualization, I.S.-R.; supervision, F.-R.L.-E. and G.B.; project administration, F.-R.L.-E. and G.B. All authors have read and agreed to the published version of the manuscript.

Funding: This research has been supported by the Consejo Nacional de Ciencia y Tecnología (CONA-CyT) and by Tecnológico Nacional de México under the program *Proyectos de Investigación Científica y Desarrollo Tecnológico e Innovación 2022*.

References

1. Zaman, D.; Tiwari, M.K.; Gupta, A.K.; Sen, D. A review of leakage detection strategies for pressurised pipeline in steady-state. *Eng. Fail. Anal.* **2020**, *109*, 104264. [CrossRef]
2. Kilinç, Y.; Özdemir, Ö.; Orhan, C.; Firat, M. Evaluation of technical performance of pipes in water distribution systems by analytic hierarchy process. *Sustain. Cities Soc.* **2018**, *42*, 13–21. [CrossRef]
3. OECD. *Water Governance in Cities*; OECD Publishing: Paris, France, 2016.
4. Samir, N.; Kansoh, R.; Elbarki, W.; Fleifle, A. Pressure control for minimizing leakage in water distribution systems. *Alex. Eng. J.* **2017**, *56*, 601–612. [CrossRef]
5. Berardi, L.; Simone, A.; Laucelli, D.B.; Ugarelli, R.M.; Giustolisi, O. Relevance of hydraulic modelling in planning and operating real-time pressure control: Case of Oppegård municipality. *J. Hydroinformatics* **2018**, *20*, 535–550. [CrossRef]
6. Do, N.; Simpson, A.; Deuerlein, J.; Piller, O. Demand estimation in water distribution systems: Solving underdetermined problems using genetic algorithms. *Procedia Eng.* **2017**, *186*, 193–201. [CrossRef]
7. Ares-Milián, M.J.; Quiñones-Grueiro, M.; Verde, C.; Llanes-Santiago, O. A Leak Zone Location Approach in Water Distribution Networks Combining Data-Driven and Model-Based Methods. *Water* **2021**, *13*, 2924. [CrossRef]
8. De Paola, F.; Giugni, M.; Portolano, D. Pressure management through optimal location and setting of valves in water distribution networks using a music-inspired approach. *Water Resour. Manag.* **2017**, *31*, 1517–1533. [CrossRef]
9. Mazumder, R.K.; Salman, A.M.; Li, Y.; Yu, X. Performance evaluation of water distribution systems and asset management. *J. Infrastruct. Syst.* **2018**, *24*, 03118001. [CrossRef]
10. Parra, S.; Krause, S.; Krönlein, F.; Günthert, F.; Klunke, T. Intelligent pressure management by pumps as turbines in water distribution systems: Results of experimentation. *Water Sci. Technol. Water Supply* **2018**, *18*, 778–789. [CrossRef]
11. García-Ávila, F.; Aviles-Anazco, A.; Ordonez-Jara, J.; Guanuchi-Quezada, C.; del Pino, L.F.; Ramos-Fernández, L. Pressure management for leakage reduction using pressure reducing valves. Case study in an Andean city. *Alex. Eng. J.* **2019**, *58*, 1313–1326. [CrossRef]
12. Dini, M.; Asadi, A. Optimal operational scheduling of available partially closed valves for pressure management in water distribution networks. *Water Resour. Manag.* **2020**, *34*, 2571–2583. [CrossRef]
13. Hernández, J.; Galaviz, D.; Torres, L.; Palacio-Pérez, A.; Rodríguez-Valdés, A.; Guzmán, J. Evolution of high-viscosity gas–liquid flows as viewed through a detrended fluctuation characterization. *Processes* **2019**, *7*, 822. [CrossRef]
14. Navarro, A.; Delgado-Aguiñaga, J.; Sánchez-Torres, J.; Begovich, O.; Besançon, G. Evolutionary observer ensemble for leak diagnosis in water pipelines. *Processes* **2019**, *7*, 913. [CrossRef]

15. Jara-Arriagada, C.; Stoianov, I. Pipe breaks and estimating the impact of pressure control in water supply networks. *Reliab. Eng. Syst. Saf.* **2021**, *210*, 107525. [CrossRef]
16. Mosetlhe, T.C.; Hamam, Y.; Du, S.; Monacelli, E. A survey of pressure control approaches in water supply systems. *Water* **2020**, *12*, 1732. [CrossRef]
17. Mathye, R.P.; Scholz, M.; Nyende-Byakika, S. Optimal Pressure Management in Water Distribution Systems: Efficiency Indexes for Volumetric Cost Performance, Consumption and Linear Leakage Measurements. *Water* **2022**, *14*, 805. [CrossRef]
18. Chaudhry, M.H. *Applied Hydraulic Transients*, 3rd ed.; Springer: New York, NY, USA, 2014.
19. Besançon, G.; Georges, G.; Begovich, O.; Verde, C.; Aldana, C. Direct observer design for leak detection and estimation in pipelines. In Proceedings of the European Control Conference, Kos, Greece, 2–5 July 2007.
20. Dulhoste, J.; Besançon, G.; Torres, L.; Begovich, O.; Navarro, A. About Friction Modeling For Observer Based Leak Estimation In Pipelines. In Proceedings of the 50th IEEE Conference Decision & Control and European Control Conference, Orlando, FL, USA, 12–15 December 2011.
21. Prescott, S.; Ulanicki, B. Improved control of pressure reducing valves in water distribution networks. *J. Hydraul. Eng.* **2008**, *134*, 56–65. [CrossRef]
22. De Persis, C.; Kallesoe, C.S. Pressure regulation in nonlinear hydraulic networks by positive and quantized controls. *IEEE Trans. Control. Syst. Technol.* **2011**, *19*, 1371–1383. [CrossRef]
23. Bermúdez, J.R.; López-Estrada, F.R.; Besançon, G.; Valencia-Palomo, G.; Torres, L.; Hernández, H.R. Modeling and simulation of a hydraulic network for leak diagnosis. *Math. Comput. Appl.* **2018**, *23*, 70. [CrossRef]
24. Santos-Ruiz, I.d.l.; Bermúdez, J.R.; López-Estrada, F.R.; Puig, V.; Torres, L.; Delgado-Aguiñaga, J. Online leak diagnosis in pipelines using an EKF-based and steady-state mixed approach. *Control Eng. Pract.* **2018**, *81*, 55–64. [CrossRef]
25. Delgado-Aguiñaga, J.; Besançon, G. EKF-based leak diagnosis schemes for pipeline networks. *IFAC-PapersOnLine* **2018**, *51*, 723–729. [CrossRef]
26. De Paola, F.; Fontana, N.; Giugni, M.; Marini, G.; Pugliese, F. An application of the Harmony-Search Multi-Objective (HSMO) optimization algorithm for the solution of pump scheduling problem. *Procedia Eng.* **2016**, *162*, 494–502. [CrossRef]
27. Delgado-Aguiñaga, J.; Santos-Ruiz, I.; Besançon, G.; López-Estrada, F.; Puig, V. EKF-based observers for multi-leak diagnosis in branched pipeline systems. *Mech. Syst. Signal Process.* **2022**, *178*, 109198. [CrossRef]
28. Valencia-Palomo, G.; Hilton, K.; Rossiter, J.A. Predictive control implementation in a PLC using the IEC 1131.3 programming standard. In Proceedings of the 2009 European Control Conference (ECC), Budapest, Hungary, 23–26 August 2009; pp. 1317–1322.
29. Valencia-Palomo, G.; Rossiter, J.A. Auto-tuned predictive control based on minimal plant information. *IFAC Proc. Vol.* **2009**, *42*, 554–559. [CrossRef]
30. Khan, B.; Rossiter, J.A.; Valencia-Palomo, G. Exploiting Kautz functions to improve feasibility in MPC. *IFAC Proc. Vol.* **2011**, *44*, 6777–6782. [CrossRef]
31. Pannocchia, G. Offset-free tracking MPC: A tutorial review and comparison of different formulations. In Proceedings of the 2015 European control conference (ECC), Linz, Austria, 15–17 July 2015; pp. 527–532.
32. Pérez-Pérez, E.; López-Estrada, F.R.; Valencia-Palomo, G.; Torres, L.; Puig, V.; Mina-Antonio, J.D. Leak diagnosis in pipelines using a combined artificial neural network approach. *Control Eng. Pract.* **2021**, *107*, 104677. [CrossRef]
33. Garriga, J.L.; Soroush, M. Model predictive control tuning methods: A review. *Ind. Eng. Chem. Res.* **2010**, *49*, 3505–3515. [CrossRef]

Article

Diagnostics and Control of Pumping Stations in Water Supply Systems: Hybrid Model for Fault Operating Modes

Marko Milošević [1,*], Milan Radić [1], Milica Rašić-Amon [1], Dragan Litričin [2] and Zoran Stajić [1]

[1] Faculty of Electronic Engineering, University of Niš, 18000 Niš, Serbia; milan.radic@elfak.ni.ac.rs (M.R.); milica.rasicamon@elfak.ni.ac.rs (M.R.-A.); zoran.stajic@elfak.ni.ac.rs (Z.S.)

[2] Siemens, Siemensova 1, 155 00 Prague, Czech Republic; dragan.litricin@siemens.com

* Correspondence: marko.milosevic@elfak.ni.ac.rs

Abstract: This paper indicates the importance and advantages of the application of hybrid models in the control of water supply systems. A range of possibilities provided by this scientific approach is presented in the practical examples related to the fault diagnostics and fault-tolerant control in the pumping station (PS) control systems. It is presented that continuous monitoring and recording of the data of the pumping stations operation processes (electrical parameters such as electrical power, pressure or flow in the pipelines, water levels in the tanks, changes in various discrete states, etc.) could represent a significant resource that can be used to develop various hybrid models using the appropriate "data-driven" techniques. During this process, data are transformed into information, and thereafter, information into knowledge. Based on this knowledge, the control of PS operation can be significantly improved and a significant increase in the user's satisfaction can be achieved while the maintenance and operation costs can be reduced.

Keywords: water supply systems control; pumping stations operation; hybrid models; "data-driven" models; fault diagnostics; fault-tolerant control

Citation: Milošević, M.; Radić, M.; Rašić-Amon, M.; Litričin, D.; Stajić, Z. Diagnostics and Control of Pumping Stations in Water Supply Systems: Hybrid Model for Fault Operating Modes. *Processes* **2022**, *10*, 1475. https://doi.org/10.3390/pr10081475

Academic Editors: Francisco Ronay López-Estrada and Guillermo Valencia-Palomo

Received: 18 April 2022
Accepted: 24 May 2022
Published: 27 July 2022

Publisher's Note: MDPI stays neutral with regard to jurisdictional claims in published maps and institutional affiliations.

1. Introduction

A hybrid model is a combination of two or more different modeling techniques, which aims to improve the modeling adequacy of complex systems [1–9]. In the processes of system design or optimization, the principles of mathematical modeling are usually used. These modeling principles include the representation of physical models using appropriate systems of equations (differential, algebraic) that describe the relations between the system's variables and the system's parameters. These models are known as parametric models [10–17]. Due to known relationships between the model parameters and variables, they are usually presented as "white boxes" ("white box models"). Despite the complexity of these models, they are not able to present the ideal behavior of the real system in all operation conditions during its exploitation. This is especially difficult for systems that are characterized by certain unexpected occurrences, irregular conditions, or failures that are difficult to predict in the systems' (re)designing phase. In situations where certain data recorded during such systems' operation are available, it is possible to explore and discover various phenomena related to the systems' processes that have not been previously noticed or sufficiently researched. It is also possible to recognize some additional relationships between the process parameters and process variables. There are also examples where it is not possible to mathematically describe these relationships. These hidden relations in data are mostly extracted through some machine learning systems [9,18–22], which are used in gaining a better understanding of the process or to help in the decision making [9,15,23,24]. These models are known as non-parametric models, presented as "black boxes" ("black box models"), and they are classified in the "data-driven" models [2,8,9,25–28]. The combination and integration of classical parametric models with one or more non-parametric models, connected into the serial, parallel, or serial–parallel structures is known as the hybrid model in the literature. An example of the serial–parallel hybrid model structure is presented in Figure 1.

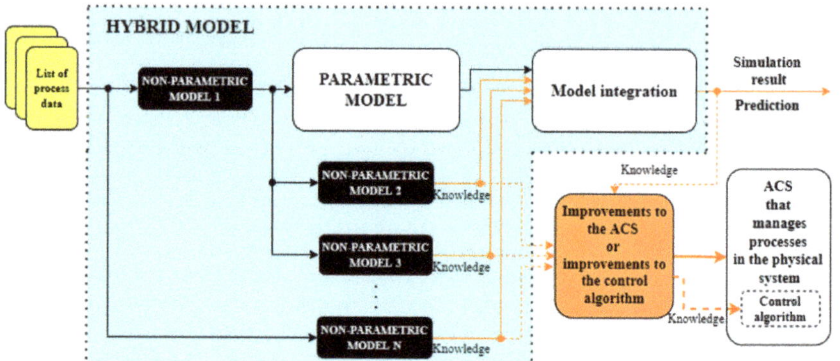

Figure 1. Serial-parallel hybrid model structure.

This modeling method allows that all of the available knowledge can be integrated into a single model without compromising the adequacy of the basic parametric model [2,8,9]. On the other hand, the quality of the model increases and processes from the reality are better modeled [6,9,13]. In addition, the information and knowledge that can be obtained from the non-parametric models or simulation results from hybrid models can be used to improve the quality of the system control. Thereafter, these models represent significantly better solutions for the modeling of the complex processes and systems in comparison to the models that use only one source of knowledge (only one modeling method) [1–9,29].

Some authors in this area recognize that a "data-driven" approach can also be applied in situations where there is a lack of information about the functioning of the real system, where hybrid models can be expanded by adding data from different sources, and that hybrid models are especially suitable for a range of applied targeted research [2,9,30]. Other authors have concluded that hybrid modeling techniques are not unknown and that they are mainly applied in scientific (academic) research, but they are insufficiently applied in practice [9].

Therefore, this paper presents the applied research in the area of pumping station control, and the main goal was to demonstrate the advantage of using hybrid models in the field of diagnostics and the tolerant control of the irregular operation modes as well as the fault-tolerant control in the water supply systems. The authors chose the concept with a number of simpler non-parametric models, because it allows for "step-by-step" expansion and an increase in the complexity of the hybrid model and facilitates its application in practice.

2. Hybrid Models of the Pump Stations Automatic Control Systems

In the process of creating a hybrid model of a certain PS, it is necessary to start from its physical model and form an appropriate parametric model based on it.

As an example of this procedure, in Figure 2, the physical model of PS with two pumping units (PU) is presented. One of the units is in operation and the other is spare, which is a common situation for a smaller PS in water supply systems. For easier understanding, in Figure 2, the water flows are presented in blue, the power flows that feed the drive motors are shown in red, the control commands are in orange, and the reference value according to which the control is performed (control signal) is in green.

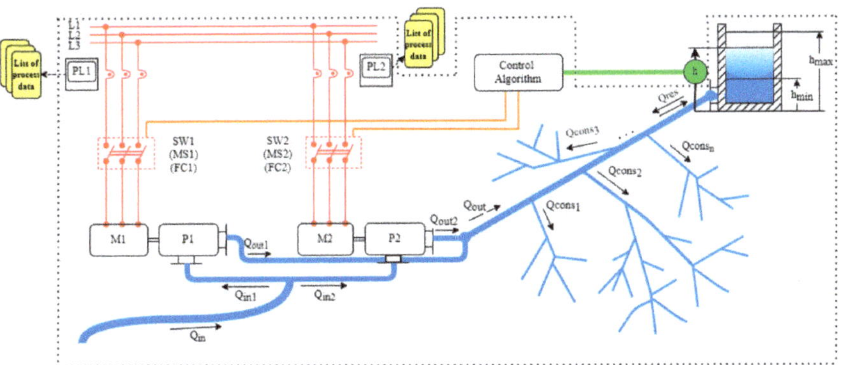

Figure 2. The physical model of a PS with two PUs pumping water into the discharge–distribution piping system with one counter-tank.

It is assumed that it was a booster PS (i.e., that PUs pump water from the common suction line that branches to the suctions of both pumps). Pump discharge lines are also connected within the PS into the single discharge–distribution pipeline that represents the situation regularly presented in practice. This pipeline supplies the connected customers and excess water is usually transferred to a counter-tank located at the appropriate location above the PS to ensure the stability of pressure in the distribution system and reduce its variations during PU switching on and off and/or major changes in the water consumption of the system.

Depending on the water consumption and water flows in the system, it is possible at certain periods that the total water flow to consumers ($\sum_{i=1}^{n} Q_{\text{cons}_i}$) is higher than the flow provided by the PU in operation (Q_{out}), which leads to the decrease in water levels in the tank. This is presented in Equation (1):

$$Q_{res} = Q_{out} - \sum_{i=1}^{n} Q_{\text{cons}_i} < 0. \tag{1}$$

However, such systems are usually designed so that in regimes when one PU operates, the total water flow to consumers is less than the flow provided by PU in operation for much longer periods, so that excess water fills the tank and increases its water levels (Equation (2)):

$$Q_{res} = Q_{out} - \sum_{i=1}^{n} Q_{\text{cons}_i} > 0. \tag{2}$$

All examples presented in this paper refer to the PS Knez Selo, whose operation has been monitored since 8 June 2009. During that period, there were several PU replacements and reconstructions of parts of the electrical installations belonging to the PS:

- At the beginning of the observed period, the PU drive motors started connecting to the three-phase electrical network via the star-delta starters. Despite the fact that the star-delta starter for a single PU consists of three switches, this situation is symbolically represented in Figure 2 with a single switch (SW1 and SW2, respectively);
- As the failure occurred in March 2010, PU2 was replaced with the repaired PU of the same type;
- During the period from May 2011 to October 2017, the start of the Pus' drive motors was performed using a motor starter (MS), where one MS was used to start both PUs (Figure 2 shows a more general case where each PU was started by a separate MS);
- After October 2017, a complete reconstruction of the PS was performed, where pump units of 15 kW were replaced by new pump units of 9 kW, and their start was performed using frequency converters FC1 and FC2, respectively.

For the analyses presented in this paper, it is important to point out that in all of the presented examples, there was no regulation of water flow in the system. All flow changes were exclusively a consequence of the process of self-regulation due to natural changes in the system pressure in the suction and discharge pipelines of the PU (changes in the PU head). Changes in the discharge pipeline pressures come as a consequence of changes in the water consumption, or changes in the counter-tank water level. On the other hand, pressure in the suction pipeline changes due to the switching of the PUs in the PS supplying water to consumers in lower altitude zones (PS Matejevac 2), and variations in the water consumption in that part of the system. Even in the presented examples in which the PUs were started via frequency converters, their output frequencies were constant and there was no regulation of the speed of the pump impellers. Any small changes in the speed of rotation of the pump impellers were due to changes in the load on the motor shaft, which was also a consequence of the natural change in the pump head. In all illustrative cases, the valves located on the suction and discharge branches of each PU were fully open and no flow control was performed with them. For this reason, their positions are not shown in the physical model presented in Figure 2.

2.1. Parametric Models of PS

There are a large number of scientific papers that have described the different parametric models of PS, which can be used to simulate their operation [10–17,26,31]. Bearing in mind the topic of this paper, the focus of research was not placed on the well-known parametric models of the PS. For the investigation covered in this paper, the parametric model of PS presented by the authors in [12] can be used as the basis.

2.1.1. A Simple Parametric Model of PU Convenient for the Analysis of PS Operation

This model indicates that relations between electrical (phase and line voltages and currents, active, reactive, and apparent powers, power factors, frequency) and non-electrical (flow, pressure, and pumP's efficiency) parameters can be established. This can be achieved by providing appropriate $H(Q)$ and $P'(Q)$ characteristics [12,15–17,32]. These characteristics can be represented by the corresponding polynomials of the fourth- and third-order, respectively:

$$H(Q) = H_0 + H_1 * Q + H_2 * Q^2 + H_3 * Q^3 + H_4 * Q^4; \tag{3}$$

$$P'(Q) = P_0 + P_1 * Q + P_2 * Q^2 + P_3 * Q^3. \tag{4}$$

In certain situations, it is assumed that the characteristic $P'(Q)$ is linear, meaning that $P_2 = 0$ and $P_3 = 0$, which will be adopted in the following examples.

The PU efficiency can be calculated based on the following formula:

$$\eta(Q) \, [\%] = \frac{P(Q)}{P'(Q)} * 100 = \frac{0.981 * Q * H(Q)}{P'(Q)}. \tag{5}$$

In Equations (3)–(5), P' is given in [kW], H in [m], and Q in [l/s].

2.1.2. The Parametric Model Application—Obtaining the Simulation Results

Knowing the parametric models of PU, historical data of the measured electrical active power P' can be used to indirectly determine the flow and pressure of the pump, and the PU efficiency in the same historic period. Furthermore, the results of modeling obtained in this way can be used for the estimation of the total volume of water pumped into the system in a certain period, or for energy efficiency analyses. Additionally, the basic idea of using the historical data could be to find out what really happened during the PS operation in the past. Fixed data sequences used as data input in the parametric model enable the detection of different types of irregularities, faults, or failures that have appeared during the PS operation as well as to explore the dependencies between the different system variables and corresponding system parameters. This is an effective way in which to transform the data into valuable information that can help in diagnosing different technical problems and in finding appropriate solutions for them.

Values of all of the electrical quantities measured by power loggers PL1 and PL2 (Figure 2) were recorded with a sampling period of 12 s. Some of them are shown in Figures 3–6 as the daily diagrams of changes in the recorded electrical quantities on 30 December 2009, but the principle of using the parametric model is identical for any other period.

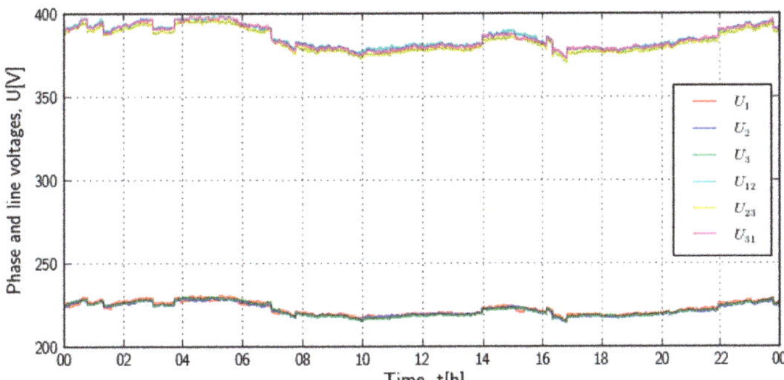

Figure 3. The diagrams of the phase and line voltage changes U(t) at PS Knez Selo on 30 December 2009.

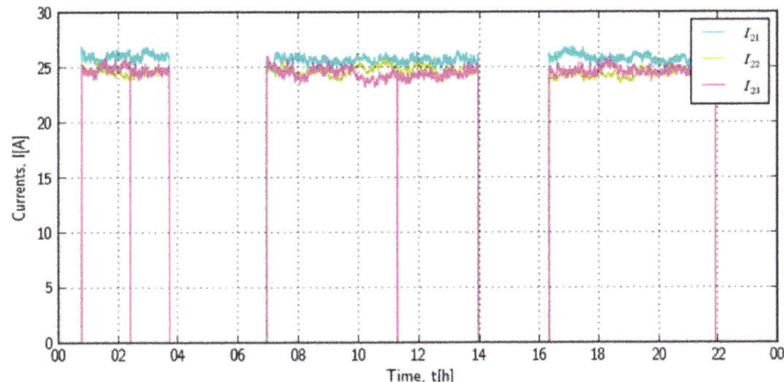

Figure 4. The diagrams of the PU2 load current changes I(t) on 30 December 2009.

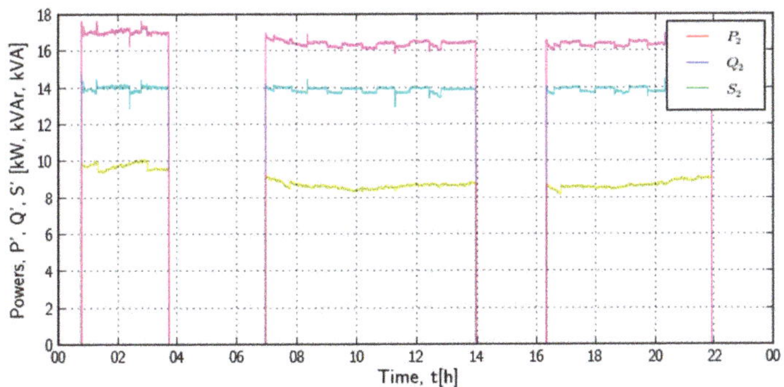

Figure 5. The diagrams of the PU2 active, reactive, and apparent power changes on 30 December 2009.

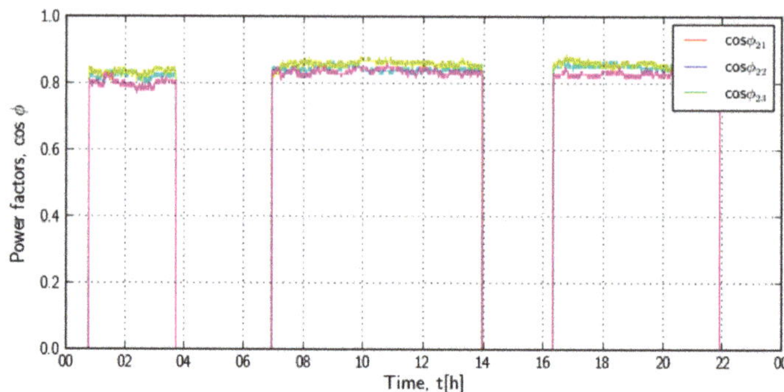

Figure 6. The diagrams of the PU2 power factor changes on 30 December 2009.

It should be pointed out that the only electrical quantity needed for the presented parametric model application is active power P'. In order to obtain the corresponding simulation results, only this diagram is actually needed (turquoise line in Figure 5). All of the other diagrams presented in Figures 3–6 can be used to explore the dependencies between the different system variables and the corresponding system parameters.

The following values of the coefficients in Equations (3) and (4) were used when calculating the non-electric quantities: $H_0 = 144$, $H_1 = -4.7273$, $H_2 = 0.3413$, $H_3 = -0.07267$, $H_4 = 0.00102$, $P_0 = 9.95$, $P_1 = 0.620448$. The same coefficients were used for both PUs, because the experimental tests performed in the period presented in the figures showed that they had approximately the same operating characteristics.

The calculation procedure was as follows: Based on the information on the electrical active power P' that the pump drive motor draws from the supplying electrical system at a certain time moment, the corresponding flow value Q can be determined using Equation (4). For the presented linear function, it is not complicated at all. By substituting this value of the flow in Equation (3), the value of the head H is determined, and by substituting the same value in Equation (5), the corresponding value of the PU efficiency η at that time moment is obtained. This calculation procedure is repeated, so that for each subsequent recorded value of the active power P', the values of the corresponding quantities Q, H, and η are obtained. Presenting these obtained results in the time moments in which there are available data on the active power P' (every 12 s), the diagrams of the Q, H, and η changes in time are obtained (Figures 7–9).

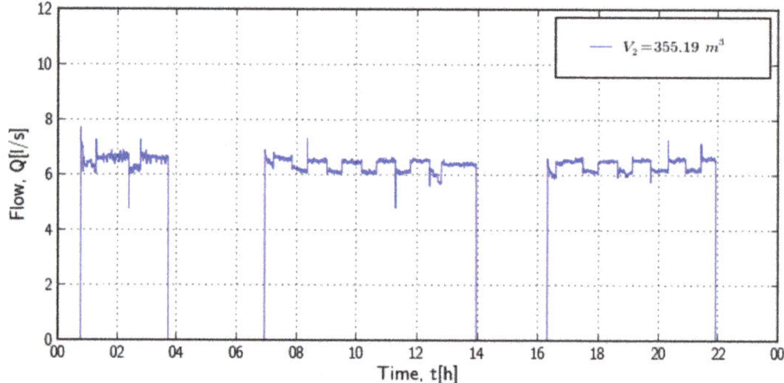

Figure 7. The PU2 flow change diagram Q(t) on 30 December 2009.

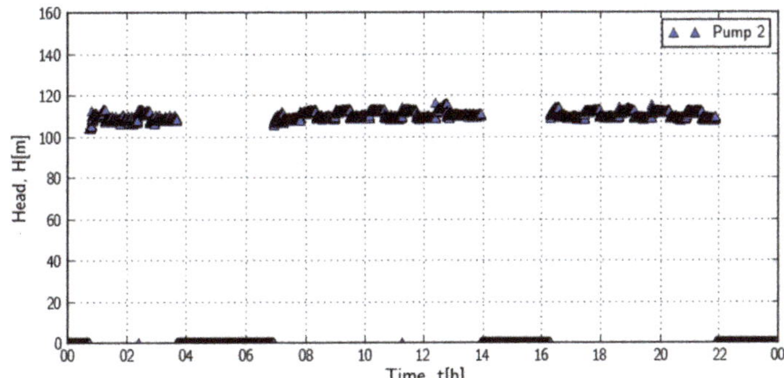

Figure 8. The PU2 head change diagram H(t) on 30 December 2009.

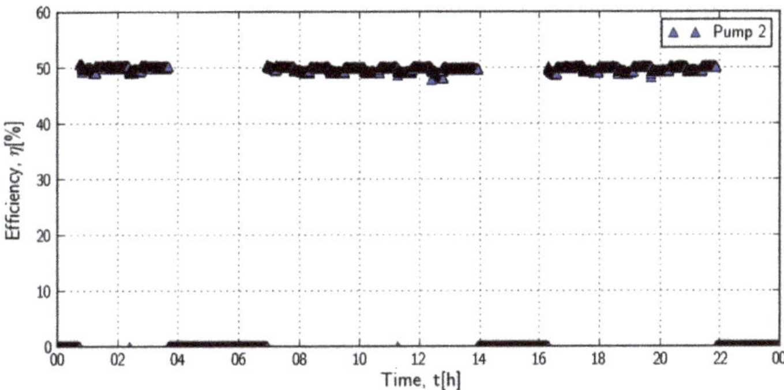

Figure 9. The PU2 efficiency change diagram η(t) on 30 December 2009.

Bearing in mind that the values of all of these quantities are known at each of the observed moments, their interdependencies can also be shown. Figures 10–12 show the functional dependencies H(Q), P′(Q), and η(Q). Unlike the diagrams presented in Figures 3–9, which show the daily changes in the different process variables on 30 December 2009, the images in Figures 10–12 refer to the entire month (December 2009).

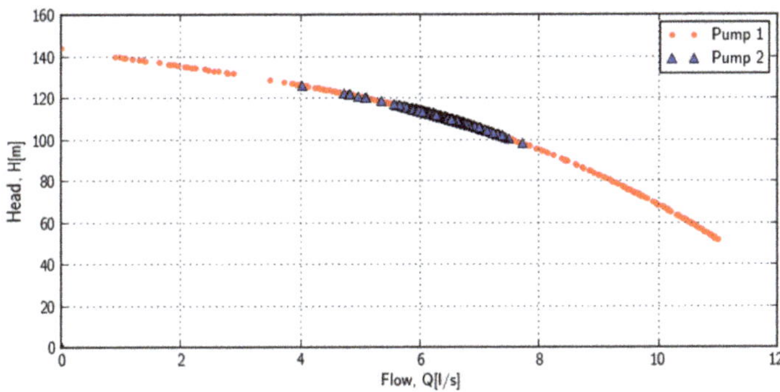

Figure 10. The operating point positions on the PU1 and PU2 H(Q) characteristics in December 2009.

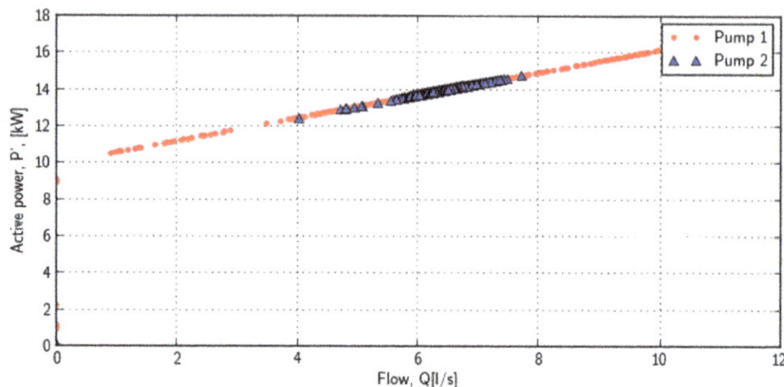

Figure 11. The operating point positions on the PU1 and PU2 $P'(Q)$ characteristics in December 2009.

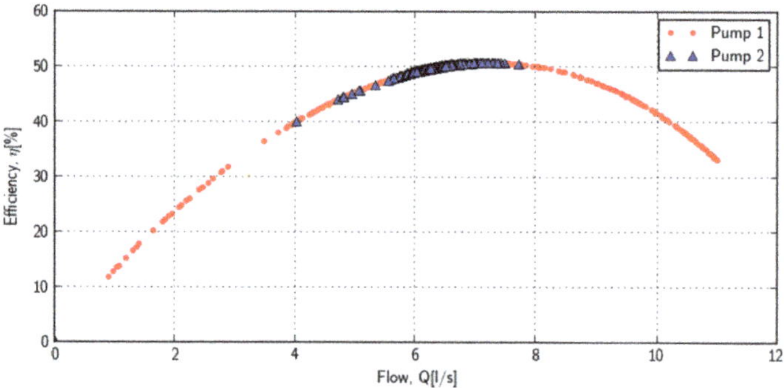

Figure 12. The operating point positions on the PU1 and PU2 $\eta(Q)$ characteristics in December 2009.

2.1.3. Practical Considerations and Simulation Analysis

When observing the time diagrams or functional dependencies in the previously presented figures, the following considerations must be taken into account:

- There was no regulation of the flow in the system shown and the valves on the suction and discharge side of the pump units were fully open in all of the presented modes. All changes that can be seen in the presented electrical quantities diagrams (except for certain changes in the supply voltage of drive motors) were due solely to the natural behavior of the system and the changes in the operating conditions occurring in it (changes in voltage, possible frequency changes, pressure differences in the discharge and suction pipelines of PU), on which the pump head and the position of the operating points on the operating characteristics of the PU directly depend.

- On some time diagrams, certain abrupt changes can be noticed between parts of the diagram with significantly less pronounced changes. Less pronounced changes that can be seen, for example, in the active power diagram (turquoise line in Figure 5) are due to small changes in the pressures in the suction and discharge pipeline of the PU, which happen due to changes in consumption in the front and behind of the PS. On the other hand, the abrupt changes in active power that are clearly seen in this diagram are due to the switching on/off PU in PS, which is located in the part of consumption in front of this PS and supplies water consumers in lower altitude zones as well as to the booster PS Knez Selo (in this case, it was PS Matejevac 2). In this sense, each PU switching on in PS Matejevac 2 causes a sharp increase in the pressure in the suction branches of the PU in PS Knez Selo, which results in a sharp decrease in the pump head (Figure 8), and a sharp increase in the water flow (Figure 7). Similarly, each shutdown of the PU at PS Matejevac 2 caused a decrease in pressure in the suction branches of the PU at PS Knez Selo, which resulted in an increased pump head (Figure 8) and decreased water flow (Figure 7). Changes in the active power diagram shown in Figure 5 are also direct consequences of these phenomena. Precisely, such recognized relations make it possible to identify certain phenomena occurring in the larger part of the hydromechanical system, based on the recorded electrical quantities changes.

- The abrupt changes observed in the phase and line voltage diagrams (Figure 3) are a consequence of the operation of automatic voltage regulators in TS 110/X kV through which this PS is supplied with electricity. These changes coincide in time with changes in the diagrams of reactive power that PU motors draw from the network (olive curve in Figure 5). These changes directly affect the reactive power consumption, the load currents of the motors, and the value of the apparent power, but their influence on the phenomena in the electromechanical part of the system can be practically neglected.

- The presented parametric model, based on the operating characteristics of the PU, is valid only for the steady-state, when all of the electrical and non-electric quantities in the system obtain constant values. For the analysis of most of the transient processes in the system, it is necessary to use dynamic models that are significantly more complex. From the point of view of the presented model's practical applicability, it would be especially problematic to use values of electrical quantities measured during electromagnetic and electromechanical transient processes such as the starting of PU drive motors. During these processes, motors draw inrush currents (powers) several times higher than rated. Electromagnetic transient processes are completed during 3–5 periods of supply voltage and do not last longer than 100 ms, while electromechanical transients are somewhat longer, lasting up to several seconds in smaller PS such as PS Knez Selo. If a measuring device samples and records the effective electrical values during the transient period (using the last 20 ms between the two consecutive current zero passing before the exact moment of recording), erroneous information on the water flow in the system will appear. Usually, a value for flow obtained from the parametric model will be extremely large, which is physically impossible. On the other hand, hydraulic transients that occur in the hydraulic part of the system have much higher time constants, and they can be observed through recorded electrical values, if the PUs have previously reached the steady state.

- Better understanding of this can be obtained from Figures 10–12, which show the variation in the operating points for both the PUs during December 2009. It can be seen that PU2 had more stable operating conditions than PU1 because its operating points were placed in much narrower ranges (points colored in blue). One reason for this is that PU2 was in operation significantly shorter than PU1. According to the data from the report generated by the decision-support system created by the authors, PU1 was in operation for a total of 14 days and 20:29:10 during the month of December, and the total number of registered on/off cycles was 117. On the other hand, the total operating time of PU2 was 2 days and 7:58:56, and the total number of 20 on/off cycles was registered. Another reason for the differences was that the tank had not been emptied during the operation of PU2. It should be noted that the differences that occurred in the area of higher flows were usually either as a consequence of the last mentioned phenomenon, or due to the registration of higher values of the starting currents (power) during the starting of the PUs' drive motors. On the other hand, the differences that could be noticed in the area of lower flow were the direct consequences of the lower values of the currents (power) recorded during the hydraulic transients (e.g., the pressure head was higher and the flow was lower due to the reflected pressure wave at the moment of recording).

- All values of the electrical quantities resulting from the electromagnetic transients can be identified and eliminated from appropriate analyses by applying certain methods of the preprocessing of input data, which could be conducted in the block, which was labeled in Figure 1 as non-parametric model 1. Bearing in mind that this paper deals with another topic, this model will not be further developed, and the original recorded data will be used in further analyses.

- It should be kept in mind that the operating characteristics of the PU degrade during its operation, so it is recommended that periodically, they should be experimentally tested and checked (e.g., once a year). When a new shape of the operating characteristic curve is determined after a certain period of time, appropriate recalculations can be performed using interpolation methods, which can take this effect into account by increasing the number of operating hours spent by the PU. Additionally, during any major reconstruction of the PS, it is recommended to re-record the operating characteristics of the PU and determine the new coefficients of the parametric model. The same procedure is mandatory after each overhaul or substitution of the PU by a new one.

2.1.4. Control Algorithms and Automatic Control System Functioning

In order to achieve a better understanding of the phenomena being observed in some PSs and to be able to form a parametric model of PS operation, it is necessary to understand the logic (i.e., the control algorithm applied during the implementation of the automatic control system (ACS) in the PS, the reference value used for the control, and the applied equipment).

In pump stations of the same type as the pump station presented in Figure 2, the "on–off" model of control is usually provided. The control is based on the information on the actual water level in the tank (h). The main goal of the regulation is the requirement that the water level (h) always remains between the previously defined minimum (h_{min}) and maximum (h_{max}).

The Knez Selo water tank has a total volume of 100 m^3, and the changes in the water level was monitored using five conductive probes placed in the tank at different heights. The system monitors and records changes that occur when any of the probes becomes submerged (which is signaled with the blue dot) or when it remains dry (which is signaled with the red x sign). Figure 13 shows the reactions of the upper- and the lower-limit probe in the Knez Selo water tank on 30 December 2009. Based on the time schedule of these signals, the change in the water level in the tank can be modeled. Comparing Figure 13 with the diagrams of changes in the quantities shown in Figures 3–9, it is easy to note that the limit probes controlled the operation of PU2 in the PS. PU1 was not put into operation on 30 December 2009.

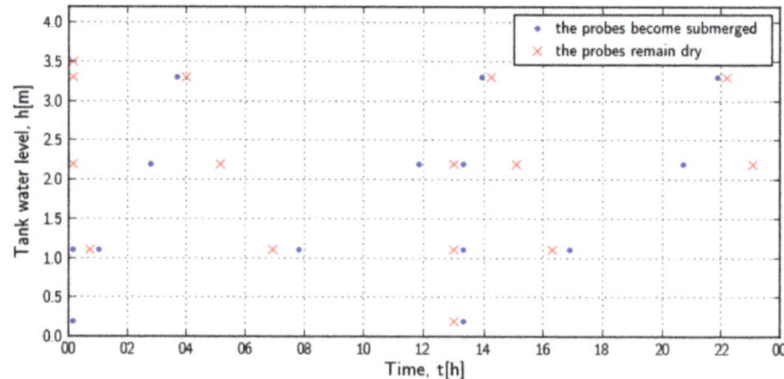

Figure 13. The diagram of the water level change h(t) in the Knez Selo water tank on 30 December 2009.

Observing Figure 13, one can see that only three of the existing five probes detected a change in the state during 30 December 2009. Two of them were used as limit probes (the upper one was set at $h_{max} = 3.3$ m and the lower one at $h_{min} = 1.1$ m height from the bottom of the tank). The third one was placed in the middle between them (i.e., at a height of 2.2 m). There were two more probes in the tank. One of them signaled the appearance of an overflow in the tank and was placed at a 3.5 m height, between the upper limit probe and the overflow pipe, which was physically located in the tank at a 3.6 m height (maximal water level in the tank). The other one was placed at a height of 0.2 m from the bottom of the tank, and when it remains dry, it means that the water tank is almost empty. During 30 December 2009, there was no change in the condition of these two probes.

With this arrangement of probes, it was chosen to turn off a PU when the volume of water in the tank reached a value of 91.6 m^3, and to turn it on when the volume of water in the tank dropped to about 30 m^3. Setting the lower limit probe to a lower level is not recommended because a certain volume of water in the tank must serve as a reserve to supply consumers from the moment of any problem in the PS until the arrival of the operator tasked with troubleshooting. A water volume of 30 m^3 allows for a water supply autonomy of at least 1 h. In most situations, it should be sufficient for the operator to be notified and to attend to the PS in order to fix the problem.

Considering the position of the upper limit probe, it might make sense to raise it by 0.1 m, which could increase the total available volume of water in the tank to about 94.4 m^3, but this could result in more frequent overflows in the tank.

In some cases, such as in the examples shown in the previous figures, the choice of PU to be in operation was selected manually. By switching the appropriate switch, the operator determines which PU will be in operation during the next period, and until the next switching, the local automation turns on/off the selected PU only.

2.1.5. The PU Parametric Model Application in More Complex Cases

After the last PS Knez Selo reengineering, performed in October 2017, two power loggers that were previously used to measure and record the electrical parameters of each PU were dismounted. Instead, a single power logger was installed in order to measure and record the parameters related to the electrical energy consumption of the complete PS. On that occasion, the pump units and the complete ACS were replaced. Two completely new PUs run by the drive motors with a rated power of 9 kW were installed. During this period, the PUs were started via appropriate frequency converters, but there was no frequency control (their output frequency was set to 50 Hz).

Relating to the parametric model application, it is a significantly more complex case. Namely, the main advantage of the previously applied concept represents the possibility of direct measurements on each PU (direct model application is possible), but there was not any information on the consumption of other electronic devices or appliances installed in the PS. The second concept, applied after October 2017, provides information about the total PS's electrical energy consumption and all the relevant parameters (Figure 14). However, the parameters related to the consumption of one or another PU cannot be obtained directly. This information has to be extracted from the originally recorded data using some kind of machine learning approach. Due to the single-phase consumers installed in the PS, it is hard to perform without the simultaneous monitoring of the parameters in all three phases.

During the implementation of the new ACS, the requirement of the cyclic operation was set—at every following achievement of the h_{min}, the command for the start of the PU that in the previous cycle was off is given. A part of a simplified algorithm that meets the control requirements is presented in Figure 15. The practical result of its application in the normal operation of the PS is presented in Figure 14. Actually, in Figure 14, the daily diagrams of the load current changes at PS "Knez Selo" are presented. These diagrams were recorded on 3 January 2022 with the sampling period of 10 s. The order of PUs starting is indicated by the appropriate ordinal number in Figure 14.

According to the goals of the research, in this paper, the focus will be on the "data-driven" non-parametric models, algorithms that can be used to implement certain hybrid models, and/or knowledge that can be used to improve the ACS or its control algorithms [7–9,15,33–35].

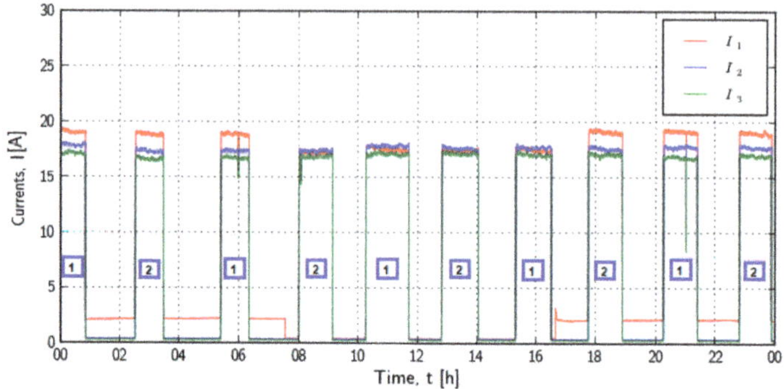

Figure 14. The diagrams of the PS load current changes recorded on 3 January 2022. 1: the PU1 is in operation, 2: the PU2 is in operation.

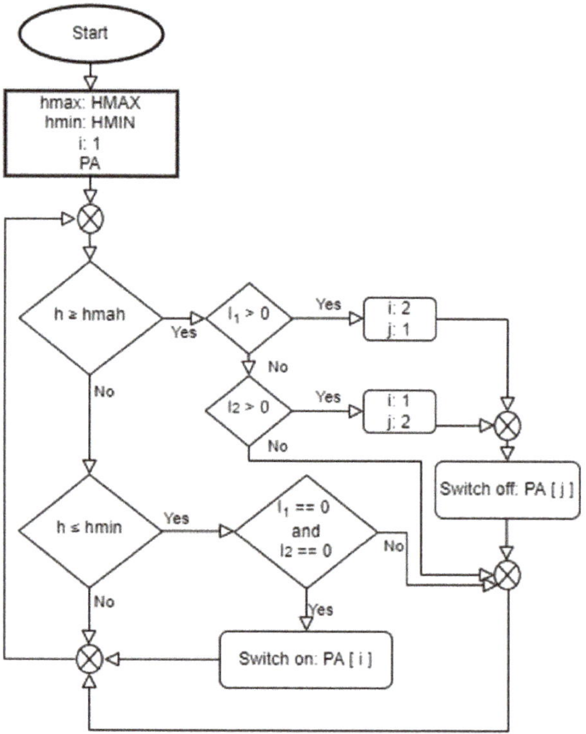

Figure 15. A part of the control algorithm allowing cyclic PU operation.

2.2. What Is the Purpose of Non-Parametric Models in the Simulations of PS Operation?

All predictable and expected situations and phenomena that occur in the PS operation can be modeled by some of many parametric models available in the literature. In all these cases, based on the simulation results, it is possible to (re)design the complete PS including appropriate automatic control and protection systems.

However, there are some specific situations and phenomena that appear in the PS operation that are not easily predictable in the system-design phase such as many of the ones belonging to the field of fault diagnostics and fault-tolerant control. These phenomena could have many negative consequences on the PS operation such as installed equipment damage, reducing the operating life, increase in the operating and maintenance costs, etc. In the hybrid modeling theory, the combined use of physics-based and data-driven models is of great importance. Some authors call this approach hybrid analysis and modeling (HAM), and consider it the third monumental step in the development of science [24].

As one of the examples showing that this is indeed an approach with huge potential, it can be stated that the considerations given in Section 2.1.3 have shown that the leaps in active power observed in the corresponding diagram in Figure 5 (in periods when some of the PU works) can be converted into useful information about the on/off of the PUs installed at PS Matejevac 2, supplying the lower altitude zone as well as PS Knez Selo. For example, this can be used to signal possible disturbances in the operation of PS Matejevac 2 (e.g., longer periods in which the PUs in this PS are not switched off, which could be due to overflows in some tanks, burst pipelines, or leaching in parts of the water supply network located in lower altitude zones). Generally speaking, the possibility of obtaining information on the operation of other PSs in the system or other parts of the system, based on data on the operation of subjected PUs, provides enormous opportunities for various innovations and improvements in the quality of system operation in regular or irregular conditions, malfunctions, or various emergency situations.

In this sense, for all of the data-driven models, the data recorded during the real system exploitation are incomparably more important for this area than any of other data (e.g., data from different simulation models). For this purpose, data recorded during the real system's operation were selected as the subject of the research in this paper. Based on these data, the appropriate "data-driven" parametric model was created in Section 2.1.

The task of the parametric models that were the subject of this paper was to process the available historical data on the system's operation and to identify the different unexpected phenomena, irregular modes, or faults that occurred in the PS. Many of them could not be described by some of the new parametric models, and would be treated as "black boxes". However, the aim should be to provide enough information about them for the appropriate analyses. These analyses can be used to obtain other useful information and knowledge based on the changes in the values or frequencies of the changes in certain process data and their interrelationships. The obtained information and knowledge could be used to create appropriate non-parametric models that will, in combination with the existing parametric model, enable better modeling and simulation of that phenomena. On the other hand, it can be expected that the obtained information and knowledge can be used to improve the existing automatic systems' control and protection or applied control algorithms. Practical reasons for these improvements are usually to raise the reliability and quality of the customer service as well as the level of the protection and safety of the facilities and equipment, thus increasing the energy efficiency of the system and reducing the maintenance and operating costs, etc.

It was previously stated that the hybrid model consists of a parametric and a number of non-parametric models whose integration should be performed in a certain way. When hybrid models are applied for the purpose of modeling certain processes, then the result of such models should be appropriate simulation results or appropriate predictions. In order to facilitate the model integration, the authors also deliberately chose to use the experimentally recorded data of the electrical quantities related to the loads of the pump drive motors as input data for the parametric model.

3. Example of the Creation of Non-Parametric and Corresponding Hybrid Models for Specific Phenomena Detected in the PS Operation

3.1. The Outdoor Lightning Detection Based on PS Load Currents

The diagrams of the PS load current changes presented in Figure 14 can be used as a simple example that can practically verify the previously described approach to non-parametric model creation. Namely, with a more precise look at the load current diagrams, it can be noticed that at the beginning and end of the day, the load current in phase L1 (red line) was significantly higher than the other two currents, while in the middle part of the day (approximately between 7:30 and 16:40), it was very close to the load currents in phases L2 and L3 (its value was between two other currents). Knowing that this phenomenon is repeated every day, it can be concluded that the cause of its appearance was the operation of the outdoor lighting of the PS, which was turned off in the morning and turned on in the evening. This diagram shows that at the time when the external lighting was off (the electrical circuit powering the external lighting was interrupted), the currents in all phases had approximately the same values. During these periods, the electricity in the PS is mostly consumed by the PUs' three-phase induction motors (strictly speaking, part of the energy is also spent on the no-load consumption of frequency converters and elements of control and protective devices, but it can be neglected). In periods when the outdoor lighting system is on, the energy required for its operation is taken over through phase L1, to which this system is connected. The characteristic time moment of this system switching on, when the peak in the load current appears, can be clearly seen in this diagram. It is a direct consequence of the characteristic transient processes taking place in light sources during their heating from the ambient to operating temperature.

Furthermore, based on the load current change diagrams, the moments of switching off and on the external lighting could be accurately detected every day. Knowing the expected periodicity during the year and voltage changes in the power distribution network, this algorithm can be extended to detect any failure in this system (e.g., burnout of light sources or the absence of command to turn off/on at certain times of the day). Such failures, with a certain time delay, can be signaled to the competent operators and their managers. For this purpose, for example, GSM/GPRS modems can be used as a communication channel through which the system can send appropriate SMS messages of predetermined content. However, this paper will deal with more complicated situations that occur in the exploitation of PS, so this simple algorithm will not be further developed.

3.2. The Detection and Analysis of the Irregular Operation of PU

For the analysis of the irregular modes of operation of the PS, the most important period was when the monitoring of its operation began. The recording of data from the power logger started on 8 June 2009. At that time, PUs with 15 kW drive motors were installed in the PS. The control concept was still based on the "on–off" regulation of the water level in the tank, and the choice of operation of one or the other PU was performed manually, using a changeover switch with two positions. Characteristic examples of regular ACS operation that was applied in this PS at that time are shown in Figures 16 and 17. The daily diagrams of the load current changes of PU1 (27 June 2009) and PU2 (23 June 2009) are presented in these figures, respectively (data sampling period was 12 s).

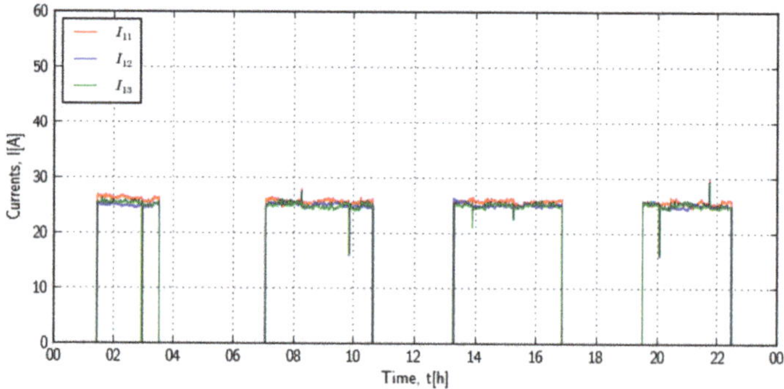

Figure 16. The diagrams of the load current changes in PU1 recorded on 27 June 2009.

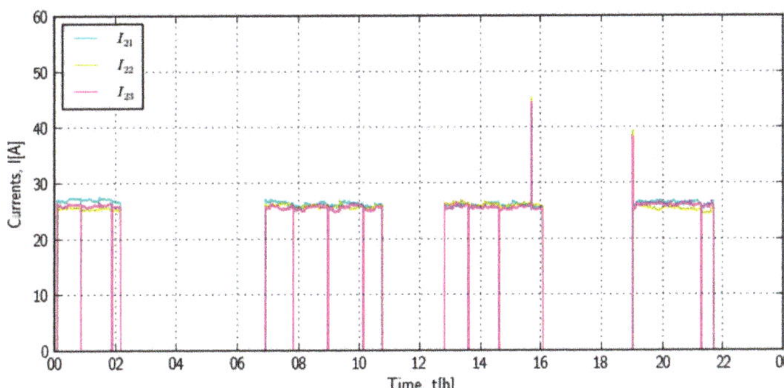

Figure 17. The diagrams of the load current changes of PU2 recorded on 23 June 2009.

In contrast, Figures 18 and 19 show that there were modes of operation that can be considered irregular and that could hardly have been predicted by any system modeling.

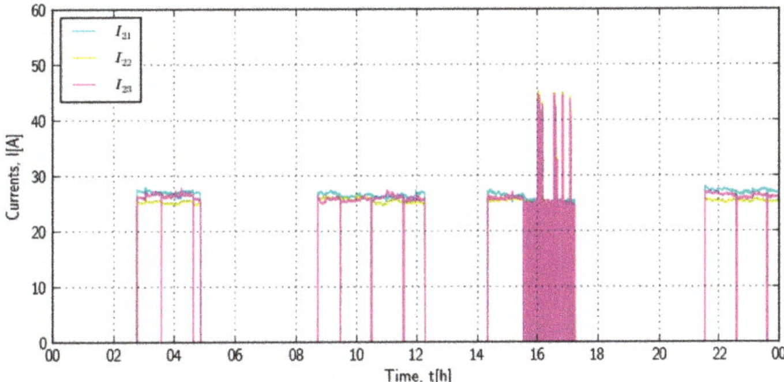

Figure 18. The diagrams of the load current changes of PU2 recorded on 21 June 2009.

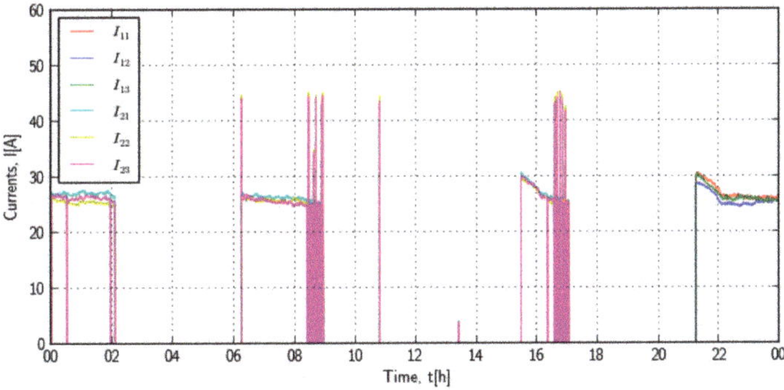

Figure 19. The diagrams of the load current changes of PU2 and PU1 recorded on 25 June 2009.

The irregularity of such a system's operation is reflected in periods in which a large number of consecutive cycles of the PU switching on/off occurred. These periods are easily noticeable in the diagrams, because, due to the large number of consecutive PU2 on/off, the diagrams are not clearly visible lines, but look like colored surfaces. Both cases shown in Figures 18 and 19 relate to the operation of PU2, but such modes of PS operation were common in both PUs.

The analysis of these phenomena revealed that in such situations, the existing ACS tries many times to turn on the PU and put it into operation, while a certain protection gives the command to turn it off, not allowing it to start. The diagrams in Figure 20, which actually represent a zoomed in part of the diagram in Figure 18, show that the automatic reconnection (AR) of the PU was set to about 30 s (two starts per minute).

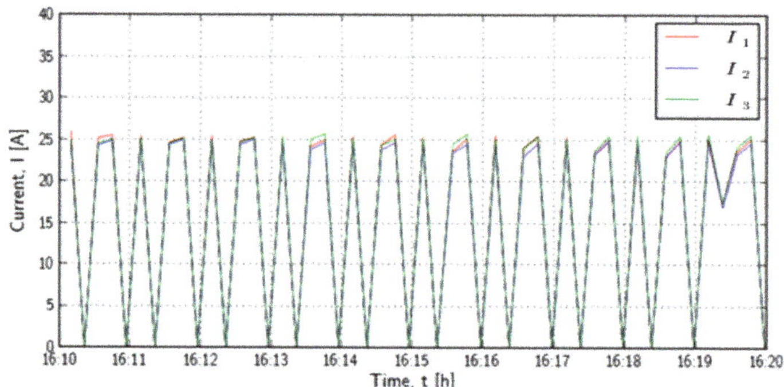

Figure 20. The diagrams of the load current changes of PU2 recorded on 21 June 2009. (Zoomed in period 16:10–16:20).

More detailed analyses have shown that the reason for this behavior of the automatic control system and protection was the dry-running protection system. At this period, this protection system consisted of a pressure switch, which under a certain pressure on the suction pipeline did not allow for the operation of the PU, giving the command to turn it off. Frequency of this phenomenon was defined by the AR system, which was set to attempt the PU reconnection about 30 s after the protection system is triggered.

Taking into account the fact that the PS Knez Selo is a booster pump station pumping water from the second to the third altitude zone, it should not be surprising that the occurrence of pressure drop on the suction pipeline that causes the pressure switch to react is relatively common.

From the diagrams shown in Figures 16–19, it can also be noticed that these phenomena are often transient (reactions of this protection system are visible, after which the local automation succeeds to turn on the PU and it continues to work for a certain period of time). However, when the pressure drop is not transient, a large number of unnecessary shutdown cycles and automatic reconnection (AR) occur. They expose the PU and additional equipment to high electro-mechanical stresses and significantly reduce their operational life. Based on the available data from June 2009, when the first data on the operation of this PU were recorded, it could be estimated that the total number of unnecessary attempts to start the PU during the month could be in the hundreds.

A Significance of Using Detailed Databases

It is very easy to show that the many phenomena described in the paper could not be detected or recognized using lower detailed databases. For example, if we try to use data from the existing SCADA (Supervisory, Control, and Data Acquisition) system (with sampling period of 15 min for all analog values), instead of the diagrams presented in Figures 18 and 19, we will obtain completely different diagram shapes (Figures 21 and 22, respectively).

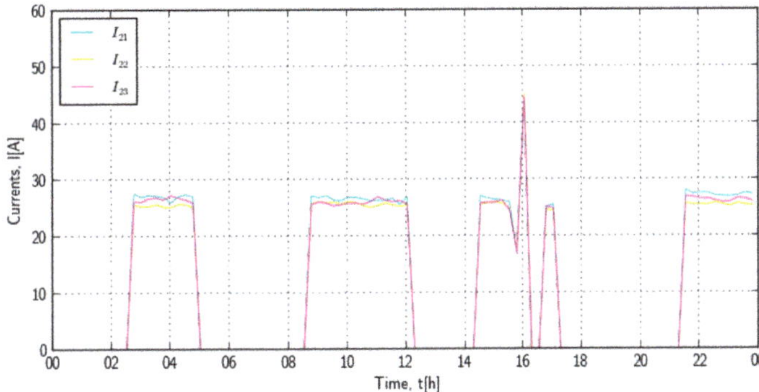

Figure 21. The diagrams of the load current changes of the PU2 recorded on 21 June 2009. (Sampling period of 15 min).

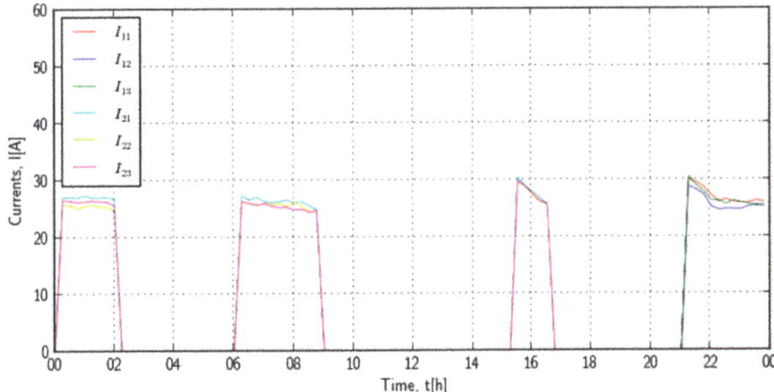

Figure 22. The diagrams of the load current changes of PU2 and PU1 recorded on 25 June 2009. (Sampling period of 15 min).

It is obvious that based on the diagrams with a sampling period of 15 min, the described irregular PU operation could not be detected at all. Knowing that the usually used control systems in the PS do not perform data acquisition (this option is eventually performed by corresponding SCADA systems), one should not be surprised by the fact that many of the important phenomena occurring in the PS remain undiscovered.

For this reason, the presented data-driven model approach, using the previously collected detailed databases, enables the detection and analyses of various phenomena, and can be considered as one of the paper's scientific contributions. The more different the data in the databases, the more qualitative analyses can be performed.

3.3. Non-Parametric Model of the Described Phenomenon (Non-Parametric Model 2)

The previously presented description and detailed analysis of the observed phenomena are sufficient to start the design of a non-parametric model, using which the described phenomena can be modeled.

The basic idea in a non-parametric model design should be to prevent unnecessary cyclical switching on and off of the PU because they can cause failure and damage to the equipment. However, in the model with the fixed input data sequences, it is not possible to simulate the mutual effects between the input data and ACS operation. For this reason, the main purpose of designing a non-parametric model should be to enable a simulation analysis, which will show how many unwanted cycles could be reduced by using the certain control logic that will be implemented in it. A corresponding software algorithm that will apply the selected control logic and that will be able to process the input data and give corresponding simulation results, can be actually treated as a non-parametric ("black box") model.

3.3.1. Practical Considerations about the Control Logic That Should Be Implemented in Non-Parametric Model Designing

The non-parametric model 2 can be formed based on the loading and processing of the original recorded data that will be used as the input data. The control logic that should be implemented should be based on the following practical considerations:

- After the dry-running protection tripping, it is necessary to allow one or at most two attempts to automatically restart the PU, because based on the diagram in Figures 16–19, it is obvious that in a number of cases, the automatic restart can be completed successfully, which is good because it eliminates the need for the handler visiting the PS in these situations and to incur additional unnecessary costs. The algorithm should be designed so that it can detect any situation when successive shutdown cycles and ARs occur in any PU and determine the limits of its duration. Then, it is necessary to define which number of cycles could be declared as a permissible number (e.g., 2, or in the general case n (n = 2, 3, 4)).

- The pressure switch used for the implementation of the dry-running protection at the PU suction reacts if a measured pressure value is less than the certain pre-set value and gives the command to switch off the PU. Since this is a PS type booster, it is logical that it will be more likely to react in periods when the PU at PS Matejevac 2, which supplies consumers in the lower altitude zone, is not working. In these regimes, lower pressure values can be expected compared to the values that would be had when these PUs pump water into the system. The most critical moments that lead to the reaction of this protection system are probably the moments of PU shutdown in the PS that belongs to the lower altitude zone when the water in that part of the system tends to return through the PU that is turned off. This is prevented by non-return valves, but the nature of the process is such that even a negative value of pressure can probably occur briefly on the suction of the PU at the subject PS. Any attempt to turn on the PU will naturally cause additional pressure drops at the suction side of the pump. As a result, short-term reductions and even possibly negative pressure values can be expected. The results of such conditions will be the reaction of the pressure switch and the shutdown of the PU, even in situations when its impeller is submerged. For these reasons, the reference value of the pressure to which the pressure switch will react should be set to the lowest possible value. It was shown that the maintenance teams set the pressure switches used in these types of protection to about 0.5 bar before detecting this phenomenon. In the experiments performed after the discovery of this phenomenon, it was shown that in this particular case, it was much more favorable to adjust the pressure to about 0.1 bar, which was the lowest value of pressure that could be adjusted on that pressure switch.

- In Section 2.1, the description of Figure 13 states that the lower limit probe in the tank was placed at a height of 1.1 m from the bottom of the tank. This means that in the case of problems with starting the PU, the water volume in the tank below this level (about 30 m^3) allows for the autonomy of the water supply to consumers of at least 1 h. With this in mind, it can be concluded that if a system that will stop further attempts to switch on after two unsuccessful cycles is implemented, it could be designed so that it has the ability to immediately notify the operator to come out to the PS and fix the problem before the consumers run out of water. However, the selected autonomy of the water supply allows, for example, to try another switch-on of the PU after 15 or 20 min. In the non-parametric model, this can be programmed in the following way: After each detection of n cycles in the original data, for cycles whose total duration is t_1 < 15 min (in the general case, the time t_1 should be determined as an input parameter (t_1 = 5,..., 15 min)), all of the other measured parameters (currents, powers, power factors), except for the input voltages and frequencies in the electrical network, should be assigned a value of 0 until the end of the cycle. If the total cycle time is greater than time t_1, after the expiration of time t_1 from the input data should be left as the originally recorded values for the first-subsequent cycle of the AR and re-shutdown, while all of the other measured values (currents, power, power factors), except for the input voltages and frequencies in the electrical network, the value of 0 should be assigned again by the end of the cycle.

- If it happens that the pressure on the suction of the PU has stabilized and that turn on is successful (which could be a consequence of the fact that in the meantime, the PU has been turned on in the PS that supplies a lower altitude zone or for some other reason), in each such case, the need for the operator to come to the PS unnecessarily would be eliminated. Due to the potential reduction in unnecessary costs, it was chosen to implement the system so that after 15 min, it would try to turn on the PU again. If this also fails, the system should block further attempts to turn on and send a message to the operator that their intervention is needed. In such situations, the fact that the time t_1 was not long enough to stabilize the hydraulic processes in the system, to eliminate the cause of the PU protection from "dry-running", and that the protection again does not allow for the start of the PU, will be enough to instruct the PS operator through an SMS message, so that they can come and fix the problem before the tank is emptied. The further operation of the system will depend on the speed of their arrival in the PS.

3.3.2. A Simulation Analysis of the Proposed Non-Parametric Model

Before the practical application of the knowledge used for the creation of the presented non-parametric model for the PS operation control, an appropriate simulation analysis should be performed.

The parametric model described in Section 2.1 was chosen to use the values of the electrical quantities related to the powers (currents) that PU drive motors take from the grid at certain moments as input data. All of these data are available and can be used as the input parameter list in the proposed non-parametric model (software designed based on the proposed control algorithm). If this model (software) uses these input data, bearing in mind that it has information on the motor load currents every 12 s, it is clear that it can detect any of the previously described cycles (protection system triggering and AR system attempt to reconnect the PU about 30 s after). The input data in the non-parametric model will be the sequences of all of the electrical parameters recorded in a certain period of time. The output data will be all of these values, changed in accordance to the previously described control logic, but only in the periods of irregular PU operation (i.e., in intervals in which all of the other measured parameters (currents, powers, power factors), except the input voltages and frequencies in the electrical network, is assigned the value 0).

The simulation results obtained by applying this model for the situations presented in Figures 18 and 19 (for the values of the parameters n = 2 and t_1 = 15) are shown in Figures 23 and 24, respectively. A simple comparison showed that now the number of unnecessary PU on/off cycles had been significantly reduced.

An overview of the exact number of switching on and off in real conditions as well as after the application of the simulation model is shown in Table 1. The first column is filled using the parametric model of the system (original data), and the second using the previously described non-parametric simulation model (simulation data).

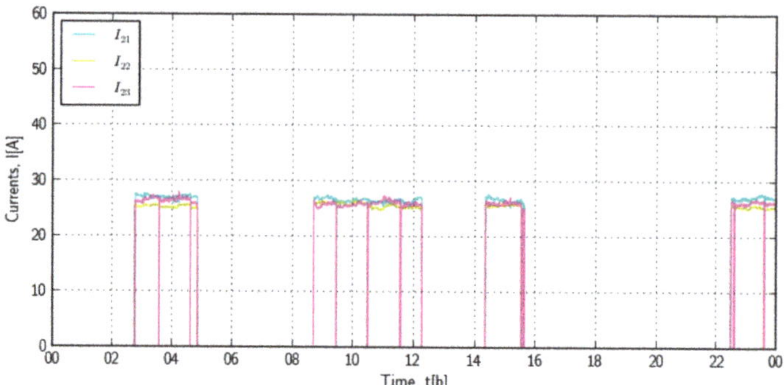

Figure 23. The simulation of the PU2 load current changes on 21 June 2009. ($n = 2$, $t_1 = 15$ min).

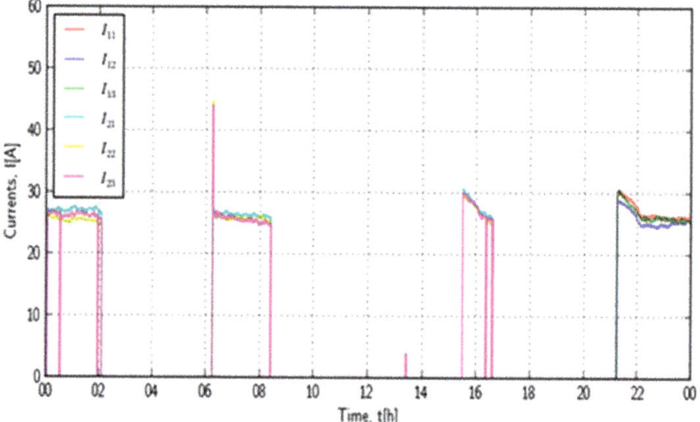

Figure 24. The simulation of the PU2 and PU1 load current changes on 25 June 2009. ($n = 2$, $t_1 = 15$ min).

From Table 1, it can be noticed that for a little more than 22 days in June 2009, the number of on/off cycles for PU1 was 209, and for PU2 425 (the system started recording the electrical energy consumption on 8 June 2009 around 18:00 h). The previously described simulation analysis showed that in this period, there should have been 138 on/off cycles for PU1 and 113 cycles for PU2 (i.e., that the unnecessary number of cycles on/off PU1 was 71, and PU2 was 312). The number of unnecessary cycles in this case was 383 and was higher than the expected number of cycles that would occur during the normal operation of PS in this period (251) by about 52.6%.

The drastic reduction in the unnecessary number of on/off cycles in both PUs, which is evident in Table 1, since July 2009, is a direct consequence of the applied knowledge obtained from previous analyses. Namely, the reference value on the pressure switch had been reduced before the end of June 2009 from 0.5 bar to a value of about 0.1 bar.

This indicates the fact that the dry-running protection system implemented by measuring the suction pressure of the PUs was not adequate and was causing shutdowns, even in situations when the pumps' impeller had been submerged.

The previously described example showed that irregular operating modes occurred in the PS operation, and that they were often unnoticed or insufficiently researched. This is especially characteristic for smaller power PSs, which are usually equipped with a simpler ACS, designed according to the expected operating modes, without connection to remote monitoring and control systems. Operators usually visit them only after certain failures appear, and leave them soon after their elimination, and this practice is common for almost all countries, regardless of their development level.

Table 1. The overview of the total number of switching off cycles and AR.

Period	PU	Original Data	Simulation Data	Difference
8–30 June 2009	PU1	209	138	71
	PU2	425	113	312
1–31 July 2009	PU1	2	1	1
	PU2	218	217	1
1–31 August 2009	PU1	267	266	1
	PU2	26	25	1
1–30 September 2009	PU1	5	5	0
	PU2	308	296	12
1–31 October 2009	PU1	0	0	0
	PU2	167	167	0
1–30 November 2009	PU1	96	95	1
	PU2	44	43	1
1–31 December 2009	PU1	115	114	1
	PU2	21	20	1
1–31 January 2010	PU1	15	14	1
	PU2	147	143	4
1–28 February 2010	PU1	151	151	0
	PU2	0	0	0

The specific example of the irregular operating mode of the PS described above is not an isolated example. The author's experience shows that identical or very similar situations occur in a large number of PS (especially in booster type PS), regardless of the dry-running protection type (pressure switch, pressure transmitter, conductive probes). Practice has shown that similar phenomena occur even in situations where the ACS is designed using modern equipment (motor-starters (MS) or frequency converters (FC) controlled by PLCs), which is also indicated next to the symbol for the corresponding switching elements on the physical model presented in Figure 2. This is evidenced by the large number of examples of PSs and the large amount of experimentally collected data that the authors possess.

3.4. Hybrid Model for More Efficient and Effective Control in PS Operation

When hybrid models are applied for the purpose of modeling certain processes, then the result of such models should be appropriate simulation results or appropriate predictions. The hybrid model consists of a parametric and a number of non-parametric models, whose integration should be performed in a certain way. In the example shown, it was very easy to perform this using software, because the parametric and non-parametric models used the experimentally recorded data on the electrical quantities related to the loads of the pump drive motors as input data (Figure 25).

The model integration box in this situation could actually have the same program lines as the parametric model. When the simulation results from a non-parametric model should be obtained, its output data (changed in accordance to the previously described control logic) are passed through the same procedure as in the parametric model. In this process, the corresponding nonelectrical values are obtained (pump flows, PU heads, and efficiencies). It should be pointed out that this procedure will not take place in parallel with the procedure that takes place in the parametric model. A certain time delay between these two procedures, needed for the non-parametric model procedure to perform, will have to be implemented.

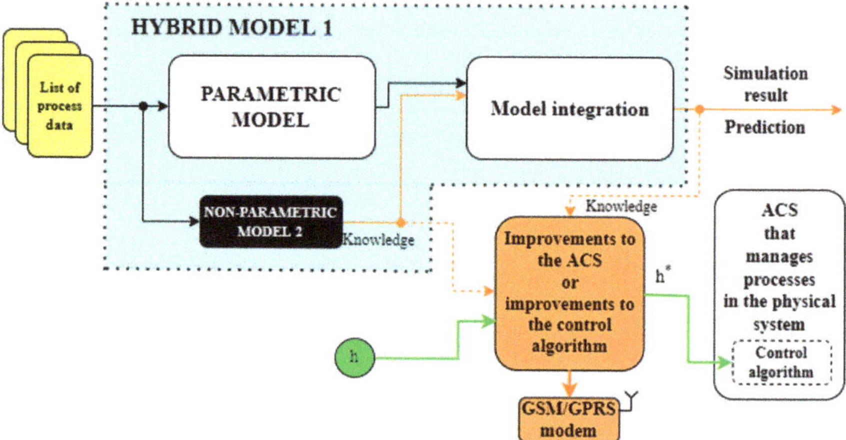

Figure 25. The hybrid model for the tolerant control of the described irregular PS operating mode.

In Section 2.1, it was clearly shown how the input data, with the help of the selected parametric model related to the measured electrical quantities, were converted into significant information about a number of non-electrical quantities in the system. On the other hand, Sections 3.2 and 3.3 described how, based on the same set of input parameters, the occurrence of irregular PS operation (resulting in numerous negative effects for electrical and hydraulic equipment in PS and for consumers) can be detected. The performed analysis explained which logic should be used to eliminate them, and the results of the simulation analysis obtained by applying the formed non-parametric model are shown. The results present what these phenomena might look like (i.e., how the appropriate electrical quantities diagrams would look like after applying the proposed algorithms). It is clear that treating the output quantities from the non-parametric model as input parameters and passing them through the parametric model would allow us to obtain the simulation values of all of the non-electrical quantities and to determine what effects the application of such control algorithms would have on the system (by how much the number of unwanted on/off cycles would be reduced, and by how much less the amount of electricity consumed would be, etc.). This could also be easily conducted in the model integration module (Figure 25).

However, the following issues that have practical significance are much more important:

- Is it possible to apply the knowledge acquired in the process of hybrid analysis and modeling in such a way that the observed negative phenomena are eliminated or that at least their negative effects are mitigated as soon as possible?
- If the results of the simulation analysis show that the proposed control algorithms used in the development of the simulation non-parametric and/or hybrid models are satisfactory, how can they be most easily applied in the automatic control system of PS operation?

It is obvious that in the presented process of the hybrid analysis and modeling, the input data (data on electrical parameters of PUs in different operating modes) are converted into knowledge (Figure 25), which allows the described situations to be reliably recognized and to create a control algorithm that allows for more efficient and effective control of the described occurrence in the irregular operation of a PS.

Part of this process has already been presented in the previous section, and the knowledge obtained from the presented information was enough to design an algorithm for the more efficient and effective management of PS, without the need of the existing ACS to be replaced (Figure 26).

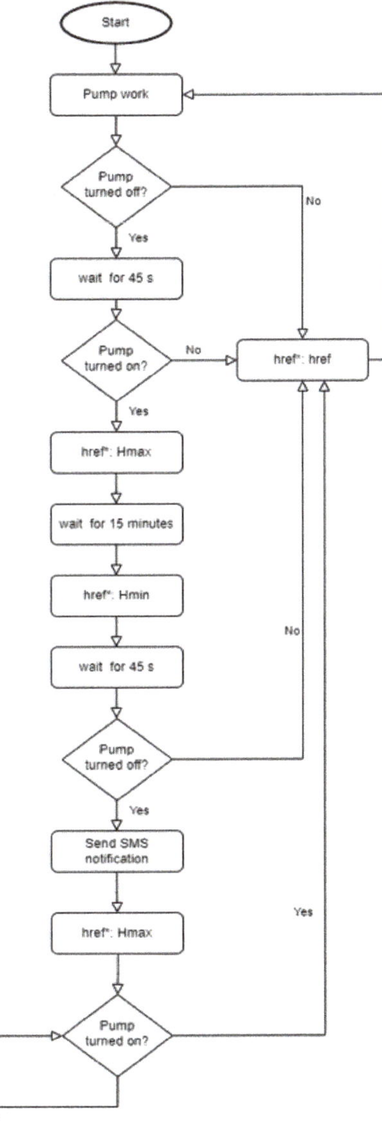

Figure 26. The algorithm for the tolerant control in the irregular PS operating modes.

From the control algorithm shown in Figure 26, it can be noticed that the basic idea for upgrading and improving the control function consists of the following:

- As soon as two consecutive shutdown cycles and the AR of the pump set are detected, regardless of the current value for reference value h, send the information h* to the existing PS control algorithm (i.e., the corresponding signal that is present in a situation where the upper limit probe is submerged ($h^* > h_{max}$)). The ACS is designed so that this signal will automatically turn off the PU in operation (it does not matter if the shutdown signal arrives faster than the PU dry protection system) and eliminate further switching until the water level falls below the set minimum value ($h^* < h_{min}$).

- After a period of 15 min, which is left as an option to stabilize the pressure in the system, try to turn on the PU again. Again, regardless of the current value of the reference value h, this will be achieved by sending information about the reference value h* to the existing control algorithm PS (i.e., the corresponding signal that is present in the situation when the lower limit probe has remained dry (h* < hmin)).

- If the PU is not switched off again within 45 s, it will mean that the suction pressure drop has been transient and that the system can continue with normal operation, for which the ACS is enough to send the real value signal of the water level in the tank as reference quantities (h* = h). In that case, the ACS will continue to operate normally until the next response of the protection system, and the integration algorithm will continue to periodically send information about the real value of the water level in the tank (h* = h), which will not affect the operation of the ACS, until it does not detect two consecutive shutdown cycles and the AR of the pump set again. On the other hand, if the PU stays switched off again within 45 s, this will mean that the suction pressure drop was not transient and that there is a possibility that the pump impeller is not submerged, which means that the cycle of unnecessary shutdowns and ARs would continue. In order to prevent this situation, the signal should be sent to the ACS again (h* > h$_{max}$), and immediately afterward, an appropriate SMS message is sent to the operator in charge of this PS, so that they can intervene in time and eliminate the problem before the tank emptying occurrs. In such a situation, the signal value (h* > h$_{max}$) should be kept until the operator reaches the PS and eliminates the cause of the protection response.

- After arrival at the PS, the operator will first deactivate the ACS by switching the appropriate ridge switch to a position that allows for manual control (which is a common procedure in such situations). After eliminating the cause of the protection response, which usually comes down to immersing the pump impeller, they will manually turn on the PU to check whether their intervention was sufficient for the PS to be able to continue its operation. The integration algorithm will detect this PU activation by the increase in the load currents in all three phases, after which the ACS will immediately transmit a signal about the real value of the water level in the tank (h* = h), which is a necessary condition for the ACS to resume with normal operation. This will also happen as soon as the operator again switches the ridge switch to the PS automatic operation position.

4. An Example of Hybrid Model Extension: Modeling New Phenomena in the PS Operation

4.1. Consequences of Long-Term Irregular PU Operation

A large number of the dry-running pump protection system activations led to a new, more serious problem in the PS operation, which also required an adequate solution. Namely, due to a large number of protection system tripping during the month as a result of material fatigue, the pressure switch relay failures occurred, after which this protection system was out of order and the PU entered into the "dry-running" operation. This state often lasted for several hours (Figure 27). Due to worsening cooling conditions, the pump bearings are often damaged in such situations, or other mechanical damage occurs due to the excessive thermal stresses, after which the pumps must be repaired. Figure 28 clearly shows that PU2 was replaced on 8 March 2010. It is obvious that before the replacement of the pump unit PU2, which was performed on 8 March 2010, PU1 and PU2 had very similar operating characteristics, which can be seen from the fact that in that period, in normal operating modes, both PUs drew approximately equal currents (the largest number of points on the red (PU1) and blue (PU2) curves are grouped around the values of the currents in the range between 24 and 25 A). On the other hand, after the replacement of PU2, the newly installed PU drew slightly higher load currents from the network, which in operating modes was around 28 A, which indicates that it had different operating characteristics from the previous one. Additional information confirmed that the old PU was replaced by a PU of the same type that had been repaired in the meantime. After the replacement of the PU, its performance characteristics should be changed in the parametric model, according to the procedure explained in [12]. On the other hand, the diagram of the change in the current's average value in Figure 27 shows that only 3 days after the replacement (on 11 March 2010), the repaired PU2 entered into a "dry-running" operation that had lasted for more than 3 h. This is the operating mode of pump units that should be avoided by applying an adequate protection system.

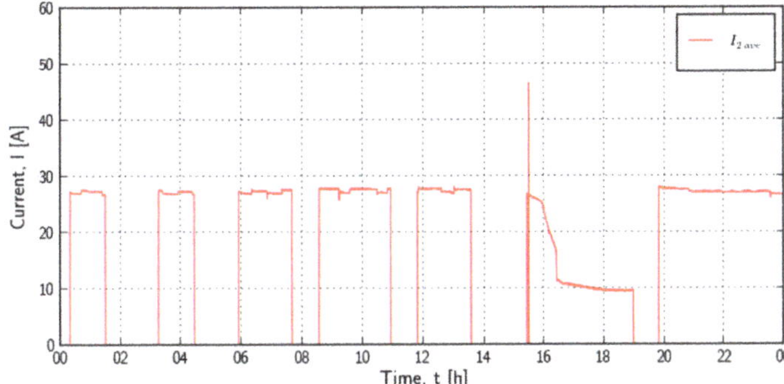

Figure 27. An example of the dry-running pump operation, visible through the PU2 load current change diagram.

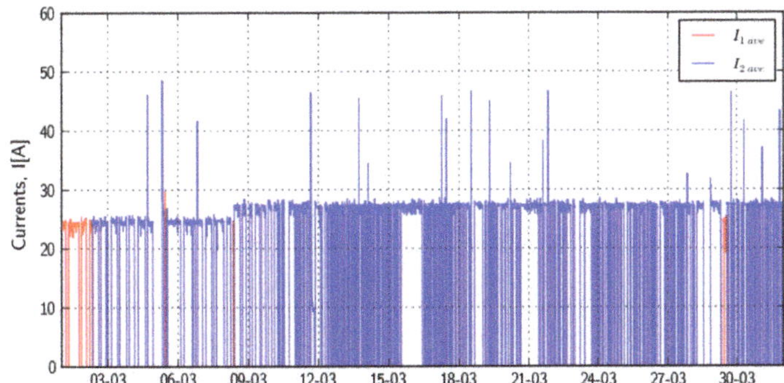

Figure 28. An example of the PU2 replacement performed on 8 March 2010, visible through the load current change diagram.

The diagram in Figure 27 shows that this mode is not problematic for the pump drive motor, because the PU load currents in such modes are significantly lower than the load currents in normal operating modes. As the PU drive motor operates at a relatively low load, such operating modes cannot in any way adversely affect its operation. The main problem that prohibits the operation of pumps in such modes is that, due to a lack of water in the impeller, the cooling of the pump was much worse compared to normal operating modes. Prolonged operation in such modes can lead to increased heating of the pump shaft bearings, which in situations where they are not in good condition can lead to thermal stresses and possible damage to their housings.

Based on the results of the conducted experiments, the authors concluded that when running dry, the PUs enter into a mode where their motors draw less current from the grid, even less than the PUs' no-load currents. Measurements performed on the PU led to the conclusion that the average value of the current that the drive motors of both the PUs draw from the grid at no-load (with the submerged impeller and maximally closed valves on the thrust) was greater than 23 A.

4.2. An Extended Hybrid Model for More Efficient and More Effective Control of PS Operation

This idea can be used to form another non-parametric model that can be used to simulate the operating modes of the PS in which the failure of the PU dry-running protection system fails, which are obviously situations that can realistically occur in practice. However, it is much more practical based on this information and knowledge to design a backup protection system that will, even in the event of the failure of basic protection systems, prevent the further operation of PUs in such regimes. Through the formation of a new hybrid model and simulation analysis, it is possible to simulate its operation.

The basic idea for such a backup protection system can be described as follows: As soon as the system detects that the average load current of the drive motor is less than 23 A at several consecutive points (e.g., 5 points, which corresponds to a period of 60 s), it gives a signal to turn off the PU after this period. This command in the system has the treatment of an identical command issued by a pressure switch under normal circumstances. This is enough to form the non-parametric model 3, an extended hybrid model corresponding to this situation (Figure 29).

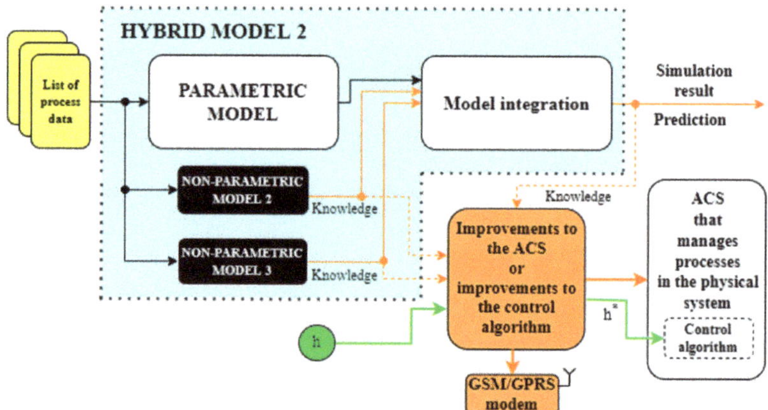

Figure 29. The extended hybrid model of the serial-parallel structure.

As in the previously described case, this non-parametric model would use the same input parameters and its integration into a more advanced hybrid model would not be complicated again.

The simulation result of the load currents' average values of the PU2 drive motor, obtained after the application of the extended hybrid model, is shown in Figure 30.

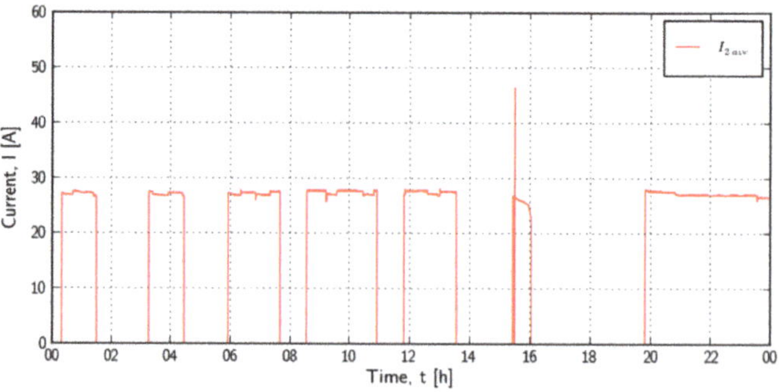

Figure 30. The simulation of the PU2 load current average values obtained by the application of the extended hybrid model.

The same idea was used to develop an algorithm for the tolerant control after pressure switch failure (Figure 31), and to implement it in a backup dry-running pump protection system.

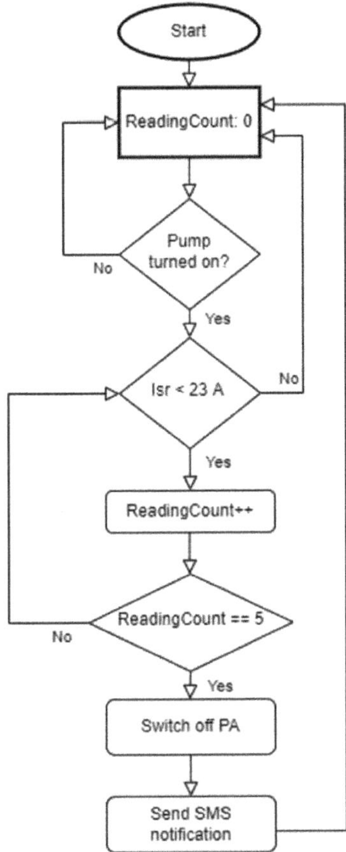

Figure 31. The algorithm for the tolerant control after a pressure switch failure.

The presented model is just an obvious example of a tolerant system control after a fault occurs. In this case, the application of non-parametric (i.e., hybrid models) enabled the realization of a more advanced ACS that has the possibility of tolerant management in the case of the failure of one protective component (in this case, the pressure switch, which served to protect the PU from dry-running).

In a similar way, non-parametric and hybrid models can be realized that will enable the tolerant control of PS in the case of the failure of other protective devices (voltage protection, current protection, overload protection, frequency protection, etc.). The presented model can be applied in all PS booster types because the described phenomena occurred in all similar PS.

The authors designed and suggested the implementation of this backup protection system in March 2010. The system was put into operation in September 2013 at PS Knez Selo after obtaining their permission for its practical implementation. Since then, the backup protection system has been tested in the operation of PS Knez Selo for many years, and its efficiency has been confirmed in the conducted experiments as well as in practice.

5. Conclusions

In the hybrid modeling theory, the combined use of physics-based and data-driven models is of great importance. In this paper, some of the numerous advantages of data-driven models and the hybrid analysis and modeling approach have been demonstrated with the example of one booster PS (the PS Knez Selo).

The authors used and processed the available historical data on the system's operation, in order to discover their mutual interrelationships and to identify different unexpected phenomena, irregular modes, or faults that occurred in the PS, especially those that previously could not be detected. The presented results pointed to a significance in the detailed database usage and showed that many phenomena described in the paper could not be either recognized or detected at all when using lower detailed databases. This was practically demonstrated by the corresponding comparison between the results taken from the existing SCADA database (with a sampling period of 15 min for all of the analog values) and the database previously collected by the authors (with the applied sampling period of 12 s).

The presented comparisons demonstrated that using diagrams with a sampling period of 15 min, the described irregular PU operation could not be detected at all. Knowing that the usually used control systems in PS do not perform data acquisition (this option is eventually performed by corresponding SCADA systems), one should not be surprised by the fact that many of the important phenomena occurring at the PS remain undiscovered. For this reason, the presented data-driven modeling approach, using previously collected detailed databases, enables the detection and analyses of various phenomena that could not be detected before, and that can be considered as one of the paper's scientific contributions. In several examples presented in the paper, it has also been shown that the more different the data in databases that are recorded, the more qualitative the analyses that can be performed.

Relating to the phenomena that could not be detected before, two of them were investigated in the paper.

The first case was the irregular PU operation that was reflected in many unnecessary on/off cycles. The analysis of these phenomena revealed that in such situations, the existing ACS tries many times to turn on the PU and put it into operation, while the dry-running protection system gives the command to turn it off, not allowing it to start. The frequency of this phenomenon is defined by the AR system, which is the part of the ACS and that was set to attempt PU reconnection about 30 s after the protection system was triggered.

The simulation results obtained by the corresponding hybrid model showed that for a little more than 22 days in June 2009, the total number of unnecessary and unwanted on/off cycles for both PUs was 383, which was higher than the expected number of cycles that would occur during the normal operation of PS in this period (251) by about 52.6%. The simulation results also showed a drastic reduction in the unnecessary number of on/off cycles in both PUs since July 2009. This was a direct consequence of the applied knowledge obtained from the presented analyses. Namely, before the end of June 2009, the authors suggested that the reference value on the pressure switch should be reduced from 0.5 bar to a value of about 0.1 bar, which was implemented.

This analysis directly pointed out the fact that the dry-running protection system based on measuring the suction pressure of PUs was not adequate and was causing shutdowns even in situations when the pumps' impeller had been submerged.

Without the information and knowledge acquired during the presented research, the authors supposed that this specific phenomenon could not be discovered. The best evidence that this assumption is real represents the fact that even today, many PS designers apply a dry-running protection system based on pressure monitoring. The results of the research presented in this paper undoubtedly confirm that this is not an adequate solution at all.

The second specific case analyzed in the paper was the failure of the dry-running protection system (caused by the actuator relay failure due to the material fatigue) occurred in March 2010. After this failure, the protection system was completely out of order, and the PU entered into the dry-running operation modes.

Based on the analyses presented in the paper, the authors designed and suggested the implementation of a backup protection system in March 2010. This protection system was based on monitoring the drive motor load currents, which became lower than some of the pre-set reference values only in the PU dry-running operation modes. This backup protection system was put into operation in September 2013 at PS Knez Selo after obtaining their permission for their practical implementation. Since then, it has been tested in the operation of PS Knez Selo for many years, and its efficiency has been confirmed in the conducted experiments as well as in practice.

According to the authors' opinions, it should be pointed out that the examples presented in the paper undoubtedly show that a data-driven approach to hybrid analysis and modeling is indeed an approach that has huge potential. Generally speaking, it is shown that it is possible to obtain information on the operation of other PSs in the system or other parts of the system based on the data on the operation of the subjected PUs. In the related example is shown that the leaps in the active power observed in the corresponding diagram in periods when some of the PUs in the PS Knez Selo operate can be converted into useful information about the ons/offs of the PUs installed at PS Matejevac 2, which supplies the lower altitude zone as well as PS Knez Selo. These possibilities provide enormous opportunities for various innovations and improvements in the quality of the system operation in regular or irregular conditions, malfunctions, or various emergency situations.

Although the complete analysis presented in this paper refers to one PS, the conclusions are of general importance because similar phenomena occur in all booster PSs.

According to the authors, one of the most important conclusions, indirectly indicated by this paper, is the high importance of continuous data recording on the operation of controlled facilities (in this case, the PU in the PS). The possibilities of data recording with a sampling period of a few seconds are especially important because the data collected in this manner enables the detection of numerous specific phenomena during the operation of the system.

Hybrid modeling techniques enable the application of various innovations that lead to the reduction in system exploitation and maintenance costs.

These types of applications could also have importance for the future development of "smart grid" and "smart city" projects.

Author Contributions: Conceptualization, M.M., M.R., M.R.-A., D.L. and Z.S.; Methodology, Z.S.; Software, M.M.; Validation, M.R., M.R.-A. and Z.S.; Formal analysis, D.L.; Investigation, M.R.-A.; Resources, Z.S.; Data curation, M.M.; Writing—original draft preparation, M.R.-A. and Z.S.; Writing—review and editing, M.M.; Visualization, M.M.; Supervision, D.L. and Z.S.; Project administration, M.R. and Z.S.; Funding acquisition, Z.S. All authors have read and agreed to the published version of the manuscript.

Funding: This paper was supported by the Ministry of Education, Science, and Technological Development of the Republic of Serbia through the following projects and/or contracts (Projects: NP EE413-118A, NPV-19.B, 401-00-00144/2008-01-IP, III 44006; Contract No. 451-03-68/2022-14/200102, 4 February 2022).

Data Availability Statement: The data presented in this study are available upon request from the submitting author.

Conflicts of Interest: The authors declare no conflict of interest.

References

1. Mosterman, P.J.; Biswas, G. A comprehensive methodology for building hybrid models of physical systems. *Artif. Intell.* **2000**, *121*, 171–209. [CrossRef]
2. Stosch, M.; Oliveira, R.; Peres, J.; Feyo de Azevedo, S. Hybrid semi-parametric modeling in process systems engineering: Past, present and future. *Comput. Chem. Eng.* **2014**, *60*, 86–101. [CrossRef]
3. Li, K.; Xu, W.; Han, Y.; Ge, F.; Wang, Y. A hybrid modeling method for interval time prediction of the intermittent pumping well based on IBSO-KELM. *Measurement* **2019**, *151*, 107214. [CrossRef]

4. Xie, Y.; Hu, P.; Zhu, N.; Lei, F.; Xing, L.; Xu, L.; Sun, Q. A hybrid short-term load forecasting model and its application in ground source heat pump with cooling storage system. *Renew. Energy* **2020**, *161*, 1244–1259. [CrossRef]

5. Han, Q.; Meng, F.; Hu, T.; Chu, F. Non-parametric hybrid models for wind speed forecasting. *Energy Convers. Manag.* **2017**, *148*, 554–568. [CrossRef]

6. Lee, D.S.; Vanrolleghem, P.A.; Park, J.M. Parallel hybrid modeling methods for a full-scale cokes wastewater treatment plant. *J. Biotechnol.* **2005**, *115*, 317–328. [CrossRef] [PubMed]

7. Mustafee, N.; Harper, A.; Onggo, B.S. Hybrid modelling and simulation (M&S): Driving innovation in the theory and practice of M&S. In Proceedings of the IEEE Winter Simulation Conference (WSC), Orlando, FL, USA, 14–18 December 2020; pp. 3140–3151. [CrossRef]

8. Stosch, M. Novel Strategies for Process Control Based on Hybrid Semi-Parametric Mathematical Systems. Ph.D. Thesis, Universidade do Porto, Porto, Portugal, 2011.

9. Rasheed, A.; San, O.; Kvamsdal, T. Digital Twin: Values, Challenges and Enablers From a Modelling Perspective. *IEEE Access* **2020**, *8*, 21980–22012. [CrossRef]

10. Klempous, R.; Kotowski, J.; Nikodem, J.; Olesiak, M.; Ułasiewicz, J. Some models for water distribution systems. *J. Comput. Appl. Math.* **2015**, *21*, 257–269. [CrossRef]

11. Seifollahi-Aghmiuni, S.; Bozorg, H.; Omid, M.; Miguel, A. Generalized Mathematical Simulation Formulation for Reservoir Systems. *J. Water Resour. Plan. Manag.* **2016**, *142*, 04016004. [CrossRef]

12. Rašić-Amon, M.; Radić, M.; Stajić, Z.; Brkić, D.; Praks, P. Simplified Indirect Estimation of Pump Flow Discharge: An Example from Serbia. *Water* **2021**, *13*, 796. [CrossRef]

13. Alves, Z.; Muranho, J.; Albuquerque, T.; Ferreira, A. Water Distribution Network's Modeling and Calibration. A Case Study based on Scarce Inventory Data. *Procedia Eng.* **2014**, *70*, 31–40. [CrossRef]

14. Luo, Y.; Yuan, S.; Tang, Y.; Yuan, J.; Zhang, J. Modeling optimal scheduling for pumping system to minimize operation cost and enhance operation reliability. *J. Appl. Math.* **2012**, *2012*, 370502. [CrossRef]

15. Stajić, D.; Milošević, M.M.; Milošević, M. Water Supply System Management in Presence of Major Disturbances. *Facta Univ. Ser. Autom. Control. Robot.* **2012**, *11*, 153–160, UDC 681.5.01.628.29.

16. Radić, M.; Rašić, M.; Stajić, Z. Practical Example of Significant Energy Savings Achieved by Optimal Reengineering in Pump Stations. In Proceedings of the 3rd Virtual International Conference on Science, Technology and Management in Energy, eNergetics 2017, Niš, Serbia, 22–23 October 2017; pp. 157–161, ISBN 978-86-80616-02-5.

17. Radić, M.; Stajić, Z.; Rašić, M. Experimental identification of actual performance curves in pump stations. In Proceedings of the XIII International Conference on Systems, Automatic Control, and Measurements—SAUM2016, Niš, Serbia, 9–11 November 2016; pp. 55–58, ISBN 978-86-6125-170-2.

18. Camarinha-Matos, L.M.; Martinelli, F.J. Application of Machine Learning in Water Distribution Networks Assisted by Domain Experts. *J. Intell. Robot. Syst.* **1999**, *26*, 325–352. [CrossRef]

19. Filipović, F.; Stajić, Z.; Banković, B.; Radić, M. Application of Deep Learning for Electrical Appliance Classification. In Proceedings of the 3rd Virtual International Conference on Science, Technology and Management in Energy, eNergetics 2017, Niš, Serbia, 22–23 October 2017; pp. 101–106, ISBN 978-86-80616-02-5.

20. Sun, C. Multi-Layer Model Predictive Control of Complex Water Systems. Ph.D. Thesis, Universitat Politecnica de Catalunya, Barcelona, Spain, 2015.

21. Okeya, I.; Kapelan, Z.; Hutton, C.; Naga, D. Online Modelling of Water Distribution System Using Data Assimilation. *Procedia Eng.* **2014**, *70*, 1261–1270. [CrossRef]

22. Thompson, K.; Kadiyala, R. Leveraging Big Data to Improve Water System Operations. In Proceedings of the 16th Conference on Water Distribution System Analysis, WDSA 2014, Procedia Engineering, Bari, Italy, 14–17 July 2014; Volume 89, pp. 467–472. [CrossRef]

23. Vegas Niño, O.T.; Martínez Alzamora, F.; Tzatchkov, V.G. A Decision Support Tool for Water Supply System Decentralization via Distribution Network Sectorization. *Processes* **2021**, *9*, 642. [CrossRef]

24. San, O.; Rasheed, A.; Kvamsdal, T. Hybrid analysis and modeling, eclecticism, and multifidelity computing toward digital twin revolution. *GAMM-Mitteilungen* **2021**, *44*, e202100007. [CrossRef]

25. Monshizadeh, N. Amidst data-driven model reduction and control. *IEEE Control Syst. Lett.* **2020**, *4*, 833–838. [CrossRef]

26. Filipe, J.; Bessa, R.J.; Reis, M.; Alves, R.; Póvoa, P. Data-driven predictive energy optimization in a wastewater pumping station. *Appl. Energy* **2019**, *252*, 113423. [CrossRef]

27. Brkić, D.; Čojbašić, Ž. Intelligent flow friction estimation. *Comput. Intell. Neurosci.* **2016**, *2016*, 5242596. [CrossRef]

28. Floranović, N.; Radić, M.; Stajić, Z. Smart Metering and Power Loggers as Key Element of Effective Energy Efficiency Improvement. In Proceedings of the XII International Conference on Systems, Automatic Control and Measurements—SAUM 2014, Niš, Serbia, 12–14 November 2014; pp. 372–375, ISBN 978-86-6125-117-7.

29. Booske, J.H.; Converse, M.C.; Kory, C.L.; Chevalier, C.T.; Gallagher, D.A.; Kreischer, K.E.; Heinen, V.O.; Bhattacharjee, S. Accurate parametric modeling of folded waveguide circuits for millimeter-wave traveling wave tubes. *IEEE Trans. Electron Devices* **2005**, *52*, 685–694. [CrossRef]

30. Mendoza, P.A.; Clark, M.P.; Barlage, M.; Rajagopalan, B.; Samaniego, L.; Abramowitz, G.; Gupta, H. Are we unnecessarily constraining the agility of complex process-based models? *Water Resour. Res.* **2014**, *51*, 716–728. [CrossRef]

31. Olszewski, P.; Arafeh, J. Parametric analysis of pumping station with parallel-configured centrifugal pumps towards self-learning applications. *Appl. Energy* **2018**, *231*, 1146–1158. [CrossRef]
32. Brkić, D.; Praks, P. Air-Forced Flow in Proton Exchange Membrane Fuel Cells: Calculation of Fan-Induced Friction in Open-Cathode Conduits with Virtual Roughness. *Processes* **2020**, *8*, 686. [CrossRef]
33. Skovmose Kallesøe, C.; Cocquempot, V.; Izadi-Zamanabadi, R. Model Based Fault Detection in a Centrifugal Pump Application. *IEEE Trans. Control. Syst. Technol.* **2006**, *14*, 204–215. [CrossRef]
34. Sunela, M.I.; Puust, R. Modeling water supply system control system algorithms, Science Direct. *Procedia Eng.* **2015**, *119*, 734–743. [CrossRef]
35. Sun, J.W.; Xi, L.F.; Du, S.C.; Ju, B. Reliability modeling and analysis of serial-parallel hybrid multi-operational manufacturing system considering dimensional quality, tool degradation and system configuration. *Int. J. Prod. Econ.* **2008**, *114*, 149–164. [CrossRef]

 *processes*

Article

Research on the Optimization Method of Safety Input Structure in Coal Mine Enterprise

Xiu-Zhi Shi, Jin-Yun Zhu and Shu Zhang *

School of Resources and Safety Engineering, Central South University, Changsha 410083, China
* Correspondence: shuzhang303@gmail.com

Abstract: In order to study the application of the Cobb-Douglas production function on the optimization of safety inputs and further reduce accident losses, two safety input structures of a coal mine enterprise were constructed using literature, and the weight order of each safety input indicator was determined by the entropy weight method (EWM) and the analytical hierarchy process (AHP). The Cobb-Douglas production function was used to calculate the accident loss function of the safety input structure, and the accident loss function was obtained by multiple regression analysis. The optimal configuration of safety inputs was obtained by fitting the accident loss function. Finally, the optimal loss and mean squared error (MSE) of the corresponding functions of the two safety input structures were compared. The results show that the optimal configuration of Safety Input Structure 2 is better than that of Safety Input Structure 1, and the MSE of Safety Input Structure 2 is less than that of Safety Input Structure 1. The research results demonstrate that coal enterprises can find more significant indicators by refining the safety input structure and increasing monetary resources for more crucial indicators of safety input to effectively minimize accident loss and boost economic benefits, and to test the quality of safety input structures' regression function using MSE.

Keywords: safety input; Cobb-Douglas production function; structure optimization; safety input structure; comprehensive empowerment

Citation: Shi, X.-Z.; Zhu, J.-Y.; Zhang, S. Research on the Optimization Method of Safety Input Structure in Coal Mine Enterprise. *Processes* **2022**, *10*, 2497. https://doi.org/10.3390/pr10122497

Academic Editors: Francisco Ronay López-Estrada, Guillermo Valencia-Palomo and Anthony Rossiter

Received: 14 October 2022
Accepted: 15 November 2022
Published: 24 November 2022

Publisher's Note: MDPI stays neutral with regard to jurisdictional claims in published maps and institutional affiliations.

1. Introduction

Due to the continued expansion of China's economy and society, personal safety, particularly the life and health of enterprise workers, has received increasing attention in recent years. Coal, which accounts for more than half of China's total energy supply, will continue to be the primary energy source for a long time. Various safety accidents occur occasionally in large-scale coal mines. According to statistics, the total number of coal mine accident fatalities in China is 1–2 times higher than that of other coal-producing countries. The coal mine fatality rate per million tons in 2017 was 0.06, which is twice that of the United States and 1.5 times that of Germany [1,2]. One of the most essential approaches to managing the frequency of accidents successfully is to ensure sufficient safety input. Because the development of businesses is always focused on lowering costs and increasing benefits, the money businesses invest in safety is limited. Safety input is characterized by latency and invisibility [3]. As a result of limited safety funds, how to enhance the utilization rate of safety funds, maximize the protection of employees' lives and maintain normal corporate production is a subject to be explored.

Numerous researchers have conducted extensive studies of the utilization of safety input funds. Some researchers established a safety input model and used data analysis of the safety input–output function or the safety input–output relative outcomes of each scheme to determine the ideal safety solution. For example, Jiang et al. [2] used a type of entropy-close to ideal solution (TOPSIS) to establish a safe input scheme evaluation model and calculate the safety input's best configuration. Song et al. [4] used the order relation method and expert scoring method to determine the weight, and the

Cobb-Douglas (C-D) production function to construct the safety input–output function relationship. Guo et al. [5] comprehensively worked out the optimal scheme of safety input allocation by using grey correlation degree analysis, partial correlation analysis and the C-D production function model. Based on the literature, Ye et al. [6] identified widely used safety input indicators and used data envelopment analysis to build a safety output efficiency assessment model and evaluate the efficiency methods of various safety inputs and outputs. Various academics have also proposed some strategies for identifying safety input indicators: Wang et al. [7] used cluster analysis to pick the primary indications of safety input. By studying the rationale and accuracy of the assessment indicators of safety input, Li et al. [8] created an evaluation indicator system comprising three Level Two and 16 Level Three indicators. From the standpoint of a safety system, Zhang et al. [9] built a safety input assessment indicator system. Xin [10] converted the safety decision in engineering to a mathematical problem, establishing three robust optimization models using robust optimization technology, thereby providing safety input decision-making models for different enterprise risk scenarios. Chao [11] presented the economic optimization model and relative optimization algorithm based on CBA and CEA for the process industry. Roy and Gupta [12] created a nonlinear optimization model for safety input. Son et al. [13] established a safety input optimization model that aims to minimize the overall safety input expenditure of coal mining enterprises. Compared with foreign countries' research on industry enterprises, many studies of coal mine enterprises' safety input have been conducted in China. Although research on safety input has reached a relatively mature stage in China [14], China lags far behind the United States in terms of coal mine safety [15]; there is currently less relevant study of the influence of various safety input models on safety output.

To sum up, the current studies on the optimization of safety input structure have all established only one indicator structure [2,4–6,10–13] for optimization. In order to better increase the efficiency of safety inputs, this research uses two safety input structures for comparison for the first time. The two safety input structures are obtained by combining references to previous security input construction methods and models [2,16–18]. The research objectives are twofold: in contrast to the current situation, where only one indicator structure has been used, two safety input structures were used to verify that different indicator structures make different optimization assignments for safety input. While the C-D production function was used to construct a function model, the Mean Squared Error (MSE) of the two obtained regression functions were compared for the first time to illustrate that when selecting safety input indicators for modeling, the regression function that minimizes the MSE should be selected. The regression function with the lower MSE is not only closer to reality, but also obtained a better safety input assignment. The research discusses that when the C-D production function is used to model and the accident loss function equation is obtained by multiple linear regression, if the safety input indicator structure can be refined, the optimal solution of the accident loss function is lower and the coal mine enterprise's safety input efficiency also can be raised to a higher extent.

Thus, the analytic hierarchy process and the entropy weight method (AHP-EWM) is used to assign weight comprehensively to each indicator, and the C-D production function is applied to construct the accident loss function model. Finally, the optimal accident losses of functions are compared and the fitness quality of accident loss functions is examined using MSE. While constructing accident loss functions, the preventive safety input is assumed to be constant, minimizing the accident losses of the enterprise by changing the assignment of each safety input in the safety input structure. Simultaneously, two safety input structures of the same coal mine enterprise are constructed according to the various hierarch and the number of indicators of the preventive safety input, and the optimal accident loss and MSE of the two safety input structures are compared under the same preventive safety input expenditure.

2. Construction of Safety Input Structure of Coal Mine

2.1. Establishment of a Hierarchical Structure of Safety Input in Coal Mine Enterprises

At the moment, there is no consistent definition of safety input, and there are variances in the separation of various elements of safety input. Safety input [8,16] refers to the total amount of money spent by a company to ensure the safety and health of its personnel during the manufacturing process. Safety input [8,19] refers to the total of a country's or enterprise's safety-related costs, such as expenditures on safety measures, personal protective equipment and occupational illness prevention, among other things. The human, material and financial resources invested to manage hazard sources in the manufacturing process, prevent possible accidents and provide safe production conditions are referred to as safety input [8,20]. The term "safety input" [2] refers to the amount of human, material and financial resources expended during the production and operation processes to assure the safe production of enterprises, eliminate possible mishaps and lower the fatality rate. The State Administration of Work Safety has split the safety expenditures of coal enterprises into 10 categories in the Measures for the Administration of Extraction and Use of Enterprise Safety Production Expenses. According to Duan et al. [17], the safety expenditures of coal firms are divided into two categories: safety input and accident loss. In terms of enterprise safety input, coal mining businesses may be classified into a variety of safety input variables based on the current condition. Professor Mei [16] classified safety input into five categories: safety technology, industrial cleanliness, safety education, personal protective equipment, daily management and labor expenses. When researching the logic of safety input structure, Zhao et al. [18] separated safety input into first-level indicators and second-level indicators, with the first-level indicators including personnel, safety technology and safety management input.

In conjunction with existing safety input structure and references, the research constructed the evaluation indicator system of coal mine safety inputs using Jiang et al. [2] and the TOPSIS method based on the entropy weight of coal mine safety input decision analysis. The safety input of coal mining enterprises is divided into five major categories—safety science and technology, safety engineering, safety equipment, safety management and safety education and training—and each category can be further subdivided into one or more safety indicators. Enterprise safety input has an effect on an enterprise's safety output to a certain extent, which can include impairment and value-added output. The former relates to accident economic damage, while the latter refers to coal yield. To clearly demonstrate the link to the resulting safety input structure according to the nature of various kinds of safety inputs, the inputs of safety management and safety education and training were integrated into the input in people, and the input of safety technology, safety engineering and safety equipment were integrated into the input in objects.

2.2. Construction of Safety Input Structure

Based on the definition of coal mine safety input and safety cost in Duan et al. [17], given the definition of safety cost and the safety standards of the benchmark, the safety related costs incurred include two parts: preventive safety input and accident loss. The cost of safety input (which is also called total safety input) can be divided into preventive safety input and loss of safety input.

Preventive safety input, also known as safety input, refers to the safety input before the occurrence of an accident and the loss of safety input refers to the total loss caused by an enterprise accident, also known as accident loss. Total safety input includes both loss of safety input and preventive safety input. The following two safety input structures were constructed using the above hierarchical structure of safety input. Safety Input Structure 1 is more detailed than Safety Input Structure 2 and has three more safety input indicators, as shown in Figures 1 and 2.

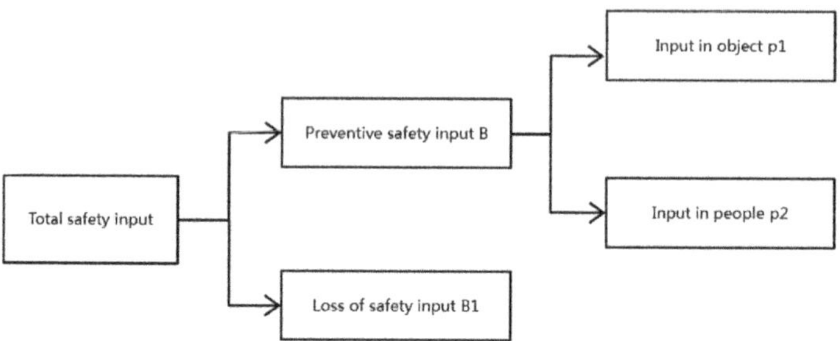

Figure 1. Safety Input Structure 1 of coal mine enterprise.

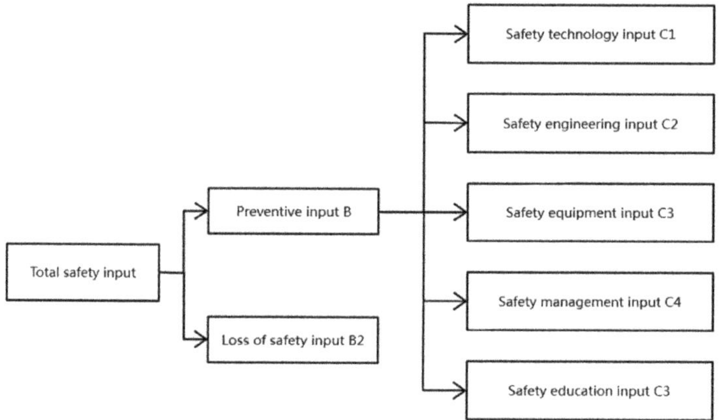

Figure 2. Safety Input Structure 2 of coal mine enterprise.

3. Method of Weight Determination of Safety Input Index

3.1. Weight Determination by the Analytic Hierarchy Process

(1) Construction of a safety input structure comparison matrix

The safety input fund allocation generates countless solutions, and each indicator of safety input is greater than 0. The evaluation indicators of safety input were determined by safety input hierarchy. For each evaluation indicator, quantitative comparison to the expert scoring method was used to establish comparison matrixes, using mutual comparisons of two indicators to determine the relative degree of importance of each evaluation indicator with a structure comparison matrix of n × n. The relative importance value is shown in Table 1.

$$a_{ij} = i\text{'s relative importance}/j\text{'s relative importance}, \tag{1}$$

Table 1. Relative importance table.

Scale a_{ij}	Implication
1	factor i is as important as factor j
3	factor i is slightly more important than factor j
5	factor i is significantly more important than factor j
7	factor i is mightily more important than factor j
9	factor i is extremely more important than factor j
2, 4, 7, 8	the median of the above two adjacent judgments
Reciprocal value	when scale a_{ij} is n, scale a_{ji} is 1/n

(2)   Indicator weight and consistency test

The normalization of the feature vector of the comparison matrix's maximum characteristic root max is designated as W, and the element W is the ordering weight of the elements at the same level to the relative importance of a component at the next level. This is known as hierarchical single ordering. The consistency indicator is calculated using the following formula:

$$CI = (\lambda_{max} - n)/(n - 1),\qquad(2)$$

where n denotes the number of matrix evaluation indicators. Based on the number of indicators in the judgment matrix, look for the average random consistency RI in Table 2.

Table 2. Average stochastic consistency indicator RI.

n	1	2	3	4	5	6	7	8	9	10	11	12	13	14	15
RI	0	0	0.52	0.89	1.12	1.26	1.36	1.41	1.46	1.49	1.52	1.54	1.56	1.58	1.59

The consistency ratio CR is calculated by the following formula:

$$CR = CI/RI,\qquad(3)$$

It is generally believed that when the consistency ratio is:

$$CR < 0.1,\qquad(4)$$

it indicates standard consistency, and its normalized feature vector can be used as the weight vector through the consistency test. Otherwise, the comparison matrix should be reconstructed and the value of a_{ij} adjusted.

3.2. Weight Determination by the Entropy Weight Method

The entropy weight method (EWM) is a thorough, objective approach to weight assignment that determines the indicator weight based on current data while minimizing the variation caused by subjective assignment. Entropy is a measure of uncertainty in information. The lower the entropy number, the more information and weight there is. The specific steps of this model are as follows:

(1)   Data standardization. Standardized processing of data into dimensionless data.

Positive indicators:

$$xij = \frac{xij - \min\{x1j \cdots xmj\}}{\max\{x1j \cdots xmj\} - \min\{x1j \cdots xmj\}}\qquad(5)$$

Matrix:

$$\begin{pmatrix} x_{11} & \cdots & x_{1n} \\ \vdots & \ddots & \vdots \\ x_{m1} & \cdots & x_{mn} \end{pmatrix}$$

where, x_{ij} is the value of the j indicator of the i sample, x'_{ij} is the value of the j indicator of the standardized i sample, n is the number of indicators and m is the number of samples.

(2)   Calculate the entropy value of the j indicator.

$$Pij = \frac{x'_{ij}}{\sum_{i=1}^{m} x'_{ij}}\qquad(6)$$

$$e_j = -k \sum_{i=1}^{m} P_{ij} \ln\left(P_{ij}\right)\qquad(7)$$

$$j = 1, \ldots, n$$

In the formula, P_{ij} is the proportion of x'_{ij} in i sample and

$$K = 1/\ln(n). \tag{8}$$

(3)   Calculate the information entropy of j indicator.

$$dj = 1 - ej \tag{9}$$

$$j = 1, \ldots, n$$

In the formula, d_j is the information entropy of j indicator.

(4)   Determine the weight of each indicator β_j.

$$\beta_j = \frac{d_j}{\sum\limits_{j=1}^{n} d_j} \tag{10}$$

$$j = 1, \ldots, n$$

In the formula, β_j is the weight of j indicator.

3.3. AHP-EWM for Comprehensive Weight Assignment

The analytic hierarchy process (AHP) is highly subjective and based on specialists' practical experience. In contrast, EWM is based only on objective data, which has objective advantages but cannot be applied to the actual situation. To compensate for the lack of single weighting, the AHP-EWM are coupled to assure the dependability of evaluation outcomes. The comprehensive weight of AHP-EWM is calculated as:

$$W_j = \frac{\alpha_j \beta_j}{\sum\limits_{j=1}^{n} \alpha_j \beta_j} \tag{11}$$

In the formula, α_j is the weight of the j indicator that was calculated by the analytic hierarchy process; β_j is the weight of the j indicator that was calculated by the entropy weight method.

4. Establishment of the Accident Loss Model Based on the C-D Production Function

4.1. Cobb-Douglas Production Function

The C-D production function, proposed by American mathematician C.W. Cobb and American economist Paul H. Douglas in the 1930s, is widely used in economic quantitative analysis and can be used to analyze the relationship between input and output under certain conditions of time.

The original form of the C-D production function is as follows:

$$Y = AK^{\alpha} L^{\beta} \tag{12}$$

In the formula, Y is the total output, K is the total capital input, L is the total labor input, A is the constant determined by the technical conditions of production in a certain era, α is the capital elasticity coefficient and β is the labor elasticity coefficient.

4.2. Establishment of Accident Loss Models

When preventive safety input remains constant in a safety input model, total safety input depends on loss of safety input, and there is a certain functional relation between loss of safety input and each preventive safety input's safety input indicator. The word "model" means a set of relationships between two or more variables. These relationships can be expressed in terms of mathematical equations (34). The C-D production function [4,21,22] is used to simulate the functional relationship between safety input indicators and loss of safety input.

(1) According to the Safety Input Structure 1 of coal mine enterprises in Figure 1 and C-D production function, the accident loss multiple regression model S_1 was constructed [4,21]:

$$B_1 = F(p1, p2) = A_1 P_1^{\alpha 1} P_2^{\alpha 2} \tag{13}$$

In the above formula, p_1 was the input in people and p_2 was the input in objects; α_1 and α_2 were the capital elasticity coefficient corresponding to p_1 and p_2, respectively; and B_1 was the loss of safety input (accident loss) of Structure 1. At this point, Safety Input Structure 1's total safety input was $B + B_1$.

(2) According to the Safety Input Structure 2 of coal mine enterprises in Figure 2 and C-D production function, the accident loss multiple regression model S2 was constructed [4,21]:

$$B_2 = F(x1, x2, x3, x4, x5) = A_2 x_1^{\beta 1} x_2^{\beta 2} x_3^{\beta 3} x_4^{\beta 4} x_5^{\beta 5} \tag{14}$$

In the formula above, x_1 represented the input of safety technology, x_2 represented the input of safety engineering, x_3 represented the input of safety equipment, x_4 represented the input of safety management and x_5 represented the input of safety education. β_1, β_2, β_3, β_4 and β_5 corresponded to the capital elasticity coefficient of safety science and technology input, safety engineering input, safety equipment input, safety management input and safety education input, respectively. B_2 was the loss of safety input (accident loss) of Structure 2. At this time, the total safety input of Structure 2 was $B + B_2$.

4.3. Optimization of Safety Input Fund Assignmentf

First, the data of the enterprise safety input and loss of safety input were processed into logarithmic form using Excel, and then multiple linear regression was performed using Matlab to determine the accident loss models S_1 and S_2 corresponding to Safety Input Structure 1 and 2. The functional relationship between each safety input indicator and the accident loss was obtained by multiple linear regression analysis [23]. The accident loss models S_1 and S_2 were obtained using the same statistics and modelling method. Between the two models, model S_2 had more specific indicators than model S_1. We then took the mean value of the preventive safety input of the coal mine enterprise and inserted it in the obtained function to calculate the minimum accident loss under constraints, to determine the optimal assignment of safety input and to compare the minimum values of accident loss.

4.4. Comparison of Accident Loss Regression Functions by MSE

Few studies have been conducted to validate the obtained functions for modeling using the C-D production function. MSE is a common metric for testing regression functions, and its value provides a visual indication of how well the function fits the actual data.

The mean squared error is the expected value of the squared difference between the parameter estimate and the true value of the parameter, denoted as MSE.

The MSE is calculated by the following formula:

$$MSE = \frac{1}{N} \sum_{i=1}^{n} (y_i - \hat{y}_i)^2 \tag{15}$$

5. Application Analysis of Method

5.1. Safety Input Data of a Coal Mine Mnterprise

Simulated and analyzed using the safety input details of a large state-owned coal mine enterprise, Table 3 [24] shows the detailed statistical data of different safety element inputs and the accident economic loss of the state-owned coal mine enterprise from 2001 to 2010.

Table 3. Safety input and safety output statistics of a coal mine enterprise from 2001 to 2010.

Year	Input in Safety Education and Training/ 10,000 RMB	Input in Safety Science and Technology/ 10,000 RMB	Input in Safety Management/ 10,000 RMB	Input in Safety Equipment/ 10,000 RMB	Input in Things/ 10,000 RMB	Input in People/ 10,000 RMB	Accident Economic Loss/ 10,000 RMB
2001	203.12	168.08	1257.04	1687.67	2166.21	1460.16	983.88
2002	223.58	186.78	1400.71	1596.82	2127.84	1624.29	962.44
2003	310.16	205.32	1270.13	1430.32	1989.37	1580.29	1066.96
2004	362.38	311.73	1390.75	1073.97	1841.85	1753.13	1016.27
2005	385.37	344.61	1645.56	1032.21	1838.2	2030.93	985.58
2006	391.85	358.95	1859.71	1010.93	1856.83	2251.56	889.35
2007	428.67	473.06	1768.85	1052.52	2015.16	2197.52	923.68
2008	496.03	489.33	1730.91	1349.57	2356.77	2226.94	880.27
2009	502.31	490.23	1859.95	1513.58	2523.33	2362.26	913.81
2010	509.38	495.95	1895.63	1965.58	2983.92	2405.01	882.39

5.2. Weight Determination of the Safety Input Indicators

(1) According to the Safety Input Structure 1 and Safety Input Structure 2 of coal enterprises, the expert scoring method was adopted to establish the comparison matrix B_1-P_j and B_2-C_j. The obtained results are shown in Tables 4 and 5, and the consistency test was conducted. According to the expert scoring results, the weight was calculated by the geometric evaluation method.

Table 4. Comparison matrix of B_1-P_j.

B_1	P_1	P_2	W_P
P_1	1	2	2/3
P_2	1/2	1	1/3

Table 5. Comparison matrix of B_2-C_j.

B_2	C_1	C_2	C_3	C_4	C_5	W_C
C_1	1	1/3	1/5	1/7	1/2	0.0529
C_2	3	1	1/2	1/3	2	0.1547
C_3	5	2	1	1/2	3	0.2659
C_4	7	3	2	1	4	0.4322
C_5	2	1/2	1/3	1/4	1	0.0942

The consistency test revealed that:

$$CR_1 = 0, CR_2 = 0.008. \tag{16}$$

The consistency ratio CR_1 and CR_2 were less than 0.1, hence the comparison matrixes were consistent and the feature vectors equaled the weight vectors.

(2) The weight was determined by the entropy weight method. The results of the entropy weight method assignment obtained by Python programming calculations are shown in Tables 6 and 7:

Table 6. Model 1 Entropy weight method assignment results.

	Input in Objects	Input in People
The weight of entropy weight method	0.6764	0.3236

Table 7. Model 2 Entropy weight method assignment results.

	Safety Science and Technology	Safety Engineering	Safety Equipment	Safety Management	Safety Education
The weight of entropy weight method	0.2009	0.1448	0.2911	0.2085	0.1547

(3) AHP-EWM comprehensive empowerment: Python programming was used to calculate the final weights of each safety input indicator. The results are shown in Tables 8 and 9:

Table 8. Comprehensive empowerment results of Model 1.

	Input in Objects	Input in People
The comprehensive weight	0.8070	0.1930

Table 9. Comprehensive empowerment results of Model 2.

	Safety Science and Technology	Safety Engineering	Safety Equipment	Safety Management	Safety Education
The comprehensive weight	0.0494	0.1042	0.3598	0.4188	0.0677

In Safety Input Structure 1, the ranking of the relative importance of the comprehensive weight assignment was $p_1 > p_2$.

In Safety Input Structure 2, the ranking of the relative importance of the comprehensive weight assignment was safety management C_4 > safety equipment C_3 > safety engineering C_2 > safety education C_5 > safety science and technology C_1.

5.3. Analysis of the Extreme Value of the Accident Loss Function by Matlab Software

(1) Before multivariate regression of the accident loss function, logarithms of Formulas (13) and (14) corresponding to Safety Input Structure 1 and Safety Input Structure 2, respectively, were taken and the results were as follows:

Model S_1:

$$\ln B_1 = \ln A_1 + \beta_1 \ln P_1 + \beta_2 \ln P_2 \tag{17}$$

Model S_2:

$$\ln B_2 = \ln A_2 + \beta_1 \ln X_1 + \beta_2 \ln X_2 + \\ \beta_3 \ln X_3 + \beta_4 \ln X_4 + \beta_5 \ln X_5 \tag{18}$$

(2) The datasets from 2001 to 2010 were sorted into pairs by Excel, and multiple linear regression was performed on Equations (17) and (18) by Matlab programming [25]. After obtaining the parameters, the accident loss functions were obtained as follows:

Model S_1:

$$B_1 = 14338.64 P_1^{-0.2496} P_2^{-0.107368} \tag{19}$$

Model S_2:

$$B_2 = 24247.7 x_1^{-0.081823} x_2^{-0.1618} x_3^{-0.2544} x_4^{-0.3363} x_5^{-0.1191} \tag{20}$$

(3) Taking accident loss minimization as the goal, the average safety input of 4159.157 million RMB over the 10-year period 2001–2010 and the minimum economic loss of

950.463 million RMB as the condition, the Matlab software was used to calculate the extreme values of Equations (19) and (20). The minimum values and corresponding safety input indicators values of the accident loss functions were determined. The results of the two models were calculated as follows:

Minimum value of B_1 was 910.894 million RMB, P_1 was 1251.200 million RMB and P_2 was 2907.900 million RMB.

Minimum value of B_2 was 679.770 million RMB, x_1 was 356.00 million RMB, x_2 was 729.400 million RMB, x_3 was 111.140 million RMB, x_4 was 144.557 million RMB and x_5 was 516.600 million RMB.

The results indicated that B_1's minimum value > B_2's minimum value.

The optimal results of comparing the two types of safety input structure indicates that the optimal accident loss of Safety Input Structure 2 was significantly less than the optimal accident loss of Safety Input Structure 1. Although these two optimization results were the optimal solution of two safety input structures that were modeled by the C-D production function, they were not necessarily the optimal solution of safety input in coal enterprises.

5.4. MSE of Two Accident Loss Functions

The MSE of the two accident loss functions was calculated separately by the MSE formula and the results were as follows:

$$\text{Accident Loss Function 1's MSE of Model } S_1: \text{MSE}_1 = 1763.681538$$

$$\text{Accident Loss Function 2's MSE of Model } S_2: \text{MSE}_2 = 392.7999477$$

The MSE_1 of function 1 was found to be significantly larger than that of function 2. If the MSE is less, the more the function reflects the input–output situation of production; that is, the production function should minimize the MSE. Therefore, the accident loss function 2 should be chosen, and the corresponding safety input indicator structure 2.

5.5. Discussion

In the article, two safety input structures were constructed for the same coal mine, modeled using the C-D production function, and the analytic formula of the accident loss function was obtained by multiple regression analysis. The optimal outputs (least loss) and MSEs of the two accident loss functions were obtained and compared, respectively. The comparison reveals that the MSE of the loss function of the higher output (less loss) Safety Input Structure 2 is also smaller.

Generally, when constructing an accident loss function, we expect the extreme value of this function to be the least possible, which means that the company can save money on accident losses. The optimal accident loss for Safety Input Structure 2 was 679.770 million RMB and the optimal accident loss for Safety Input Structure 1 was 910.894 million RMB. When optimizing the safety input allocation of the company, we can give preference to Safety Input Structure 2. The MSE_2 of Accident Loss Function 2 was 392.7999477 and the MSE_1 of Accident Loss Function 1 was 1763.681538, yielding $\text{MSE}_2 < \text{MSE}_1$. The fitness of Accident Loss Function 2 was better than that of Accident Loss Function 1, and this result was consistent with the comparison of optimal accident losses.

To summarize: (1) When modeling with the C-D production function, the calculated optimal configuration of safety inputs will be different because of the different safety input structures constructed; (2) The quality of the fitness of the function can be checked by calculating the MSE of the function, and the function with a lower MSE value has a better fit and a better calculated optimal solution; and (3) At present, when using the C-D production function for the optimal solution, the safety input structure construction is considered less, and in the future, when using the C-D production function for modeling, it can be combined with structure optimization.

6. Conclusions

In this paper, the models S_1 and S_2 are compared. The two models use data from a coal mine enterprise to simulate the enterprise's accident loss function. In addition, AHP-EWM is employed to assign the weight of each safety input indicator of safety input structure, the C-D production function is utilized to simulate the coal enterprise accident loss function model and Matlab software is used for multiple regression analysis and solution. Due to the varied degrees of refinement of preventive safety input, the optimum solution of the accident loss model of safety input structure presents different results, hence the main conclusions can be summarized as follows.

Comparing safety input structure and two functions of the structure illustrates the more specific indicators, and more refined structures are better when modeling using the Cobb-Douglas model with limited data.

By comparing the MSE of two safety input structures and the optimal configuration of the corresponding accident loss function, we selected a better structure, which is safety input indicator structure 2. This indicates that Safety Input Structure 2 is an effective structure, so we can apply this method to identify the quality of the constructed structure.

Compared to modeling using only the C-D production function, the MSE can be used as a supplement to determine the degree of consistency of the regression function with regard to the actual situation, which provides a reference for the use of the C-D production function.

In the digital era in the future, enterprises can obtain specific internal safety input-related data and determine the best method of allocating safety funds through programming and modeling. Currently, technical conditions cannot be upgraded to improve the utilization of safety input funds effectively in order to help reduce accident losses.

The optimization of the coal firms' safety input structure still has to be strengthened. An in-depth study of the safety input structure of coal enterprises requires the use of more detailed safety input-related data. The statistics from a longer period of years can simulate more accurate accident loss function. In the future, coal enterprises can apply the refined safety input structure and the more accurate function to provide a scientific basis for the allocation and decision-making of safety input funds.

Author Contributions: Conceptualization, J.-Y.Z. and X.-Z.S.; methodology, J.-Y.Z.; validation, S.Z.; formal analysis, J.-Y.Z.; data curation, J.-Y.Z.; writing—original draft preparation, J.-Y.Z.; writing—review and editing, J.-Y.Z. and X.-Z.S.; visualization, J.-Y.Z.; supervision, S.Z. and X.-Z.S. All authors have read and agreed to the published version of the manuscript.

Funding: This research received no external funding.

Institutional Review Board Statement: Not applicable.

Informed Consent Statement: Not applicable.

Data Availability Statement: Not applicable.

Conflicts of Interest: The authors declare no conflict of interest.

References

1. Jiang, X.X.; Li, C.X. Statistical analysis on coal mine accidents in China from 2013 to 2017 and discussion on the countermeasures. *Coal Eng.* **2019**, *51*, 101–105.
2. Jiang, F.C.; Zhou, S.; Wu, Z.; Zhong, J.C.; Zhang, S.H.; Zhang, X.F.T. Analysis of coal mine safety investment decision based on entropy weight-TOPSIS method. *China Saf. Sci. J.* **2021**, *31*, 24–29.
3. Reniers, D.L.L.; Van Erp, H.R.N. *Operational Safety Economics: A Practical Approach Focused on the Chemical and Process Industies*; John Wiley & Sons, Ltd: London, UK, 2016.
4. Song, H.N.; Zhai, X.X.; Ma, J.S.; Duan, Z.W. Research on optimization of coal mine safety investment structure. *Saf. Coal Mines* **2021**, *52*, 256–259.
5. Liang, R.; Guo, H.Y.; Jiang, F. Research on distribution direction of safety investment in enterprise's safety-affecting factors. *China Saf. Sci. J.* **2010**, *20*, 128–133.

6.	Ye, W.T.; Cheng, L.H. Safety input-output efficiency evaluation of coal mining enterprises under high quality development. *J. Xi'an Univ. Sci. Technol.* **2021**, *41*, 700–707.

7.	Wang, S.M. The establishment and application of the comprehensive evaluation model for safety input of coal mines based on AHP and entropy weight. *J. Jinling Inst. Technol.* **2011**, *27*, 8–16.

8.	Li, S.G.; Cheng, L.P.; Jing, X.P.; Xu, C.X. Study on construction method of evaluation index system for safety input of coal mine. *J. Saf. Sci. Technol.* **2009**, *5*, 93–96.

9.	Zhang, X.B.; Zhu, J.F. Coal mine safety input system and its index system framework. *Saf. Coal Mines* **2009**, *40*, 117–120.

10.	Xin, C.L.; Zhang, J.W.; Wu, C.H.; Tsai, S.B. Safety Investment Decision Problem without Probability Distribution: A Robust Optimization Approach. *Math. Probl. Eng.* **2020**, *2020*, 8874057. [CrossRef]

11.	Chao, C.A.; Grab, C. Economic approaches for making prevention and safety investment decisions in the process industry. Science Direct. *Methods Chem. Process Saf.* **2020**, *4*, 355–378.

12.	Roy, S.; Gupta, A. Safety investment optimization in process industry: A risk-based approach. *J. Loss Prev. Process Ind.* **2019**, *63*, 104022. [CrossRef]

13.	Son, K.S.; Melchers, R.E.; Kal, W.M. An analysis of safety control effectiveness. *Reliab. Eng. Syst. Saf.* **2000**, *68*, 187–194. [CrossRef]

14.	Jiang, F.C.; Zhang, X.F.; Li, J.; Jia, S.M. Visual analysis on the research status of safety investment in China. *Coal Econ. Res.* **2021**, *41*, 50–55.

15.	Guo, W.C.; Wu, C. Comparative Study on Coal Mine Safety between China and the US from a Safety Sociology Perspective. *Procedia Eng.* **2011**, *26*, 2003–2011.

16.	Qiu, S.X.; Mei, Q. Analysis of the present situation of investment occupational safety of some factories. *China Saf. Sci. J.* **1995**, *5*, 50–55.

17.	Duan, H.F.; Wang, L.J.; Jin, Q.Z. Delimit safety input and safety cost in coal mine. *China Saf. Sci. J.* **2006**, *16*, 65–70+145.

18.	Zhao, P.F.; He, A.H. Reasonableness evaluation for safety investment structure in coal mine. *Saf. Coal Mines* **2016**, *47*, 227–230.

19.	Luo, Y. *Security Economic*, 3rd ed.; Publishing House of Chemical Industry: Beijing, China, 2004; p. 54.

20.	Chen, B.Z. *Safe Principle*; Metallurgical Industry Press: Beijing, China, 2002; p. 164.

21.	Vilcu, G.E. A geometric perspective on the generalized Cobb-Douglas production functions. *Appl. Math. Lett.* **2011**, *24*, 777–783. [CrossRef]

22.	Nikkhah, A.; Emadi, B.; Soltanali, H.; Firouzi, S.; Rosentrater, K.A.; Allahyari, M.S. Integration of life cycle assessment and Cobb-Douglas modeling for the environmental assessment of kiwifruit in Iran. *J. Clean. Prod.* **2017**, *137*, 843–849. [CrossRef]

23.	Ciulla, G.; D'Amico, A. Building energy performance forecasting: A multiple linear regression approach. *Appl. Energy* **2019**, *253*, 113500. [CrossRef]

24.	Gao, R. Research and application of mathematical model of optimum security investment to coal mine. *China Coal* **2012**, *38*, 105–108.

25.	Available online: https://mbounthavong.com/blog/2019/2/19/cobb-douglas-production-function-and-total-costs (accessed on 1 October 2022).

 *processes* 

Article

Failure Mode Analysis of Intelligent Ship Positioning System Considering Correlations Based on Fixed-Weight FMECA

Xiaofang Luo [1], Haolang He [2], Xu Zhang [2,3,*], Yong Ma [2,3,*] and Xu Bai [4]

[1] School of Economics & Management, Jiangsu University of Science and Technology, Zhenjiang 212003, China
[2] School of Ocean Engineering and Technology, Sun Yat-sen University, Zhuhai 519082, China
[3] Southern Marine Science and Engineering Guangdong Laboratory (Zhuhai), Zhuhai 519082, China
[4] School of Naval Architecture & Ocean Engineering, Jiangsu University of Science and Technology, Zhenjiang 212003, China
* Correspondence: zhangx798@mail.sysu.edu.cn (X.Z.); mayong3@mail.sysu.edu.cn (Y.M.)

Abstract: Currently, intelligent ships are still in the early stages of development in terms of autonomous navigation and autonomous berthing, so almost no source of fault data can be obtained. Conducting an in-depth analysis of the failure modes of intelligent ships is critical to optimizing the design of smart ships and ensuring their normal and safe navigation. In this paper, the fixed-weight Failure Mode Effects and Criticality Analysis (FMECA) is combined with the decision-making trial and evaluation laboratory (DEMATEL) method to analyze the failure modes and effects of intelligent ship positioning systems. This combined method not only overcomes the failure of traditional FMECA methods to differentiate between severity, incidence, and detection rates but also allows the correlation of failure causes to be analyzed, bringing the results of the analysis closer to reality. Through the expert scoring of failure modes, the failure modes of this system are risk-ranked, and the key failure causes of this system are identified. Correlations between the critical failure causes are then considered. According to the analysis results, the high-accuracy attitude sensor was identified as the subsystem with the highest level of risk. Unavoidable, unknown failures and environmental factors were found to be key factors in causing positioning system failures. The conclusions can provide a reference for the design of equipment safety for intelligent ship positioning systems.

Keywords: fixed-weight; FMECA; intelligent ship; positioning system; risk identification

Citation: Luo, X.; He, H.; Zhang, X.; Ma, Y.; Bai, X. Failure Mode Analysis of Intelligent Ship Positioning System Considering Correlations Based on Fixed-Weight FMECA. *Processes* **2022**, *10*, 2677. https://doi.org/10.3390/pr10122677

Academic Editors: Francisco Ronay López-Estrada and Guillermo Valencia-Palomo

Received: 31 October 2022
Accepted: 9 December 2022
Published: 12 December 2022

Publisher's Note: MDPI stays neutral with regard to jurisdictional claims in published maps and institutional affiliations.

1. Introduction

Due to the large potential for saving energy consumption on ships and reducing ship crew, the intelligent ship has become the focus of the industry and the trend in its development. According to the China Classification Society (CCS) [1], an intelligent ship requires the use of sensors, communications, the Internet of Things, and other technical means to automatically sense and obtain information and data about the ship itself, the marine environment, logistics, and ports. Intelligent operation is applied to ship navigation, management, maintenance, and cargo transportation. This makes intelligent ships safer and more environmentally friendly than traditional ships.

The technology of intelligent ships includes intelligent navigation, autonomous berthing, status detection, and fault diagnosis. Currently, research work in intelligent navigation is focused on collision avoidance and algorithmic optimization of path planning. Agnieszka [2] proposes a new deterministic approach using the concept of a trajectory database to calculate the safe, optimal path of a ship, considering its dynamic properties and static and dynamic obstacles. He et al. [3] proposed an open-water intelligent navigation decision method capable of dynamically adapting to system residual errors and random maneuvers of the target vessel. By utilizing the velocity obstacle (VO) theory and dynamic collision avoidance mechanism, the vessel is able to navigate autonomously in an open water environment with multiple static and dynamic objects. Since autonomous berthing requires high ship maneuverability, the existing research work is mainly focused on berthing control tasks and controller design. Lee et al. [4,5] used fuzzy control and LOS algorithms to experiment with a 4 m-long boat to solve the problem of small boat position and navigation accuracy to achieve side booster-assisted berthing. Mizuno et al. [6–8] solved the problem of automatic berthing under uncertainty disturbances using an artificial neural network approach.

Although the technology has received a lot of attention, it still requires development. There are still many unresolved issues, and the weaknesses of intelligent ships cannot be clarified. Intelligent ships, compared with traditional ships, involve more fields and disciplines, and their components are more numerous and complex. The harsh conditions at sea lead to extremely violent ship movements. The equipment, components, signal transmission, and structures on ships are, therefore, also extremely vulnerable to failure [9]. According to the data published by Allianz Global Corporate & Specialty (AGCS) [10], more than 1000 ships of more than 100 gross tons have been lost in the last decade. More than one-third of shipping accidents are caused by mechanical damage or failure, out of more than 20,000 reported in the last 10 years. It is conceivable that the failure of intelligent ship equipment, which relies on equipment and information transfer, can also suffer from similar serious consequences. Currently, there is little data available on the failure of intelligent ship equipment. Therefore, research on risk identification related to equipment must be conducted in order to prevent or mitigate potential hazards. Through Failure Mode Effects and Criticality Analysis (FMECA), finding the weak points of equipment and proposing targeted repair and maintenance strategies can significantly improve the reliability of intelligent ships.

An intelligent ship mainly includes a hull structure, power propulsion system, positioning and navigation system, control system, communication system, and interaction system. Among them, the positioning system plays an extremely significant role in normal operation. Its functions involve ship positioning, external environment sensing and platform state sensing, providing the necessary data sources for motion decision and control as well as its own state monitoring. The causes of the failure of certain subsystems and components in positioning systems have been investigated. Alan et al. [11] studied the effects of data reliability and human error on the Automatic Identification System (AIS) in the positioning system and showed that many input errors in the navigation state are due to personnel memory errors or negligence in performing the required operations. Pallotta [12] and Tsou [13] et al. studied the impact of data redundancy on AIS. Philipp et al. [14] investigated the effects of antenna height and environmental changes on information transmission. If the system fails, serious consequences will occur, such as intelligent navigation deviation, autonomous berthing and unberthing failure, and even collision between the ship and obstacles. Therefore, it is necessary to analyze its failure mode to ensure its safety. However, relevant risk identification research, especially for positioning system equipment, has not been reported. This paper intends to focus on equipment failure mode and its impact on the positioning system.

One of the main methods for failure mode analysis is the FMECA method. The key to failure analysis by the FMECA method is usually based on three risk parameters: severity (S), occurrence (O), and detection (D). The magnitude of the risk priority number (RPN), the product of the three, measures the severity of potential system problems. By prioritizing the high-risk failure modes and guiding maintenance management strategies. Although the traditional FMECA method is widely used, it is still criticized for its many drawbacks. Different values of S, O, and D can get the same RPN value, which is theoretically the same priority as the two, but in practice, the priority of the two risk ranks is different.

Therefore, many enhanced versions have been proposed in the literature. Some scholars have proposed measuring risk in more dimensions. For example, Carmignani [15] suggested the use of a fourth parameter, profitability, in the RPN calculation. Bevilacqua et al. [16] proposed that the RPN can consist of a weighted sum of six parameters (safety, importance of the machine to the process, maintenance cost, failure frequency, downtime, and operating conditions). Other studies combine FMECA methods with other methods. Zammori et al. [17] combined FMEA with analytical network process (ANP) techniques [18] to consider the possible interactions between the main causes of failure. Silvia et al. [19] proposed a method combining reliability analysis and a multi-criteria decision-making approach to improve the maintenance activities of complex systems. Some scholars combine FMECA with Analytic Hierarchy Process (AHP) to solve the problem that the traditional FMECA method cannot distinguish the different weights of risk factors by giving different weights to the evaluation parameters so that the risk ranking of failure modes is closer to the actual results. Braglia [20] proposed the analysis hierarchical method (AHP) [21] to compare pairs of potential failure causes by assuming the classical risk factors S, O, and D and the expected cost caused by the failure as criteria. Xiao et al. [22] proposed a weighted RPN evaluation method, which multiplied the RPN value with the weight parameter representing the importance of the fault causes in the system and then ranked them. Zhang et al. [23] proposed a new method for FMECA failure mode ranking based on incentive variable weight AHP. Li et al. [24] proposed a fixed-weight FMECA method. The method considers that the scales of S, O, and D and their weights are different and designs a normalization method to convert S, O, and D to the same scales as their weights and then generates the RPN of the cause of the failure as well as the failure mode. This method not only improves the problem of different weights of S, O, and D but also solves the sorting problems with the same RPN values. Due to the insufficient ability of traditional AHP to deal with fuzziness, many scholars use fuzzy theory to solve this problem. Many researchers have proven the effectiveness and superiority of fuzzy theory in dealing with fuzzy information. Luqman et al. [25] proposed an FMEA risk assessment technology based on TPFNs and DGMA. Akram et al. [26] proposed a mixed solution of TOPSIS and ELECTRE I with Pythagorean fuzzy information, using the Pythagorean fuzzy weighted average operator to aggregate their independent evaluations into group evaluations. Some studies also combine fuzzy theory and traditional AHP methods to manage the lack of information acquisition on complex problems, such as Liu et al. [27].

However, these FMECA methods do not consider the correlation between the failure factors, making the results obtained from the analysis somewhat one-sided. Many existing studies have proposed solutions to the problem of correlation of structural reliability, mainly involving integral methods [28,29] and numerical simulation (Monte Carlo method) [30]. The integral method could solve the multidimensional integration problem, but the procedure is complicated and not practical when the system composition is large. The Monte Carlo method uses a huge number of samples to simulate the variables obeying the desired distribution, and the more simulations there are, the higher the accuracy. However, it requires a lot of time and computing capacity and is less efficient [31]. Unlike structural faults, equipment faults do not allow for the construction of limit state equations. It is, therefore, difficult to apply reliable indicator vector methods for accurate correlation assessment. In this paper, the decision-making trial and evaluation laboratory (DEMATEL) method is used for the computational analysis of equipment failure correlations. The DEMATEL method was first proposed by Gabus and Fontela [32] of the Battelle Memorial Association in Geneva and aimed to analyze the causal relationships between the elements of a complex system and the degree of mutual influence [33].

Combining the above issues, considering the complex structural composition and many failure modes of the positioning system of intelligent ships, this paper uses a combination of fixed-weight FMECA and DEMATEL to study the system. On the one hand, it improves the problem of unreasonable distribution of S, O, and D weights in the traditional FMECA method, and on the other hand, it also considers the correlation between failure modes and improves the problem of mutual independence among failure causes. The results are closer to reality and increase the credibility of the failure mode analysis.

The remainder of the paper is organized as follows: Section 2 introduces the fixed-weight FMECA method. Section 3 performs FMECA analysis on the positioning system to obtain critical failure modes. Section 4 analyzes the correlation between the key failure modes, and Section 5 gives the conclusions.

2. Method

2.1. Fixed-Weight FMECA Approach to Modeling

FMECA includes Failure Mode and Effects Analysis (FMEA) and Criticality Analysis (CA), which is an analysis technique based on failure modes and targeting the effects or consequences of failures. FMECA is performed by finding all possible failures of the product, analyzing them according to the failure mode, determining the impact of each failure on the operation of the product, and identifying the hazards of the failure mode in the order of the RPN. RPN is the risk prioritization number, whose value is equal to the product of the values of severity (S), occurrence (O) and detection (D), calculated by the following formula:

$$RPN = S \times O \times D \tag{1}$$

Although the traditional FMECA method is widely used in production practice, it still has many drawbacks. For example, the same weights are assigned to severity, occurrence, and detection. However, in practice, the weights of the three in the system are not exactly equal. For some irreparable systems, the weights of factors S and O should be higher than the weights of D. In this paper, the fixed-weight FMECA method [24] is used to eliminate the above effects. This FMECA method generates the RPN of component failure caused by using the severity, incidence, and detection rates of each item and their relative weights. Considering that the scales of each factor (severity, incidence, and detection rate) and their weights are [1, 10] and [0, 1], S, O, and D are converted to the same scales as their weights before calculating the RPN.

Denote β_{Si}, β_{Oi} and β_{Di} as the average value of the severity, occurrence and detection of the fault cause, i, given by the expert. The weights of the factors: $\psi = \begin{bmatrix} K_S & K_O & K_D \end{bmatrix}$. K_S, K_O, and K_D are the weights for severity, occurrence and detection, respectively. Thus, the raw values of severity, occurrence, and detection given by the experts can be expressed as follows:

$$\begin{bmatrix} \beta_{S1} & \beta_{S2} & \cdots & \beta_{Si} & \cdots & \beta_{Sn} \\ \beta_{O1} & \beta_{O2} & \cdots & \beta_{Oi} & \cdots & \beta_{On} \\ \beta_{D1} & \beta_{D2} & \cdots & \beta_{Di} & \cdots & \beta_{Dn} \end{bmatrix} \tag{2}$$

Denote

$$\tilde{\varsigma}_{Kij} = \frac{\beta_{Ki}}{\beta_{Kj}} \tag{3}$$

where, K represents severity, occurrence, and detection.

Therefore, the comparison matrix is attained as:

$$\begin{bmatrix} \tilde{\varsigma}_{K11} & \tilde{\varsigma}_{K12} & \cdots & \tilde{\varsigma}_{K1n} \\ \tilde{\varsigma}_{K21} & \tilde{\varsigma}_{K22} & \cdots & \tilde{\varsigma}_{K2n} \\ \vdots & \vdots & \ddots & \vdots \\ \tilde{\varsigma}_{Kn1} & \tilde{\varsigma}_{Kn2} & \cdots & \tilde{\varsigma}_{Knn} \end{bmatrix} \tag{4}$$

The normalized matrix is defined as:

$$\begin{bmatrix} \phi_{K11} & \phi_{K12} & \cdots & \phi_{K1n} \\ \phi_{K21} & \phi_{K22} & \cdots & \phi_{K2n} \\ \vdots & \vdots & \ddots & \vdots \\ \phi_{Kn1} & \phi_{Kn2} & \cdots & \phi_{Knn} \end{bmatrix} \tag{5}$$

where

$$\phi_{Kij} = \frac{\tilde{\varsigma}_{Kij}}{\sum_{i=1}^{n} \tilde{\varsigma}_{Kij}} \tag{6}$$

The adjusted value of index K of failure cause i is defined as:

$$\gamma_{Ki} = \frac{\sum_{j=1}^{n} \phi_{Kij}}{n} \tag{7}$$

According to Equations (6) and (7), $\gamma_{Ki} \in [0, 1]$, which is the same as the scale of the weight vector of indices $\psi = \begin{bmatrix} K_S & K_O & K_D \end{bmatrix}$.

Hence, the weighted RPN of failure cause i is defined as:

$$RPN_i = \psi \times \Gamma_i = \begin{bmatrix} K_S & K_O & K_D \end{bmatrix} \begin{bmatrix} \gamma_{Si} \\ \gamma_{Oi} \\ \gamma_{Di} \end{bmatrix} \tag{8}$$

The method first transforms the absolute values of S, O, and D into the same values as their weight scales, and the S, O, and D transformed values take values in the range $[0, 1]$. The larger the original S, O, and D values will still be larger after transformation and will not affect the order of RPN. The importance of the cause of failure does not change. When using the fixed-weight FMECA method, the calculation of RPN involves the S, O, and D values as well as the weights. However, the range of the two values is different, $[1, 10]$ and $[0, 1]$, respectively. This could bias the final calculated RPN. The consistent scale conversion of both before calculating the RPN prevents each type of parameter from affecting the results more than the other.

2.2. Selection of Weights

For intelligent ships, we should pay more attention to the consequences of their failure occurrences and the value of severity. This is because failures with low incidence can still occur, and detectable failures can still occur. The occurrence of faults can affect certain functions of intelligent ships and even lead to major accidents if certain critical units are affected. Therefore, the weight value of importance should be the largest among the three. At the level of occurrence and detection, the likelihood of failure occurrence is significantly more important than detection. Whether a fault occurs or not is the combined result of the physical properties of the system and the internal and external effects, and it does not change with whether the fault is detectable or not, so the occurrence degree should be given more weight than the detection degree. The selection of risk evaluation index weights should follow the principle that the severity degree is greater than the occurrence is greater than the detection.

According to the basic guideline that the selection of risk evaluation index weights should follow that the severity degree is greater than the occurrence degree is greater than the detection degree, and according to reference [23], combined with the risk characteristics of the intelligent ship positioning system, the failure mode analysis conducted in this paper selects the following weight vector:

$$\psi = \begin{bmatrix} 0.40 & 0.35 & 0.25 \end{bmatrix} \tag{9}$$

2.3. DEMATEL Method

The reason for the correlation assessment of hazardous units is that failure modes that are closely linked to other failures may lead to more severe consequences than relatively independent failure modes. Therefore, identifying correlations between failure modes can lead to more reliable results. The main steps of the DEMATEL method are as follows:

(1) Establishing assessment criteria. The degree of correlation between the assessment elements is quantified by means of expert scoring. The assessment scale ranges from 0 to 4, in the order of no impact, very low impact, low impact, high impact, and very high impact. The scoring values are entered into a direct impact matrix (10).

$$A = \begin{bmatrix} 1 & a_{12} & a_{13} & \cdots & a_{1(n-1)} & a_{1n} \\ a_{21} & 1 & a_{23} & \cdots & a_{2(n-1)} & a_{2n} \\ \cdots & \cdots & \cdots & \cdots & \cdots & \cdots \\ \cdots & \cdots & \cdots & \cdots & \cdots & \cdots \\ a_{(n-1)1} & a_{(n-1)2} & a_{(n-1)3} & \cdots & 1 & a_{(n-1)n} \\ a_{n1} & a_{n2} & a_{n3} & \cdots & a_{n(n-1)} & 1 \end{bmatrix} \tag{10}$$

where a_{ij} indicates the degree of influence of factor i on factor j.

(2) The direct relationship matrix is normalized by Equations (11) and (12) such that each value of the matrix lies between [0, 1].

$$S = \max_{1 \leq i \leq n} \sum_{j=1}^{N} a_{ij} \tag{11}$$

$$K = \frac{A}{S} \tag{12}$$

(3) The total impact matrix M is obtained by Equation (13), where I denotes the unit array.

$$M = K(I - K)^{-1} \tag{13}$$

(4) The sum of each column and each row of the total impact matrix is calculated by Equations (14) and (15), denoted as D and R, respectively.

$$D_i = \left[\sum_{i=1}^{n} m_{ij} \right]_{1 \times n} \tag{14}$$

$$R_i = \left[\sum_{j=1}^{n} m_{ij} \right]_{n \times 1} \tag{15}$$

where, $M = m_{ij}$, $i, j = 1, 2, \ldots, n$.

(5) Perform the calculation of $Ri + Dj$, $Ri - Dj$. $Ri + Dj$ indicates the extent to which factor i plays a role in the problem, with a positive $Ri - Dj$ indicating that factor i assigns influence to other problems and a negative $Ri - Dj$ indicating that factor i receives influence from other factors.

3. Fixed-Weight FMECA of Positioning Systems

3.1. Positioning System Introduction

The intelligent ship positioning system is built for the needs of digitalization, networking, visualization, and intelligence in ship positioning. It can realize the rapid circulation of ship information and effective management of ships by using the BeiDou positioning system, automatic control, and other technologies, combined with the data update of dynamic changes in ocean climate, to conduct emergency command of ocean ships.

The positioning system includes five subsystems: a high-precision attitude sensor, the BeiDou positioning system, the electronic chart display information system (ECDIS), the automatic identification system (AIS), and a mobile communication receiver. The structure diagram of the positioning system is shown in Figure 1. The high-precision attitude sensor is used to capture dynamic reference signals. The BeiDou positioning system is used to receive satellite positioning signals and, in conjunction with ECDIS, to precisely locate the ship's position at sea. The AIS and mobile communication receiver are responsible for sending and receiving to the ships and shore stations in the nearby waters so that the neighboring ships and shore stations can grasp the dynamic and static information of all the ships in the nearby sea and can immediately talk to each other for coordination. They can also calculate the voyage heading and take necessary avoidance actions to effectively ensure the safety of ship navigation.

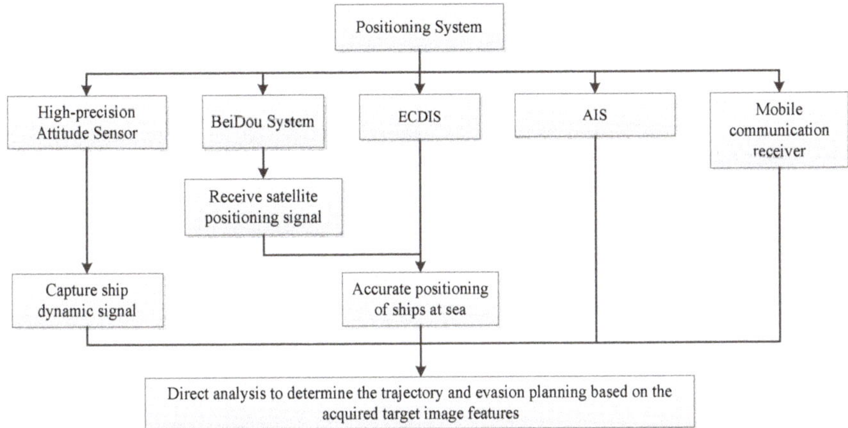

Figure 1. Schematic diagram of positioning system structure.

3.2. FMECA of Positioning Systems

Based on the composition and working principle of the positioning system and the human factors and environmental influences identified in scholarly research, this section will analyze the reliability of the intelligent ship positioning system based on the fixed-weight FMECA method [24]. The scoring of each failure mode and failure cause of the intelligent ship positioning system is based on the evaluation table, and the results are shown in Table 1. Detailed failure causes, transformed values for severity, occurrence and detection, and risk number ranking for each failure mode are also given in Table 2.

Table 1. Evaluation table of failure model risk evaluation indicators.

Score	Severity (S)	Occurrence (O)	Detection (D)
1–3	The device function is disturbed by faults but basically does not affect the overall function	The probability of occurrence is less than 1/1000	The probability that the failure can be detected is greater than 80%
4–5	The device loses some functionality and has a partial impact	The probability of occurrence is between 1/100 and 1/1000	The probability that the failure can be detected is greater than 60%
6–7	The device functionality is severely impacted	The probability of occurrence is between 1/10 and 1/100	The probability that the failure can be detected is 50%
8–10	Total loss of device function	The probability of failure is greater than 1/10	The probability of failure detection is less than 20%

Table 2. FMECA of the positioning system of intelligent ships.

Subsystems	Unit	Failure Mode	Failure Effects	Code	Cause of Failure	S	O	D	Conversion Value of S, O, and D			RPN	Rank
									S	O	D		
High-precision attitude sensor	Tri-axis gyroscope	A Drift fault	Severe lack of attitude measurement accuracy	A1	Interference torque due to friction of gyroscope elements	5	6	6	0.00684	0.01342	0.01575	0.0114	19
				A2	The gyro has a residual unbalance moment	5	3	7	0.00684	0.00671	0.01837	0.0097	44
				A3	Thermal decomposition and deformation	5	7	2	0.00684	0.01566	0.00525	0.0095	47
				A4	Component damage	5	4	5	0.00684	0.00895	0.01312	0.0091	57
				A5	Static imbalance caused by imperfect assembly	5	2	7	0.00684	0.00447	0.01837	0.0089	60
		B Vibration fault	Gyroscope vibration, attitude measurement accuracy is seriously lacking	B1	Damaged gyro motor	6	3	6	0.00821	0.00671	0.01575	0.0096	46
				B2	Rotor imbalance	6	7	6	0.00821	0.01566	0.01575	0.0127	4
				B3	Insufficient power supply voltage	6	6	3	0.00821	0.01342	0.00787	0.0099	40
				B4	Bearing wear	6	6	2	0.00821	0.01342	0.00525	0.0093	51
		C High-temperature shock failure	Affect the normal operation of the gyroscope	C1	High-temperature environment influence	6	6	2	0.00821	0.01342	0.00525	0.0093	51
		D Poor contact	Discontinuous gyroscope signal	D1	Vibration and shock factors	4	7	6	0.00547	0.01566	0.01575	0.0116	9
		E Loose and worn components	Gyroscope damage, not working properly	E1	Vibration and shock factors	7	7	4	0.00958	0.01566	0.01050	0.0119	6
		F Rotor imbalance	Attitude measurement accuracy is severely lacking	F1	High-intensity and high broadband electromagnetic interference	7	8	5	0.00958	0.01790	0.01312	0.0134	1
	Three-axis accelerometer	G Stuck gyro motor	The gyroscope does not work properly	G1	Humidity factors	8	8	4	0.01094	0.01790	0.01050	0.0133	2
		H Low sensitivity	Cannot accurately judge the ship's status	H1	Temperature effects	6	8	2	0.00821	0.01790	0.00525	0.0109	21
				H2	Hardware decline of sensors	6	6	7	0.00821	0.01342	0.01837	0.0126	5
				H3	Data connection failure	6	5	2	0.00821	0.01119	0.00525	0.0085	67
	Electronic compass	I Measurement result deviation, difficult to correct	Ship's direction deviation, unable to avoid collision	I1	Magnetic field interference	8	7	5	0.01094	0.01566	0.01312	0.0131	3
				I2	Hardware failure	8	3	6	0.01094	0.00671	0.01575	0.0107	25
		J Voltage is too low	Ship's direction deviation, unable to avoid collision	J1	Insufficient power supply	7	3	2	0.00958	0.00671	0.00525	0.0075	88
	ARM Processor	K No output of signal	Ship's direction deviation, unable to avoid collision	K1	Processor hardware damage	9	5	4	0.01231	0.01119	0.01050	0.0115	17
				K2	Circuit connection failure	9	6	3	0.01231	0.01342	0.00787	0.0116	10

Table 2. *Cont.*

Subsystems	Unit	Failure Mode	Failure Effects	Code	Cause of Failure	S	O	D	Conversion Value of S, O, and D			RPN	Rank
									S	O	D		
BeiDou System	BeiDou Navigation Satellite	L Disconnection	The entire BeiDou positioning system of a ship fails	L1	Mechanical Failure	9	2	2	0.01231	0.00447	0.00525	0.0078	84
				L2	Space wear and tear	9	2	2	0.01231	0.00447	0.00525	0.0078	84
	Master Control Station	M disconnected	The entire BeiDou positioning system of a ship fails	M1	Mechanical failure	9	2	2	0.01231	0.00447	0.00525	0.0078	84
				M2	Wear and tear	9	2	2	0.01231	0.00447	0.00525	0.0078	84
	Terminal satellite receiver	N disconnected	Vessel BeiDou positioning system failure	N1	Interference from various external factors	9	3	3	0.01231	0.00671	0.00787	0.0092	54
	ARM processor	O No signal output	Vessel's direction deviates, unable to avoid collision	O1	Processor hardware damage	9	5	4	0.01231	0.01119	0.01050	0.0115	17
				O2	Circuit connection failures	9	6	3	0.01231	0.01342	0.00787	0.0116	10
	Multi-interface terminal	P Poor contact	Poor contact or disconnection of terminal	P1	Environmental erosion	6	4	3	0.00821	0.00895	0.00787	0.0084	72
				P2	Wear and tear	6	4	3	0.00821	0.00895	0.00787	0.0084	72
	Cable	Q Poor contact or disconnection	Poor contact or disconnection of terminal	Q1	Overload by tensile stress	6	4	3	0.00821	0.00895	0.00787	0.0084	72
	Display screen	R System dead	Unable to navigate and avoid collision	R1	Software design defects	8	2	6	0.01094	0.00447	0.01575	0.0099	41
		S Partial loss of function	Unable to navigate and avoid collision	S1	Imperfect software function design	8	4	4	0.01094	0.00895	0.01050	0.0101	37
		T Signal failure	Unable to navigate and avoid collision	T1	Part of the industrial control machine board is damaged	8	7	3	0.01094	0.01566	0.00787	0.0118	8
	Display screen	U No signal output	Unable to navigate and avoid collision	U1	Control panel part of the circuit damage	9	4	2	0.01231	0.00895	0.00525	0.0094	48
				U2	Display screen damage	9	3	3	0.01231	0.00671	0.00787	0.0092	54
	Power connector	V Bad or broken contact	The whole system cannot operate without electricity	V1	Aging or external wear and tear	6	4	3	0.00821	0.00895	0.00787	0.0084	72
		W No signal output	Cannot send this radar information to external systems	W1	The interface part of the circuit is damaged	9	6	3	0.01231	0.01342	0.00787	0.0116	10
	Enclosed cover	X cracked	The terminal is susceptible to moisture or dust erosion	X1	Aging	4	6	3	0.00547	0.01342	0.00787	0.0089	61
				X2	External wear and tear	4	6	3	0.00547	0.01342	0.00787	0.0089	61
ECDIS	Image display	Y Black screen	The image cannot be displayed	Y1	Circuit Failure	6	5	3	0.00821	0.01119	0.00787	0.0092	56
				Y2	Display failure	6	6	2	0.00821	0.01342	0.00525	0.0093	51
				Y3	System program error	6	3	4	0.00821	0.00671	0.01050	0.0083	79

Table 2. *Cont.*

Subsystems	Unit	Failure Mode	Failure Effects	Code	Cause of Failure	S	O	D	Conversion Value of S, O, and D			RPN	Rank
									S	O	D		
		Z No signal output	The image cannot be displayed	Z1	Part of the control panel circuit is damaged	6	3	4	0.00821	0.00671	0.01050	0.0083	79
				Z2	Display screen damage	6	3	4	0.00821	0.00671	0.01050	0.0083	79
				AB1	Circuit failure	8	5	3	0.01094	0.01119	0.00787	0.0103	35
	Text Display	AB Black screen	Data cannot be reflected correctly	AB2	Display screen malfunction	8	6	2	0.01094	0.01342	0.00525	0.0104	30
				AB3	System program error	8	3	4	0.01094	0.00671	0.01050	0.0094	49
		AC Data error	Cannot reflect data correctly	AC1	System program error	8	3	4	0.01094	0.00671	0.01050	0.0094	49
				AC2	Data detection error	8	4	5	0.01094	0.00895	0.01312	0.0108	22
Navigation system interface		AD Poor GPS positioning accuracy	Cannot locate the vessel accurately	AD1	Poor sea conditions	7	6	3	0.00958	0.01342	0.00787	0.0105	28
				AD2	Interference from outside	7	5	4	0.00958	0.01119	0.01050	0.0104	31
Radar interface		AE Radar detection interference	Reduced detection accuracy	AE1	GPS is seriously interfered with	8	4	5	0.01094	0.00895	0.01312	0.0108	22
				AE2	Frequency synthesis module failure	8	5	5	0.01094	0.01119	0.01312	0.0116	13
		AF No signal output	Unable to send this radar information to external systems	AF1	Fuse breakage, AC/DC module damage	8	5	5	0.01094	0.01119	0.01312	0.0116	13
Radar interface				AF2	The interface part of the circuit is damaged	8	5	5	0.01094	0.01119	0.01312	0.0116	13
		AG cannot be recognized	Cannot identify tracking vessel information	AG1	Transmission data loss	9	1	2	0.01231	0.00224	0.00525	0.0070	91
				AG2	Poor wireless communication signal	9	2	3	0.01231	0.00447	0.00787	0.0085	70
Compass interface		AH Externally influenced	Failed to get heading information	AH1	Large calibration error	6	4	3	0.00821	0.00895	0.00787	0.0084	72
				AH2	Influenced by magnetic fields	6	4	4	0.00821	0.00895	0.01050	0.0090	58
Rangefinder interface		AI Failure of the speed measurement component	Unable to calculate range, speed, track	AI1	Wear of velocity measurement components	7	3	4	0.00958	0.00671	0.01050	0.0088	65
				AI2	Affected by water flow	7	5	4	0.00958	0.01119	0.01050	0.0104	31
		AJ Calculation component failure	Unable to calculate range, speed, track	AJ1	Program error	7	2	3	0.00958	0.00447	0.00787	0.0074	89
	Probe interface	AK Abnormal beam reception	Unable to measure	AK1	Voltage and power not up to standard	7	2	3	0.00958	0.00447	0.00787	0.0074	89
				AK2	Receiving probe problem	7	4	4	0.00958	0.00895	0.01050	0.0096	45
				AK3	System Handling Failure	7	3	3	0.00958	0.00671	0.00787	0.0081	82
				AL1	Unstable mobile signal	5	4	4	0.00684	0.00895	0.01050	0.0085	68
Anemometer interface		AL Data abnormality	Data cannot be collected properly	AL2	Poor sensor response	5	4	3	0.00684	0.00895	0.00787	0.0078	83
				AL3	Insufficient supply voltage	5	2	4	0.00684	0.00447	0.01050	0.0069	93

Table 2. *Cont.*

Subsystems	Unit	Failure Mode	Failure Effects	Code	Cause of Failure	S	O	D	Conversion Value of S, O, and D			RPN	Rank
									S	O	D		
	Processor	M Dead during startup, error report, black screen	The server cannot be started	AM1	Poor contact or broken pins	8	4	2	0.01094	0.00895	0.00525	0.0088	63
		AN File error during startup	The server cannot be started	AN1	Wrong working parameter setting	8	4	2	0.01094	0.00895	0.00525	0.0088	63
		AO Only boot in safe mode or command line mode	Reduced efficiency	AO1	Wrong setting of working parameters	5	4	4	0.00684	0.00895	0.01050	0.0085	68
	Data storage	AP Bad contact of memory stick	The server cannot be started	AP1	Poor contact between memory stick and motherboard	8	4	4	0.01094	0.00895	0.01050	0.0101	37
		AQ System message about memory error	Unable to operate the server efficiently	AQ1	Insufficient memory	6	4	3	0.00821	0.00895	0.00787	0.0084	72
AIS	VHF Transmitter	AR power unit failure	The power light does not light up after the host is turned on, and the whole machine has no power	AR1	No DC24V voltage output due to failure of the voltage regulator	7	4	5	0.00958	0.00895	0.01312	0.0102	36
				AR2	Host fuse blown	7	5	3	0.00958	0.01119	0.00787	0.0097	43
		AS antenna failure	Poor reception and transmission signal of VHF equipment	AS1	Antenna failure	6	6	4	0.00821	0.01342	0.01050	0.0106	26
	VHF Receiver	AT No GPS ship position signal	The fault alarm of the main unit sounds every once in a while, and the display shows no GPS ship position signal.	AT1	GPS data distributor malfunction or poor wiring contact	6	6	6	0.00821	0.01342	0.01575	0.0119	7
				AT2	Improper setting of GPS signal input mode of VHF device	6	6	5	0.00821	0.01342	0.01312	0.0113	20
				AT3	GPS signal output format change	6	2	3	0.00821	0.00447	0.00787	0.0068	94
		AU Antenna failure	Poor receiving and transmitting signal of VHF equipment	AU1	Antenna hardware or circuit failure	6	6	4	0.00821	0.01342	0.01050	0.0106	26
	Navigation system interface	AV GPS poor positioning accuracy	Cannot locate the ship accurately	AV1	Poor sea conditions	7	6	3	0.00958	0.01342	0.00787	0.0105	28
				AV2	Interference from outside	7	5	4	0.00958	0.01119	0.01050	0.0104	31
	Rangefinder interface	AW Speed measurement component failure	Cannot calculate the range, speed, track and other information	AW1	Wear of speed measurement components	7	3	4	0.00958	0.00671	0.01050	0.0088	65
				AW2	Affected by current	7	5	4	0.00958	0.01119	0.01050	0.0104	31
	Compass interface	AX External influence	Failure to obtain heading information	AX1	Large calibration error	6	4	3	0.00821	0.00895	0.00787	0.0084	72
				AX2	Affected by magnetic field	6	4	4	0.00821	0.00895	0.01050	0.0090	58

Table 2. *Cont.*

Subsystems	Unit	Failure Mode	Failure Effects	Code	Cause of Failure	S	O	D	Conversion Value of S, O, and D			RPN	Rank
									S	O	D		
	AIS Information Processor	AY Unable to process information	Cannot identify and track vessel information	AY1	Loss of transmission data	9	1	2	0.01231	0.00224	0.00525	0.0070	91
				AY2	Poor wireless communication signal	9	2	3	0.01231	0.00447	0.00787	0.0085	70
	Radar interface	AZ Interference with radar detection	Reduced detection accuracy	AZ1	Severe GPS interference	8	4	5	0.01094	0.00895	0.01312	0.0108	22
				AZ2	Frequency synthesis module failure	8	5	5	0.01094	0.01119	0.01312	0.0116	13
	ECDIS interface	BC Data loss	Failure to transmit electronic charts	BC1	Network connection error or information channel failure	8	4	4	0.01094	0.00895	0.01050	0.0101	37
Mobile Communication Receiver	Antenna	BD No signal accepted. The signal receiver does not work (no signal output)	Affect the communication signal or even lead to a short circuit by burning the components	BD1	Water in the antenna	5	2	1	0.00684	0.00447	0.00262	0.0050	105
				BD2	Sealing measures are not done	5	2	2	0.00684	0.00447	0.00525	0.0056	102
				BD3	Antenna impedance mismatch	5	2	3	0.00684	0.00447	0.00787	0.0063	98
	Antenna Switch	BE No signal or weak signal, signal not transmitting or difficult to transmit	Lead to mobile communication receiver work difficult or not work	BE1	Switch itself quality problems	3	1	1	0.00410	0.00224	0.00262	0.0031	111
				BE2	Physical damage or water ingress	3	2	1	0.00410	0.00447	0.00262	0.0039	109
	Filter	BF Low voltage fuse is blown, and the signal is not screened	The system does not work normally. The signal does not play a filtering role	BF1	Narrow passband, high loss	4	2	3	0.00547	0.00447	0.00787	0.0057	101
				BF2	Short circuit caused by the breakdown of capacitors	4	2	2	0.00547	0.00447	0.00525	0.0051	104
	Mixer	BG Signal is not mixed	Frequent automatic shutdown	BG1	Damage to the charging resistor	4	3	3	0.00547	0.00671	0.00787	0.0065	96
				BG2	The module circuit burned out	4	2	4	0.00547	0.00447	0.01050	0.0064	97
	Demodulator	BH Signal is not restored, error code appears	Results in system abnormalities. Signal lights show failure	BH1	The power supply connection line is not connected properly	4	2	1	0.00547	0.00447	0.00262	0.0044	106
				BH2	The quality of the problem itself	4	1	1	0.00547	0.00224	0.00262	0.0036	110
	Power supply	BI There is no power supply phenomenon. There is an instantaneous high- or low-voltage phenomenon	The power supply is damaged, the system stops working, or other parts of the machine are damaged	BI1	Power line aging, static electricity, power supply itself problems	4	3	2	0.00547	0.00671	0.00525	0.0059	100

Table 2. *Cont.*

Subsystems	Unit	Failure Mode	Failure Effects	Code	Cause of Failure	S	O	D	Conversion Value of S, O, and D			RPN	Rank
									S	O	D		
	CPU	BJ Work overload, and work instructions are not delivered	The system does not receive commands and cannot work properly	BJ1	Poor heat dissipation performance	6	5	4	0.00821	0.01119	0.01050	0.0098	42
	Memory Chip	BK No signal reception record, data loss	Mobile communication receiver does not have the function of storing signal information, and some modules do not work properly	BK1	Model mismatch	6	2	1	0.00821	0.00447	0.00262	0.0055	103
				BK2	Chip burned out	6	2	3	0.00821	0.00447	0.00787	0.0068	94
	Power Management Chip	BL off, not docked well, more confusion in the system	The conversion, distribution, detection and other power management functions of electrical energy in the electronic equipment system fail	BL1	Programming too high voltage	6	2	2	0.00821	0.00447	0.00525	0.0062	99
	Peripheral Circuits	BM Not energized, or the circuit is wrong	The system does not work, and the signal cannot be received and transmitted	BM1	Human error	4	2	1	0.00547	0.00447	0.00262	0.0044	106
				BM2	Circuit aging	4	2	1	0.00547	0.00447	0.00262	0.0044	106

In this section, five subsystems of the intelligent ship with a total of 111 fault causes are analyzed, as shown in Figure 2. The RPN values of each fault cause in the positioning system and its RPN share in the respective unit are given in Figure 3.

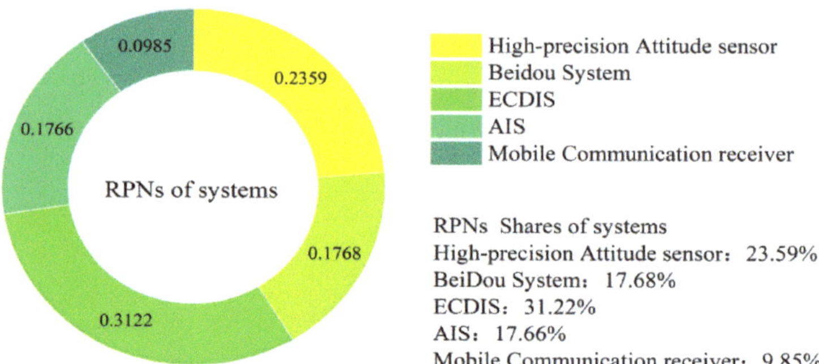

Figure 2. RPNs' share of each subsystem of the positioning system.

The FMECA table shows that, overall, the ECDIS is the most important system for positioning systems (RPN of 0.3122), followed by high-precision attitude sensor (0.2359), BeiDou positioning system (0.1768), AIS (0.1766), and mobile communication receivers (0.0985). On the one hand, the RPN values of the fixed-weight FMECA method are additive, and systems with more failure modes will have larger RPN values. In terms of RPN values of subsystems, ECDISs, and high-precision attitude sensors have numerous failure modes, both of which occupy more than half of the RPN and are important subsystems that cause the failure of positioning systems. On the other hand, to determine the critical failure mode for locating the system, it is important to consider not only the magnitude of the RPN value but also the average value of the RPN. Analyzing the importance of the failure from an average perspective can better evaluate the risk ranking of the system failure causes. The mean values of RPN for the causes of failure for these five systems were 0.0092, 0.0107, 0.0093, 0.0098, and 0.0055, respectively.

The total RPN value of the ECDIS is higher than that of the high-precision attitude sensor, while the average value of RPN is lower than that of the high-precision attitude sensor, indicating that although the failure modes of the ECDIS are many, the severity of their failure consequences is smaller compared to that of the high-precision attitude sensor. The average RPN of mobile communication receivers is much lower than other subsystems, and the risk level is low. Combining the results of the RPN values as well as the average RPN values, the high-precision attitude sensor was identified as the riskiest subsystem of the positioning system, followed by the ECDIS, AIS, and BeiDou, and finally, the mobile communication receiver.

3.3. Critical Failure Cause Analysis

The identification of critical failure causes is related to the development and planning of restorative and preventive measures. And the identification of critical failure units mainly lies in the selection of risk thresholds. Lorenzo et al. [34] proposed a new RPN threshold estimation method for FMECA. The method requires the following:

a. Calculate the RPN values for each failure mode;
b. Identify the main statistical parameters of the RPN values (maximum, 75% quantile, median, mean, 25% quantile, and minimum);
c. Draw box plots based on the main statistical parameters of the RPN values;
d. Identify critical faults. Based on the resulting box plot, critical faults (RPN above the 75th percentile) and negligible faults (RPN below the median) are identified. Faults between the median and the 75th percentile are classified by the designer, and some of the causes of failure that cannot be ignored are classified as critical failure factors.

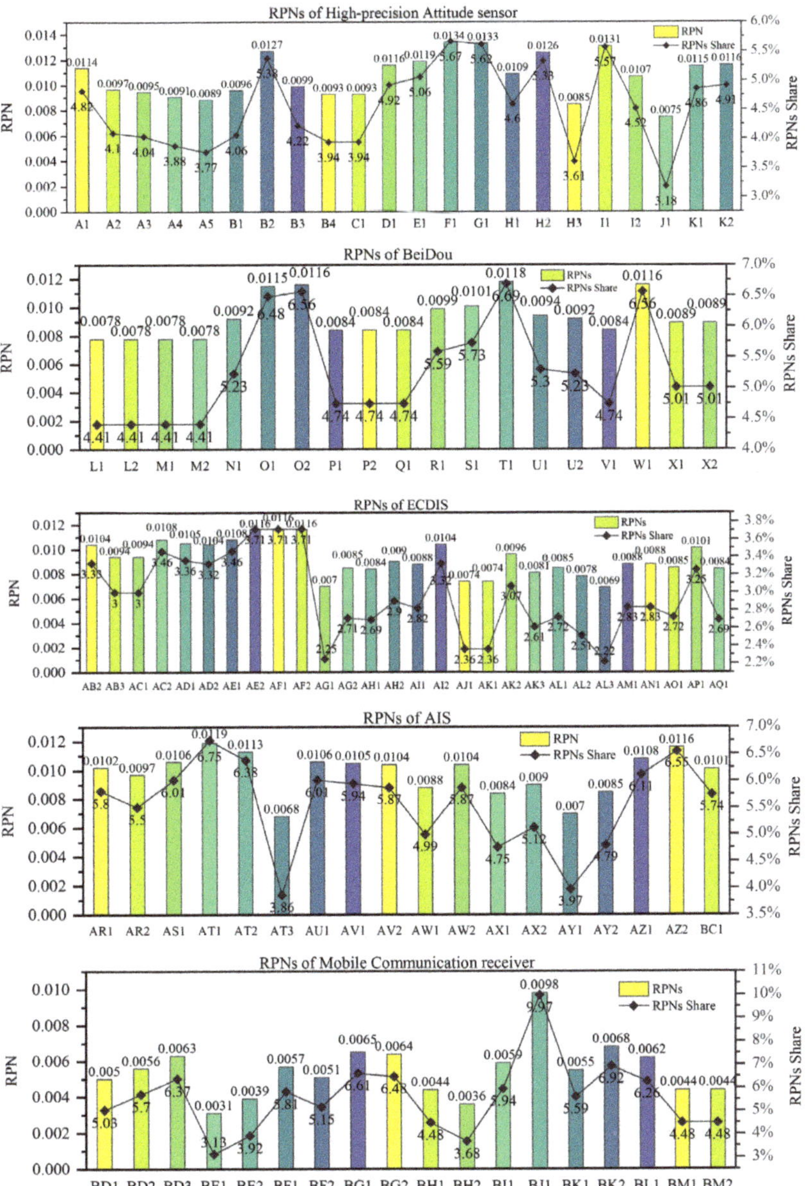

Figure 3. RPNs value and percentage of each cause of failure in the positioning system.

In this section, the critical cause identification schematic of the intelligent ship positioning system equipment is drawn according to the RPN threshold estimation method proposed by Lorenzo et al., as shown in Figure 4. The 28 failure causes in the top 25% of the RPN value ranking of the intelligent ship positioning system are identified, plus two of the pending failure causes are identified, and a total of 30 failure causes are identified as critical failure causes, as shown in Table 3.

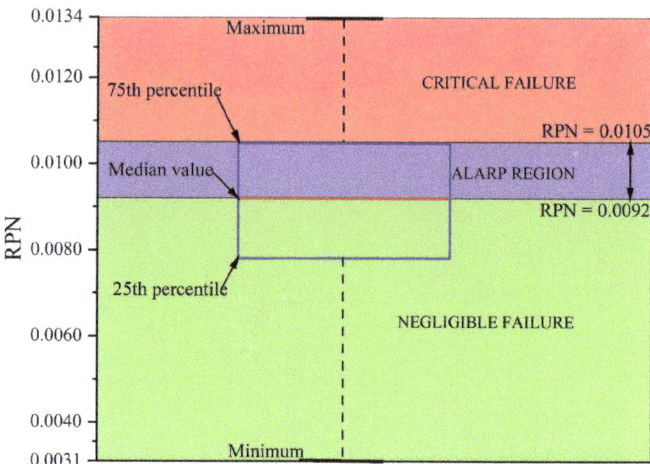

Figure 4. Box diagram of the positioning system RPN.

Table 3. Critical failure causes of the positioning system.

Subsystems	Unit	Cause of Failure	Share of RPN	Subsystems	Unit	Cause of Failure	Share of RPN
High-precision attitude sensor	Three-axis gyroscope	A1	1.14%	BeiDou System	Display	T1	1.18%
		B2	1.27%		Power connector	W1	1.16%
		D1	1.16%	ECDIS	Text Display	AB1	1.03%
		E1	1.19%			AC2	1.08%
		F1	1.34%		Navigation system interface	AD1	1.05%
		G1	1.33%		Radar interface	AE1	1.08%
	Triaxial accelerometer	H1	1.09%			AE2	1.16%
		H2	1.26%			AF1	1.16%
	Three-axis electronic compass	I1	1.31%			AF2	1.16%
		I2	1.07%	AIS	VFH transmitter	AS1	1.06%
	Arm Processor	K1	1.15%		VFH receiver	AT1	1.19%
		K2	1.16%			AT2	1.13%
BeiDou system	Terminal ARM Processor	O1	1.15%			AU1	1.06%
		O2	1.16%		Radar interface	AZ1	1.08%
	Display	S1	1.01%			AZ2	1.16%

The high-precision attitude sensor relies on multiple precision sensing units and contains 12 critical failure causes. Not only does this subsystem have many failure modes, but the consequences of failure are severe, and the risk level of the equipment is extremely high. The main causes of high precision attitude sensor failure are design-related factors (interference torque due to friction, resonance), environmental factors (magnetic field interference, temperature and humidity effects), and unavoidable unknown failures (hardware failure, device wear and tear). Serious causes of failure are interference caused by environmental factors such as F1 (electromagnetic interference, 1.34% of RPN), G1 (humidity factor, 1.33%), and I1 (magnetic field interference, 1.31%). The three-axis gyroscope, which in turn concentrates most of the key failure causes (A1–G1), is the key unit of the high-precision attitude sensor. This subsystem is related to whether the intelligent ship can accurately identify the surrounding environment. This affects the capability of the intelligent ship to complete berthing and unberthing and may even lead to the ship colliding with the shore wall. Close attention should be paid to this subsystem.

The BeiDou positioning system is one of the cores of the positioning system, which consists of BeiDou satellites, ground base stations, receivers, terminal processors, displays and other units. The system includes 5 key causes of failure. The main causes of BeiDou positioning system failures are design-related factors (poorly designed software functions) and unavoidable unknown failures (circuit damage, hardware failure). The serious causes of failure are O1 (processor hardware damage, 1.15%), O2 (circuit failure, 1.16%), and T1 (display the IPC part of the board damage, 1.18%). In general, BeiDou positioning system units are not prone to problems [35] (the BeiDou navigation system is the responsibility of the state and has a low probability of failure).

ECDIS is mainly used to accurately display the position of ships at sea. It consists of image display, text display, processor, data storage, and various data interfaces, including 7 key fault causes. The causes of system failure include unavoidable unknown faults (failure of frequency synthesis module, circuit damage), environmental factors (magnetic field interference, poor sea conditions), and human factors (wrong setting of operating parameters). The serious causes of failure are AE2 (failure of frequency synthesis module, 1.16%) and AF1 (damage to AC/AD module, 1.16%). Among them, the radar interface concentrates more critical failure causes and is the key unit of the electronic chart system. The majority of the remaining failures are minor, easy to detect and repair, and do not significantly affect the overall positioning system.

AIS consists of a VHF receiver/transmitter and an AIS information processor, and each part of the interface contains six critical failure causes. The system relies on the reception and processing of GPS data, so in practice, the failure of the working unit itself and the loss of GPS data caused by external interference are the main factors causing the system to fail. The faults with high RPN are the lack of GPS position signal caused by AT1 (GPS data distributor of VHF/TDMA receiver unit has a fault or poor connection, 1.19%). Overall, the system has a high failure rate, but failures occur less frequently, are easier to detect, and are less likely to cause very serious effects.

A mobile communication receiver is used to receive and send communication information. It consists of an antenna, filter, mixer, demodulator, CPU, and peripheral circuits and none of the failure causes are identified as critical failure causes. The unit of this subsystem was low in precision, and the basic failure modes were divided into circuit damage and capacitor breakdown due to the power supply and the aging and substandard quality of the equipment in the unit itself. In practice, receiver failures are less frequent and can be easily inspected and repaired. And temporary damage to the mobile communication receiver does not seriously affect the work of the entire positioning system and is the least dangerous in failure mode analysis.

From the analysis results, it is clear that the main reasons for the failure of the positioning system of intelligent ships are unavoidable unknown failures, environmental factors, and design-related factors.

The unavoidable unknown failures are mainly circuit failures, hardware damage, and parts damage. In order to cope with such failures, the control of key and fragile parts of intelligent ships should be strengthened, and good-quality parts should be used. At the same time, fragile parts and units should be checked regularly. It will also be beneficial to strengthen the research on condition monitoring, fault warning, and diagnosis of intelligent ships and improve the inspection system of intelligent ships to ensure the normal use of intelligent ship units and parts.

Intelligent ship positioning systems are highly susceptible to the influence of the surrounding environment. Interference from magnetic fields and other factors in the environment can easily affect the use of the positioning system and lead to deviations in positioning, which can easily lead to dangerous situations. In order to reduce the influence of the environment on intelligent ship navigation, research on extreme environmental conditions should be strengthened, especially the influence of magnetic field disturbance, strong wind, and strong waves on intelligent ship equipment.

Failures caused by design-related factors include interference torque due to gyroscope component friction, vibration, and poorly designed positioning system software functions. To avoid such failures, the design process should be improved, and the development of software should be enhanced.

4. Correlation Analysis of Critical Failure Cause

The complexity of intelligent ship systems leads to numerous failure modes; therefore, calculating the correlation between each two failure modes would take a lot of effort, and, in addition, further analysis of low-risk failure modes would be of limited significance. Therefore, only the critical failure causes obtained in Section 3.3 are subject to correlation analysis. The units analyzed are shown in Table 4. After scoring by three experts based on the steps in Section 2.3, the total relationship matrix was calculated, as shown in Table 5. The influence degree of factors is shown in Table 6.

Table 4. Analyzed units.

Subsystems	Code	Unit
High-precision attitude sensor	FM1	gyroscope
	FM2	accelerometer
	FM3	electronic compass
BeiDou system	FM4	Arm Processor
	FM5	Display
	FM6	Power connector
ECDIS	FM7	Text Display
	FM8	Navigation system interface
	FM9	Radar interface
AIS	FM10	VFH transmitter
	FM11	VFH receiver

Table 5. The matrix of total relation.

Factors	FM1	FM2	FM3	FM4	FM5	FM6	FM7	FM8	FM9	FM10	FM11
FM1	0.136	0.184	0.241	0.086	0.055	0.011	0.064	0.056	0.143	0.095	0.095
FM2	0.231	0.128	0.248	0.090	0.062	0.012	0.099	0.073	0.131	0.096	0.096
FM3	0.213	0.189	0.130	0.062	0.062	0.024	0.067	0.024	0.064	0.085	0.085
FM4	0.031	0.027	0.032	0.119	0.288	0.050	0.234	0.056	0.077	0.195	0.195
FM5	0.020	0.017	0.022	0.056	0.120	0.041	0.240	0.046	0.049	0.067	0.067
FM6	0.097	0.086	0.097	0.284	0.363	0.079	0.309	0.142	0.155	0.198	0.198
FM7	0.034	0.032	0.051	0.051	0.184	0.009	0.100	0.012	0.017	0.023	0.023
FM8	0.049	0.016	0.021	0.117	0.117	0.044	0.175	0.066	0.021	0.032	0.032
FM9	0.167	0.142	0.156	0.144	0.135	0.013	0.169	0.042	0.095	0.142	0.142
FM10	0.041	0.039	0.044	0.134	0.146	0.011	0.145	0.014	0.021	0.130	0.258
FM11	0.041	0.039	0.044	0.134	0.146	0.011	0.145	0.014	0.021	0.258	0.130

Table 6. The influence of the degree of factors.

Factor	Ri + Di	Ri-Di
FM1	2.225	0.105
FM2	2.165	0.368
FM3	2.091	−0.082
FM4	2.581	0.027
FM5	2.423	−0.934
FM6	2.312	1.704
FM7	2.283	−1.212
FM8	1.236	0.146
FM9	2.143	0.554
FM10	2.303	−0.338
FM11	2.303	−0.338

In this study, we set the threshold (k) in the total relationship matrix to 0.2. 0.2 is the most appropriate value obtained from the attempt. The causality diagram of the total relationship obtained according to k > 0.2 is shown in Figure 5.

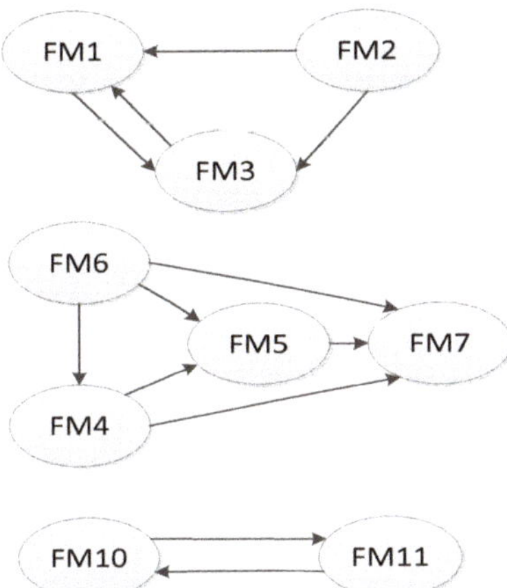

Figure 5. The causal diagram of total relation (k > 0.2).

The results of the analysis show that there is a high degree of correlation between the gyroscope, accelerometer, and electronic compass. That is, the failure of any one of the three components has the probability of leading to the failure of the other two units. All three belong to the same high-precision attitude sensor unit, which is a high-precision unit for monitoring the ship's attitude and is highly susceptible to interference from the external environment. The Arm processor, display, power interface, and text display are another group of units with a high degree of relevance. These four failure modes mainly concern the compass positioning system as well as the electronic charting system. The VFH transmitter and VHF receiver are more influenced by each other than by the other hazard units. The navigation system interface and the radar system interface weakly influenced other units and were barely influenced by other units, being two more independent units.

This section further determines that the high-precision attitude sensor is the most dangerous subsystem of the positioning system. It not only has many failure modes and serious consequences but also has a high degree of correlation between failure modes. It is easy to break down. Failure of this system will result in inadequate ship positioning accuracy, posing a serious safety hazard for intelligent navigation and autonomous berthing operations, and it should be given high priority. Regular servicing of similar high-precision components can effectively improve the reliability of smart ships.

Equipment failures on smart ships do not occur in isolation; each failure that occurs may lead to the occurrence of another. The correlation between failure modes shows that in reliability analysis, where failure modes have a cascade relationship with each other, we cannot simply consider the failure modes as independent of each other. To obtain reliable results, correlations between failure modes must be taken into account. The combined approach can be used not only for equipment failure analysis of smart ships but also for failure analysis of other marine engineering equipment.

5. Conclusions

This paper considers the shortcomings of the traditional FMECA method in that the weights of severity, occurrence, and detection are unreasonably assigned, and the correlation between failure modes is taken into account. The failure modes of the positioning system of an intelligent ship are analyzed using a combination of fixed-weight FMECA and DEMATEL, and the failure causes of the failure modes are identified. The following conclusions were reached.

(1) High-precision-attitude sensors are the most dangerous subsystem of the positioning system. It has many failure modes and serious consequences, and the correlation between failure modes is high;

(2) Unavoidably unknown failures (mechanical and component failures) and environmental factors (magnetic fields and temperature disturbances) are the key causes of positioning system failures. Regular maintenance of components and reducing environmental interference with precision components will be effective means of improving the reliability of smart ships;

(3) The critical fault units of the subsystems in the positioning system were derived. The correlation between the critical fault units was also evaluated. In order to conduct an accurate risk assessment of the entire system, it is essential to clarify the correlation between each failure mode.

The relevant conclusions can provide a reference for the maintenance of intelligent ship positioning system equipment. The safety of intelligent ships in navigation can be ensured by reducing the possibility of malfunctioning or reducing the severity of damage caused by intelligent ship equipment.

However, this paper analyzes the intelligent ship positioning system by fixed-weight FMECA using only one weight assignment method. In practice, the intelligent ship positioning system is a complex system integrated with multiple components. The failure mechanism and failure characteristics of the system itself and its components vary greatly. In the future, it is expected to consider variable and floating risk evaluation index weights, combine real data information, and select specific weights for different systems and components for FMECA analysis.

Author Contributions: Conceptualization, X.L., X.Z. and X.B.; methodology, X.L.; software, X.L.; validation, X.L. and H.H.; formal analysis, X.L.; investigation, H.H. and X.Z.; resources, X.Z. and Y.M.; data curation, H.H.; writing—original draft preparation, H.H.; writing—review and editing, X.Z.; visualization, X.B.; supervision, X.Z. and Y.M.; project administration, Y.M.; funding acquisition, X.B. and X.Z. All authors have read and agreed to the published version of the manuscript.

Funding: This research was partially supported by the Key R & D Projects in Guangdong Province (No. 2020B1111500001), the National Natural Science Foundation of China (42276225, 52001112), the Natural Science Foundation of Jiangsu Province (Grants 389 No. BK20211342), and the National Key Research and Development Program of China (No. 2022YFC2806300).

Institutional Review Board Statement: This study does not involve any institutional review issues.

Informed Consent Statement: Not applicable.

Conflicts of Interest: The authors declare no conflict of interest.

References

1. China Classification Society. *Rules for Intelligent Ships*; China Classification Society: Beijing, China, 2015.
2. Lazarowska, A. A new deterministic approach in a decision support system for ship's trajectory planning. *Expert Syst. Appl.* **2017**, *71*, 469–478. [CrossRef]
3. He, Y.; Liu, X.; Zhang, K.; Mou, J.; Liang, Y.; Zhao, X.; Wang, B.; Huang, L. Dynamic adaptive intelligent navigation decision making method for multi-object situation in open water. *Ocean. Eng.* **2022**, *253*, 111238. [CrossRef]
4. Lee, S.D.; Tzeng, C.Y.; Shu, K.Y. Design and experiment of a small boat auto-berthing control system. In Proceedings of the 2012 12th International Conference on ITS Telecommunications, Taipei, Taiwan, 5–8 November 2012; pp. 397–401.
5. Lee, S.D.; Tzeng, C.Y.; Kehr, Y.Z.; Kang, C.K.; Huang, C.C. Design and application of an image-processing-based fuzzy autopilot for small-boat approaching maneuvers. *J. Mar. Sci. Technol.* **2010**, *18*, 558–567. [CrossRef]
6. Mizuno, N.; Uchida, Y.; Okazaki, T. Quasi Real-Time Optimal Control Scheme for Automatic Berthing. *IFAC-PapersOnLine* **2015**, *48*, 305–312. [CrossRef]
7. Mizuno, N.; Tamai, Y.; Okazaki, T.; Ohtsu, K. A ship's minimum-time maneuvering system using neural networks. In Proceedings of the IEEE 2002 28th Annual Conference of the Industrial Electronics Society, IECON 02, Seville, Spain, 5–8 November 2002; pp. 1848–1853.

8. Mizuno, N.; Kakami, H.; Okazaki, T. Parallel Simulation Based Predictive Control Scheme With Application To Approaching Control For Automatic Berthing. *IFAC Proc. Vol.* **2012**, *45*, 19–24. [CrossRef]

9. Zhang, X.; Ni, W.; Sun, L. Fatigue Analysis of the Oil Offloading Lines in FPSO System under Wave and Current Loads. *J. Mar. Sci. Eng.* **2022**, *10*, 225. [CrossRef]

10. Allianz Global Corporate and Specialty Company (AGCS). *AGCS Safety Shipping Review*; Allianz Global Corporate and Specialty Company (AGCS): Munich, Germany, 2019.

11. Harati-Mokhtari, A.; Wall, A.; Brooks, P.; Wang, J. Automatic Identification System (AIS): Data Reliability and Human Error Implications. *J. Navig.* **2007**, *60*, 373–389. [CrossRef]

12. Pallotta, G.; Vespe, M.; Bryan, K. Vessel Pattern Knowledge Discovery from AIS Data: A Framework for Anomaly Detection and Route Prediction. *Entropy* **2013**, *15*, 2218–2245. [CrossRef]

13. Tsou, M.C. Discovering Knowledge from AIS Database for Application in VTS. *J. Navig.* **2010**, *63*, 449–469. [CrossRef]

14. Last, P.; Hering-Bertram, M.; Linsen, L. How automatic identification system (AIS) antenna setup affects AIS signal quality. *Ocean. Eng.* **2015**, *100*, 83–89. [CrossRef]

15. Carmignani, G. An integrated structural framework to cost-based FMECA: The priority-cost FMECA. *Reliab. Eng. Syst. Saf.* **2009**, *94*, 861–871. [CrossRef]

16. Bevilacqua, M.; Braglia, M.; Gabbrielli, R. Monte Carlo simulation approach for a modified FMECA in a power plant. *Qual. Reliab. Eng. Int.* **2000**, *16*, 313–324. [CrossRef]

17. Zammori, F.; Gabbrielli, R. ANP/RPN: A multi criteria evaluation of the Risk Priority Number. *Qual. Reliab. Eng. Int.* **2012**, *28*, 85–104. [CrossRef]

18. Saaty, T.L.; Ozdemir, M.S. *The Encyclicon: A Dictionary of Decisions with Dependence and Feedback Based on Analytic Network Process*; RWS Publications: Pittsburgh, PA, USA, 2005.

19. Carpitella, S.; Certa, A.; Izquierdo, J.; La Fata, C.M. A combined multi-criteria approach to support FMECA analyses: A real-world case. *Reliab. Eng. Syst. Saf.* **2018**, *169*, 394–402. [CrossRef]

20. Braglia, M. MAFMA: Multi-attribute failure mode analysis. *Int. J. Qual. Reliab. Manag.* **2000**, *17*, 1017–1033. [CrossRef]

21. Saaty, T.L. *The Analytic Hierarchy Process*; McGraw Hill Company: New York, NY, USA, 1994.

22. Xiao, N.; Huang, H.Z.; Li, Y.; He, L.; Jin, T. Multiple failure modes analysis and weighted risk priority number evaluation in FMEA. *Eng. Fail. Anal.* **2011**, *18*, 1162–1170. [CrossRef]

23. Zhang, S.Z.; Zeng, Q.D.; Zhang, G. A New Approach for Prioritization of Failure Mode in FMECA Using Encouragement Variable Weight AHP. *Appl. Mech. Mater.* **2013**, *289*, 93–98. [CrossRef]

24. Li, H.; Diaz, H.; Soares, C.G. A developed failure mode and effect analysis for floating offshore wind turbine support structures. *Renew. Energy* **2021**, *164*, 133–145.

25. Luqman, A.; Akram, M.; Alcantud, J.C.R. Digraph and matrix approach for risk evaluations under Pythagorean fuzzy information. *Expert Syst. Appl.* **2021**, *170*, 114518. [CrossRef]

26. Akram, M.; Luqman, A.; Alcantud, J.C.R. Risk evaluation in failure modes and effects analysis: Hybrid TOPSIS and ELECTRE I solutions with Pythagorean fuzzy information. *Neural Comput. Appl.* **2020**, *33*, 5675–5703. [CrossRef]

27. Liu, Y.; Eckert, C.M.; Earl, C. A review of fuzzy AHP methods for decision-making with subjective judgements. *Expert Syst. Appl.* **2020**, *161*, 113738. [CrossRef]

28. Song, B.F. A numerical integration method in affine space and a method with high accuracy for computing structural system reliability. *Comput. Struct.* **1992**, *42*, 255–262. [CrossRef]

29. Drezner, Z. Computation of the Multivariate Normal Integral. *ACM Trans. Math. Softw.* **1992**, *18*, 470–480. [CrossRef]

30. Da Silva, A.M.L.; Fernandez, R.A.; Singh, C. Generating Capacity Reliability Evaluation Based on Monte Carlo Simulation and Cross-Entropy Methods. *IEEE Trans. Power Syst.* **2010**, *25*, 129–137. [CrossRef]

31. Ni, W.; Zhang, X.; Zhang, W. Modified approximation method for structural failure probability analysis of high-dimensional systems. *Ocean. Eng.* **2021**, *237*, 109486. [CrossRef]

32. Gabus, A.; Fontela, E. *Perceptions of the World Problematique: Communication Procedure, Communicating with those Bearing Collective Responsibility*; DEMATEL Report No.1; Battelle Geneva Research Centre: Geneva, Switzerland, 1973.

33. Tzeng, G.H.; Chiang, C.H.; Li, C.W. Evaluating intertwined effects in e-learning programs: A novel hybrid MCDM model based on factor analysis and DEMATEL. *Expert Syst. Appl.* **2007**, *32*, 1028–1044. [CrossRef]

34. Catelani, M.; Ciani, L.; Galar, D.; Patrizi, G. Risk Assessment of a Wind Turbine: A New FMECA-Based Tool With RPN Threshold Estimation. *IEEE Access* **2020**, *8*, 20181–20190. [CrossRef]

35. Zhang, X.; Ni, W.; Liao, H.; Pohl, E.; Xu, P.; Zhang, W. Improved condition monitoring for an FPSO system with multiple correlated components. *Measurement* **2020**, *166*, 223–237. [CrossRef]

 *processes*

Article

Validation of a Classical Sliding Mode Control Applied to a Physical Robotic Arm with Six Degrees of Freedom

Andres González-Rodríguez, Rogelio E. Baray-Arana *, Abraham Efraím Rodríguez-Mata, Isidro Robledo-Vega and Pedro Rafael Acosta Cano de los Ríos

Tecnólógico Nacional de México, IT de Chihuahua, Ave. Tecnólógico #2909, Chihuahua 31310, Mexico
* Correspondence: rogelio.ba@chihuahua.tecnm.mx

Abstract: The control of robotic manipulators has become increasingly difficult over recent years due to their high accuracy, performance, speed, and reliability in a variety of applications, such as industry, medicine, research, etc. These serial manipulator systems are extremely complex because their dynamic models include perturbations, parametric variations, coupled nonlinear dynamics, and non-modular dynamics, all of which require robust control for trajectory tracking. This paper compares two control techniques: computational torque control (CTC) and sliding mode control (SMC). In this study, the latter was used for a physical robotic arm with six degrees of freedom (DOF) and online experiments were conducted, which have received little attention in the literature. As a result, the contribution of this work was based on the real-time application of this controller via a self-developing interface. The great resilience of sliding mode controllers to disturbances was also demonstrated in this study.

Keywords: sliding mode control; six degrees of freedom; manipulator; real-time application

Citation: González-Rodríguez, A.; Baray-Arana, R.E.; Rodríguez-Mata, A.E.; Robledo-Vega, I.; Acosta Cano de los Ríos, P.R. Validation of a Classical Sliding Mode Control Applied to a Physical Robotic Arm with Six Degrees of Freedom. *Processes* **2022**, *10*, 2699. https://doi.org/10.3390/pr10122699

Academic Editors: Francisco Ronay López-Estrada and Guillermo Valencia-Palomo

Received: 19 March 2022
Accepted: 24 April 2022
Published: 14 December 2022

Publisher's Note: MDPI stays neutral with regard to jurisdictional claims in published maps and institutional affiliations.

1. Introduction

Serial manipulators can perform complex tasks that human beings cannot perform because they could be harmed or because they do not have the precision or force, or either, that are required to perform those tasks. However, to fully fulfill this ability, serial manipulators require robust control. Thus, the design of manipulators needs to consider nonlinear dynamical models, parameter uncertainties, perturbations, and non-model dynamics. PID and PD controllers are some of the most widely applied control schemes for robot manipulators due to the simplicity of their implementation. However, these types of controllers have some drawbacks. A PD controller needs a gravitational term in the control law for nonplanar manipulators [1]. Meanwhile, a PID controller requires a tuning procedure to allow good performance [2]. These controllers also lack precision for non-modeled dynamics, such as friction or unknown external torques. Finally, they can only be used to track set points [3].

Trajectory tracking and motion control refer to when a manipulator follows a proposed path, which is usually obtained using inverse kinematics. One of the common solutions is the CTC (which is a feedback linearization method). This method has the disadvantages of knowing the dynamical model a priory and being not robust. To counter these problems, control researchers have combined CTC with other techniques, such as adaptive control [4], fuzzy logic [5], and neural networks [6]. Although these methods can estimate the dynamics of the proposed model, they are not robust to perturbations.

SMC is a robust control method that has been extensively studied by control researchers and demonstrates qualities such as parametric in-variance, dynamic collapse, and asymptotic convergence in the presence of perturbations [7,8]. One way to design an SMC is to use the equivalent control method (ECM) [9]. This method introduces a known part of the model system into the control law to reduce the chattering phenomena, which are a well-known drawback of SMC systems. By doing this, the SMC only deals with the nonmodeled dynamics and the estimation errors of the parameters [10–12].

In the field of robotics, the known part of an ECM is the dynamic model of a robotic arm [13]. These dynamic models are complex due to the coupling of the chain of masses and inertia and the large number of terms that appear for arms with more than three DOFs [14]. For this reason, in the literature, various authors have combined different control techniques and computer algorithms to estimate the known part of the ECM. In [15], Bailey and Arapostathis used a classical SMC surface with the known part of the ECM in a 2DOF manipulator. Kumar and Prasad [16–18] compared the use of CTC and SMC to the known part of the ECM, which showed the advantage of using SMC over the ECM in a 3DOF manipulator. The paper [19] proposed a novel sliding mode control (NSMC) that was based on an extended gray wolf optimizer (EGWO). The NSMC employed a PD surface with an exponential term that was combined with the ECM in a 2DOF manipulator and surface gains were selected using the EGWO. However, as in CTC, it is necessary to know the dynamic model in the ECM. In [20], Thuan et al. combined intelligent control with second-order sliding mode control (SOSMC). A radial basis function network (RBFN) calculated the dynamic model and nonmodeled dynamics of a 4DOF dual arm (DAM) (2DOF in each arm). Furthermore, Thuan et al. in [21] used model reference adaptive control (MRAC) instead of RBFN with SOSMC for the same DAM. An optimal sliding mode control approach, which was used in [22], combined optimal control with SMC for optimal robustness properties. In this approach, an observer of the disturbance was in charge of the adaptive part of the control and applied this control to the 2DOF planar manipulator.

There are two main problems with the techniques that were presented in [9–14]. One is that SMC is a simple control law that depends only on the selection of a gain and a surface, which, in turn, depends on the state of the system [23]. The use of this technique makes the control law and its application more difficult. Second, they only simulate the control strategy. For implementation, ref. [24] proposed an SMC with a sliding perturbation observer (SPO). The SPO estimated the reaction force of a 6DOF DAM (3DOFs in each arm). Jeong et al. [25] applied a super-twisting algorithm (STA) with an adaptive law to an industrial 4DOF robot using only 2DOF and an XY planar manipulator. Paper [26] presented an adaptive sliding mode neural network control (ASMNN). This control approach coupled the SMC with a radial basis function neural network (RBFNN), which was implemented in a three-link robot manipulator (3DOF). Similarly, ref. [27] used the same control strategy for the same manipulator, taking into account the dead zone.

Control techniques for other types of system can also be found in the literature. In [28], an NN in combination with a fractional order SMC was used to control an active power filter. The NN was in charge of the estimation of the uncertainties and nonlinearities, while the fractional order SMC improved the precision and performance of the control. The controller was implemented in real time and produced excellent results.

Ref. [29] combined a pole placement control, time delay estimation (TDE), and adaptive sliding mode control. The authors implemented the controller with a 6DOF Mitsubishi robot that used only 2DOF. To increase robustness in the reaching phase of the SMC, article [30] proposed an adaptive integral sliding mode control (AISMC) with a TDE, which was implemented with 3DOF of a 6DOF PUMA robot. The articles [15–20] only implemented controllers with a few DOFs of the manipulator. Hee et al. in [31] implemented a PID controller for the gravitational term of a physical 7DOF DAM. An SPO was used to estimate the evaluation force of the manipulator. However, as mentioned above, the PID controller could not be used for trajectory tracking.

The literature has proposed a variety of control techniques. Most of them involve the simulation of the controllers. For the implemented controllers, the majority used fewer DOFs than were available in the corresponding system. One example was an implementation of a 7DOF controller, but only for set-point tracking. Another point to address is the gain selection of the sliding-mode control. For a robot driven by DC motors, the limit gain is the maximum voltage in the power converter. When the controller needs more energy, the robot cannot reach or stay on the surface. The main focus of this paper was to find the voltage (SMC gain) that is needed for trajectory tracking. The novelty of this paper was the procedure that was used to validate the SMC gain in an n-DOF manipulator driven by DC motors. To the authors' knowledge, there has not yet been a discussion regarding the verification of the above.

In [32], a suggested strategy was compared to achieve finite-time convergence, flicker-free control input, superior tracking performance, and resilience of the robotic manipulator. The difficulty of this type of method is implementation in real time, as in this project, since the programming of fractional control algorithms is quite complex to carry out in an embedded way. In [33], an unmanned aerial vehicle (UAV) system with 6 degrees of freedom (6DOF) and external disturbances that corresponded to sensor failure was discussed. This form of pure SMC controller is still employed in robotics, and we utilized it in this paper to control the manipulator robot in a robust trajectory, following our proposed work. Using Lyapunov's theory, see Ref. [34] showed that a well-designed control could ensure that transnational and rotational tracking errors converge at the origin in a finite amount of time. However, only numerical simulations were carried out to demonstrate that the developed control scheme had a high level of robustness and a quick convergence time and demonstrated elimination of entry saturation and suppression of chattering. In this study, we proposed the use of real-time control through an SMC.

In this paper, a classical SMC dealt with the nonmodeled dynamics for the error estimation of the parameters and perturbations of a 6DOF manipulator that was driven by DC motors. The dynamical model was not applied to the control law (the known part of the ECM), which made the control law of the SMC simpler. When the numerical simulations were compared to those of CTC and SMC, SMC showed better robustness with easy implementation. The SMC experiments were performed using self-developed hardware that sent and received data between MATLAB and the robot. SMC gain was validated using Lyapunov stability analysis and actuator dynamics.

The paper is organized as follows. The mathematical model of the actuator manipulator is presented in Section 2. Section 3 deals with the CTC and the SMC. Numerical simulations of the CTC and SMC are presented in Section 4. Section 5 details the experimental results of the obtained SMC gain. Finally, Section 6 discusses the conclusions and areas for further development.

2. Mathematical Problem and Model

In this section, we introduce the mathematical model that was used to construct the robust control that was based on sliding modes. We rely on the most popular mathematical models from the literature, since this type of dynamic modeling is still used for the design of automatic control technology [11]. Therefore, the classical dynamic model of an n-DOF manipulator is expressed as follows.

$$D(\mathbf{q}(t))\ddot{\mathbf{q}}(t) + C(\mathbf{q}(t), \dot{\mathbf{q}}(t))\dot{\mathbf{q}}(t) + \mathbf{g}(\mathbf{q}(t)) = \mathbf{u}(t) \tag{1}$$

The inertia, Coriolis, and centrifugal matrices and the gravity vector are given by $D(\mathbf{q})$, $C(\mathbf{q}, \dot{\mathbf{q}})$, and $\mathbf{g}$, respectively, and $\mathbf{u(t)}$ is the control input of the system. By representing (1) as a separation of equations for each DOF, it yielded the following.

$$\sum_{j=1}^{n} d_{kj}(\mathbf{q})\ddot{q} + \sum_{i,j}^{n} c_{ijk}(\mathbf{q})\dot{q}_i\dot{q}_j + g_k(\mathbf{q}) = u_k \tag{2}$$

where k is the DOF of the manipulator.

Much of the previous research has focused on electrical positions, which was helpful as it allowed us to model more accurately and helped us develop our controller. The control input of the dynamic Equation (1) was the torque that was produced by the actuators in each DOF. In this case, the manipulator was driven by DC motors. The system was the input for the voltage in the motor terminals. For this reason, we could obtain the input of the model in voltage terms by combining the dynamical models of the manipulator and the actuators. The linear model of a DC motor was given by:

$$v_a(t) = L_a \frac{di_a(t)}{dt} + R_a i_a(t) + K_e \frac{d\theta(t)}{dt}$$

$$K_t i_a(t) = J \frac{d^2\theta(t)}{dt^2} + B \frac{d\theta(t)}{dt} + n\tau(t)$$

(3)

where the inductance L_a, electrical resistance R_a, and electrical constant K_e are the electrical parameters and the rotor inertia J, viscous friction B, and mechanical constant K_t are the mechanical parameters. The current and voltage are $v_a = v_a(t)$ and $i_a = i_a(t)$, the motor shaft angular displacement is $\theta = \theta(t)$, n is the transmission ratio, and $\tau = \tau(t)$ is the toque load, which is also known as the disturbance.

When we coupled Equations (1) and (3), there were three state variables (angular position, angular velocity, and current). In the literature, we found that Equation (3) could be reduced to one state variable. To achieve this, we compared the magnitude of the electrical time constant (ETC) with the mechanical time constant (MTC) by defining these constants as:

$ETC = \frac{L_a}{R_a}$

$MTC = \frac{J}{B}$

For the manipulator used in this article, the MTC was 168 times higher than the ETC in the arm motor and 40 times higher in the wrist motors. Dividing Equation (3) by R_a, i.e., $L_a/R_a = 0$, and writing Equation (3) according to its components yielded:

$$J_k \ddot{\theta}_k + (B + \frac{K_{ek}K_{tk}}{R_{ak}})\dot{\theta}_k = \frac{K_{tk}}{R_{ak}}v_{ak} - n_k\tau_k$$

(4)

By dividing Equation (4) by n_k and expressing $B = (K_{ek}K_{tk}/R_{ak})$, we obtained:

$$\frac{J_k}{n_k}\ddot{\theta}_k + \frac{B}{n_k}\dot{\theta}_k = \frac{K_{tk}}{R_{ak}n_k}v_{ak} - \tau_k$$

(5)

where θ_k is the angular displacement that the motor saw. When we wanted to combine (5) and (2), θ_k was necessary in terms of the manipulator. By substituting $\theta_k = \frac{q_k}{n_k}$ into Equation (5), it yielded:

$$\frac{J_k}{n_k^2}\ddot{q}_k + \frac{B}{n_k^2}\dot{q}_k = \frac{K_{tk}}{R_{ak}n_k}v_{ak} - \tau_k$$

(6)

When comparing Equation (2) with (6), the control input of (2) was the load torque of (6). By combining these equations, we obtained a model for the manipulator that was driven by a DC motor, which was presented as follows:

$$\frac{J_k}{n_k^2}\ddot{q}_k + \sum_{j=1}^{n} d_{kj}(\mathbf{q})\ddot{q} + \sum_{i,j}^{n} c_{ijk}(\mathbf{q})\dot{q}_i\dot{q}_j + \frac{B}{n_k^2}\dot{q}_k + g_k(\mathbf{q}) = \frac{K_{tk}}{R_{ak}n_k}v_{ak}$$

or as matrices:

$$(D(\mathbf{q}) + J)\ddot{\mathbf{q}} + C(\mathbf{q},\dot{\mathbf{q}})\dot{\mathbf{q}} + B\dot{\mathbf{q}} + \mathbf{g}(\mathbf{q}) = \mathbf{u}$$

(7)

where

$D(\mathbf{q}), C(\mathbf{q},\dot{\mathbf{q}}), B, J \in \mathbb{R}^{n \times n}$

$\mathbf{q}, \dot{\mathbf{q}}, \ddot{\mathbf{q}}, \mathbf{g}, \mathbf{u} \in \mathbb{R}^{n \times 1}$

where J and B are diagonal matrices with terms J_k/n_k^2 and B_k/n_k^2, respectively. The control input was given by a vector with terms:

$$u_k = \frac{K_k}{R_{ak}n_k}v_{ak}$$

(8)

As in the analysis that was performed in [1], we could compute the voltage input that was needed for the control.

3. Computed Torque Control and Sliding Mode Control

3.1. CTC

CTC is a feedback linearization technique that uses the dynamic model of a manipulator in the control law. To achieve this, we write Equation (7) as follows:

$$M(\mathbf{q})\ddot{\mathbf{q}} + \mathbf{h}(\mathbf{q}, \dot{\mathbf{q}}) = \mathbf{u} \tag{9}$$

where

$$M(\mathbf{q}) = (D(\mathbf{q}) + J)$$
$$\mathbf{h}(\mathbf{q}, \dot{\mathbf{q}}) = C(\mathbf{q}, \dot{\mathbf{q}})\dot{\mathbf{q}} + B\dot{\mathbf{q}} + \mathbf{g}(\mathbf{q})$$

As $M(\mathbf{q})$ was full-rank and square, its inverse also existed. Thus, we selected a control law as follows:

$$\mathbf{u} = M(\mathbf{q})\mathbf{v} + \mathbf{h}(\mathbf{q}, \dot{\mathbf{q}}) \tag{10}$$

where

$$\mathbf{v} = -\alpha\mathbf{q} - \beta\dot{\mathbf{q}} + \mathbf{r(t)} \tag{11}$$

where α and β are diagonal matrices and $\mathbf{r(t)}$ is the trajectory of each DOF, which was defined as:

$$\mathbf{r(t)} = \ddot{\mathbf{q}}_d(t) + \alpha\dot{\mathbf{q}}_d(t) + \beta\mathbf{q}_d(t) \tag{12}$$

Equation (11) contained the position ($\mathbf{q}_d(t)$), velocity ($\dot{\mathbf{q}}_d(t)$), and acceleration ($\ddot{\mathbf{q}}_d(t)$) of the trajectory that was proposed. Substituting Equation (11) into (10) yielded the following:

$$\mathbf{v} = \ddot{\mathbf{q}}_d(t) + \alpha\dot{\mathbf{e}}(t) + \beta\mathbf{e}(t) \tag{13}$$

where $\mathbf{e}(t) = \mathbf{q}_d(t) - \mathbf{q}$ is the error of the trajectory tracking.

Applying the control (10) to the dynamic model (9), we obtained the following.

$$\ddot{\mathbf{e}}(t) + \alpha\dot{\mathbf{e}}(t) + \beta\mathbf{e}(t) = \mathbf{0} \tag{14}$$

Equation (14) showed a homogeneous second-order linear differential equation. Another property exhibited by Equation (14) was that the set of equations did not lump together. In this sense, we could treat each differential equation separately. With this in mind, we added a disturbance to Equation (9) and, using the same control law as (10), we obtained the following:

$$\ddot{\mathbf{e}}(t) + \alpha\dot{\mathbf{e}}(t) + \beta\mathbf{e}(t) = \mathbf{f}(t) \tag{15}$$

where $\mathbf{f}(t) \in \mathbb{R}^{n x 1}$

As the equations were not lumped, we could use one equation for the Lyapunov stability analysis. To do this, we passed the first equation into state-space. By selecting $x_1(t) = e_1(t)$ and $x_2(t) = \dot{x}_1(t) = \dot{e}_1(t)$, we wrote the state-space equation as:

$$\begin{aligned} \dot{x}_1(t) &= x_2(t) \\ \dot{x}_2(t) &= -\alpha_{11}x_2(t) - \beta_{11}x_1(t) + f_1(t) \end{aligned} \tag{16}$$

Then, using the following Lyapunov candidate function:

$$V = \frac{1}{2}\alpha_{11}x_1^2(t) + \frac{1}{2}x_2^2(t) \tag{17}$$

taking the time derivative of the Lyapuniov function, it yielded:

$$\dot{V} = \begin{bmatrix} \frac{\partial V}{\partial x_1} & \frac{\partial V}{\partial x_2} \end{bmatrix} \begin{bmatrix} x_1(t) \\ x_2(t) \end{bmatrix} \tag{18}$$

$$\dot{V} = -\alpha_{11}x_2^2(t) + x_2 f_1(t) \tag{19}$$

The function presented in Equation (19) was not definitely negative because we could not use a static negative sign in the term $x_2 f_1(t)$. When we did not take the disturbance into account, Equation (19) became:

$$\dot{V} = -\alpha_{11}x_2(t)^2 \tag{20}$$

The above derivative was always negative, but only in the state $x_2(t)$. To prove total stability, we used La Salle's theorem with $\dot{V} = 0$:

$$0 = -\alpha_{11}x_2^2(t) \tag{21}$$

This meant that $x_2(t) = 0$. Thus, its derivative also equaled zero: $\dot{x}_2(t) = 0$. Substituting $x_2(t)$ and $\dot{x}_2(t)$ into the state-space Equation (16) yielded the following result:

$$\dot{x}_1(t) = 0$$
$$0 = -\beta_{11}x_1(t) \tag{22}$$

where $x_1(t) = 0$. In this manner, we show that both states become zero when following the desired trajectory with no disturbances in the system.

Numerical simulations are addressed in Section 4, which describes the control with and without disturbances.

3.2. SMC

We have seen that CTC reduced the dynamic model of the manipulator to a set of homogeneous second-order differential equations that were not lumped. However, to achieve this, the robot parameters need to be fully known. Sometimes the parameters are not available or the system has disturbances in one or various DOFs, which affects the performance of CTC in trajectory tracking.

To deal with the problems mentioned above, SMC is a highly robust control that has been discussed in the literature. To apply this control scheme, we selected a surface as follows:

$$\mathbf{s} = \dot{\mathbf{e}}(t) + c\mathbf{e}(t) \tag{23}$$

where vector $\mathbf{s}$ contains all of the sliding surfaces in each DOF and c is a diagonal matrix with the constant c_k.

Using the following Lyapunov candidate function:

$$V = \frac{1}{2}\mathbf{s}^T\mathbf{s} \tag{24}$$

we rewrite Equation (9) as follows, including a distrubance $\mathbf{f}(t)$:

$$\ddot{\mathbf{q}} = M(\mathbf{q})^{-1}(\mathbf{u} - h(\mathbf{q},\dot{\mathbf{q}}) + \mathbf{f}(t)) \tag{25}$$

The time derivative of the Lyapunov function was given by:

$$\dot{V} = \mathbf{s}^T\dot{\mathbf{s}} \tag{26}$$

Substituting the time derivative of $\mathbf{s}$ into (26) yielded:

$$\dot{V} = \mathbf{s}^T\dot{\mathbf{s}}$$
$$= \mathbf{s}^T[\ddot{\mathbf{q}}_d(t) - M(\mathbf{q})^{-1}(\mathbf{u} - h(\mathbf{q},\dot{\mathbf{q}}) + \mathbf{f}(t)) + c\dot{\mathbf{e}}(t)] \tag{27}$$

Applying the control $\mathbf{u} = \tau_o sign(\mathbf{s})$, where τ_o is a diagonal matrix of gains τ_{ok}, and substituting the control into (27) produced:

$$\begin{aligned}
\dot{V} &= \mathbf{s}^T \dot{\mathbf{s}} \\
&= \mathbf{s}^T[\ddot{\mathbf{q}}_d(t) + c\dot{\mathbf{e}}(t) + M(\mathbf{q})^{-1}(h(\mathbf{q}, \dot{\mathbf{q}}) + \mathbf{f}(t))] \\
&\quad - \mathbf{s}^T M(\mathbf{q})^{-1}\tau_o sign(\mathbf{s})
\end{aligned} \tag{28}$$

hence:

$$abs(\mathbf{s}^T) = \mathbf{s}^T sign(\mathbf{s}) \tag{29}$$

with:

$$abs(\mathbf{s}^T) = [|s_1||s_2|\dots|s_n|]^T \tag{30}$$

and:

$$abs(\mathbf{f}(t)) < \mathbf{L} \tag{31}$$

Equation (29) stated that each $f_i(t)$ had an upper bound constant of L_i. By combining Equations (29)–(31) into (28), we obtained the following inequality:

$$\begin{aligned}
\dot{V} &\leq abs(\mathbf{s}^T)[abs(\ddot{\mathbf{q}}_d(t) + c\dot{\mathbf{e}}(t) + M(\mathbf{q})^{-1}(h(\mathbf{q}, \dot{\mathbf{q}}) + \mathbf{L}) \\
&\quad - M(\mathbf{q})^{-1}\tau_0] < 0
\end{aligned} \tag{32}$$

Equation (32) was definitely negative when the following was always true:

$$\tau_0 > abs[M(\mathbf{q})(\ddot{\mathbf{q}}_d(t) + c\dot{\mathbf{e}}(t)) + (h(\mathbf{q}, \dot{\mathbf{q}}) + \mathbf{L})] \tag{33}$$

In other words, when all the gains along the diagonal of the matrix τ_0 fulfilled the inequality, the system reached the surface and stayed there for the entire period of time t. As with CTC, the simulations are addressed in Section 4 to validate the robustness of the SMC.

4. Numerical Simulations of CTC and SMC

We showed in the previous section that we considered two types of control scheme (CTC and SMC). We showed via Lyapunov stability analysis that CTC was not robust against disturbances and that the dynamic model had to be completely known in order to achieve good performance. Meanwhile, the SMC could handle trajectory tracking without knowing the dynamic model in the control law. In addition to this, the SMC was robust against disturbances in the system.

This section presents the numerical simulations of the two control laws to better understand their behaviour. The actuator parameters are shown in Table 1.

Table 1. Actuator parameters.

Parameters	Actuators at the Arm	Actuators at the Wrist
$R_a(\Omega)$	5.54	5.3
$L_a(\mu H)$	821	1052
$K_t(\frac{Nm}{A})$	0.0208	0.00363
$K_e(\frac{Vs}{rad})$	0.02076	0.00363
$J(grcm^2)$	19.4	0.0621
$B(mNmms)$	0.875	0.7971

The parameters and a diagram of the manipulator are shown in Table 2 and Figure 1, respectively. The procedure that was used to obtain the dynamic model of the manipulator can be seen in [14].

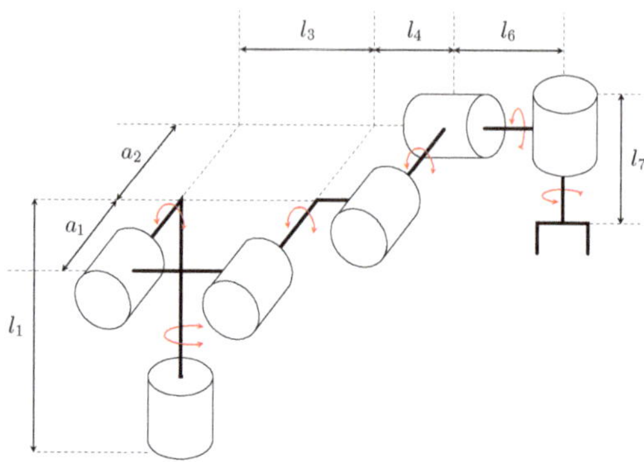

Figure 1. Diagram of the 6DOF manipulator.

Table 2. Manipulator parameters.

First Articulation	Second Articulation	Third Articulation
$a_1 = 6$ cm $a_{c1} = 3.036$ cm $m_1 = 0.11$ kg	$l_3 = 23$ cm $l_{c3} = 10.513$ cm $m_2 = 0.2125$ kg	$a_1 = 6$ cm $a_{c1} = 3.036$ cm $m_s = 0.5574$ kg
$I_{zz1} = 2.6 \times 10^{-6}$ kgm^2	$I_{xx2} = 9.2 \times 10^{-4}$ kgm^2 $I_{yy2} = 2.0515 \times 10^{-3}$ kgm^2 $I_{zz2} = 2.9224 \times 10^{-3}$ kgm^2	$I_{xxs} = 2.922 \times 10^{-6}$ kgm^2 $I_{yys} = 2.922 \times 10^{-6}$ kgm^2 $I_{zzs} = 5.615 \times 10^{-6}$ kgm^2
		$l_4 = 24$ cm $l_{c4} = 4.310$ cm $m_3 = 0.215$ kg
		$I_{xx3} = 1.9229 \times 10^{-5}$ kgm^2 $I_{yy3} = 2.1720 \times 10^{-4}$ kgm^2 $I_{zz3} = 2.3177 \times 10^{-4}$ kgm^2
Fourth Articulation	**Fifth Articulation**	**Sixth Articulation**
$a_2 = 10$ cm $a_{c2} = 5$ cm $m_4 = 0.0803$ kg	$l_6 = 4$ cm $l_{c6} = 2$ cm $m_5 = 0.1412$ kg	$l_7 = 10$ cm $l_{c7} = 5$ cm $m_5 = 0.0712$ kg
$I_{xx4} = 4.7373 \times 10^{-5}$ kgm^2 $I_{yy4} = 1.8376 \times 10^{-5}$ kgm^2 $I_{zz4} = 4.7373 \times 10^{-5}$ kgm^2	$I_{xx5} = 2.6153 \times 10^{-5}$ kgm^2 $I_{yy5} = 3.9774 \times 10^{-5}$ kgm^2 $I_{zz5} = 4.1205 \times 10^{-5}$ kgm^2	$I_{xx6} = 7.3043 \times 10^{-5}$ kgm^2 $I_{yy6} = 7.4434 \times 10^{-5}$ kgm^2 $I_{zz6} = 6.86 \times 10^{-6}$ kgm^2

For trajectory tracking, a Linear Segment Parabola Blending (LSPB) was used. This function was defined as follows:

$$
qd = \begin{cases}
q_0 + \frac{a}{2}t^2 & 0 < t \le t_b \\
\frac{q_f + q_0 + v t_f}{2} + vt & t_b < t \le t_f - tb \\
q_f - \frac{a t_f^2 t}{2} + a t_f t - \frac{a}{2}t^2 & t_f - t_b < t \le t_f
\end{cases}
\tag{34}
$$

where q_d is the trajectory LSPB, q_0 and q_f are the initial and final values of the trajectory, t_b and t_f are the mixing and final times, and v and a are the velocity and acceleration of the trajectory, respectively. We could compute v and a using the values of q_0, q_f, t_f, and $t_b = \frac{1}{3}t_f$. The LSPB trajectory with $t_f = 4$, $q_0 = 0$, and $q_f = \pi$ is shown in Figure 2. In both the CTC and STM control simulations, a step perturbation and an LSPB trajectory were used. The t_f of each DOF was 4 s and the final degree of each DOF was:

$$\mathbf{q}_f = \begin{bmatrix} \frac{\pi}{2} \\ \frac{\pi}{3} \\ -\frac{\pi}{3} \\ \frac{\pi}{4} \\ \frac{\pi}{3} \\ -\frac{\pi}{6} \end{bmatrix}$$

For CTC, a simulation without perturbations is shown in Figure 3 and a simulation with perturbations is shown in Figure 4.

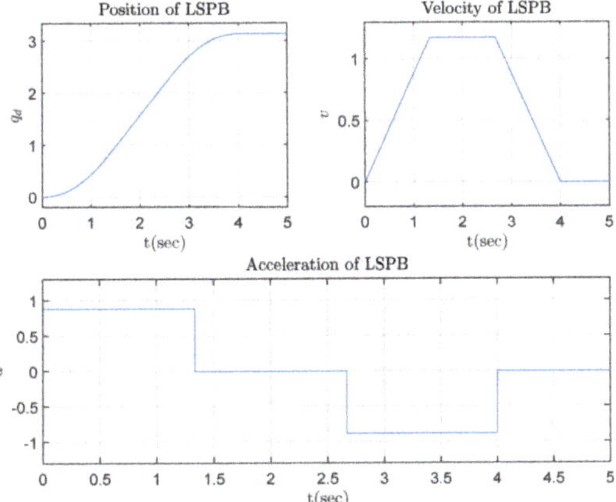

Figure 2. LSPB trajectory.

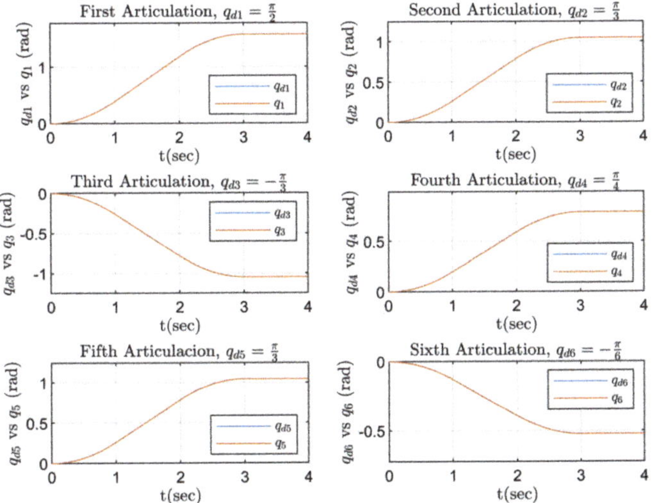

Figure 3. CTC numerical simulation with no disturbances.

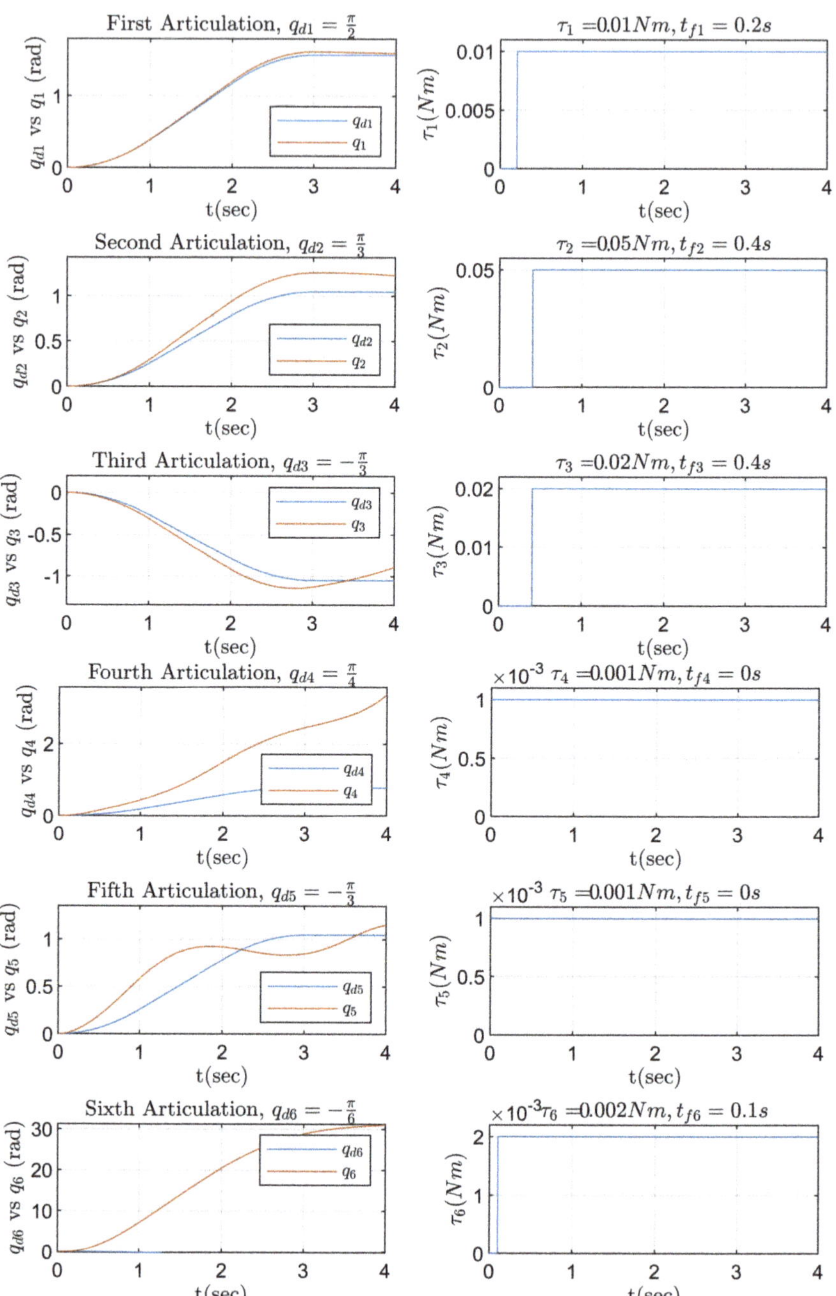

Figure 4. Numerical simulation of CTC with perturbations.

Each DOF on the manipulator followed the desired trajectory when there were no perturbations, as shown in Figure 3, but the robot lost the trajectory when we applied a perturbation in each DOF. As we showed in Section 3, the Lyapunov function could not have a definite sign when a perturbation was applied to the CTC, so there was no guarantee that stability existed. This can be seen in Figure 4. The SMC and its control are shown in Figures 5 and 6, respectively.

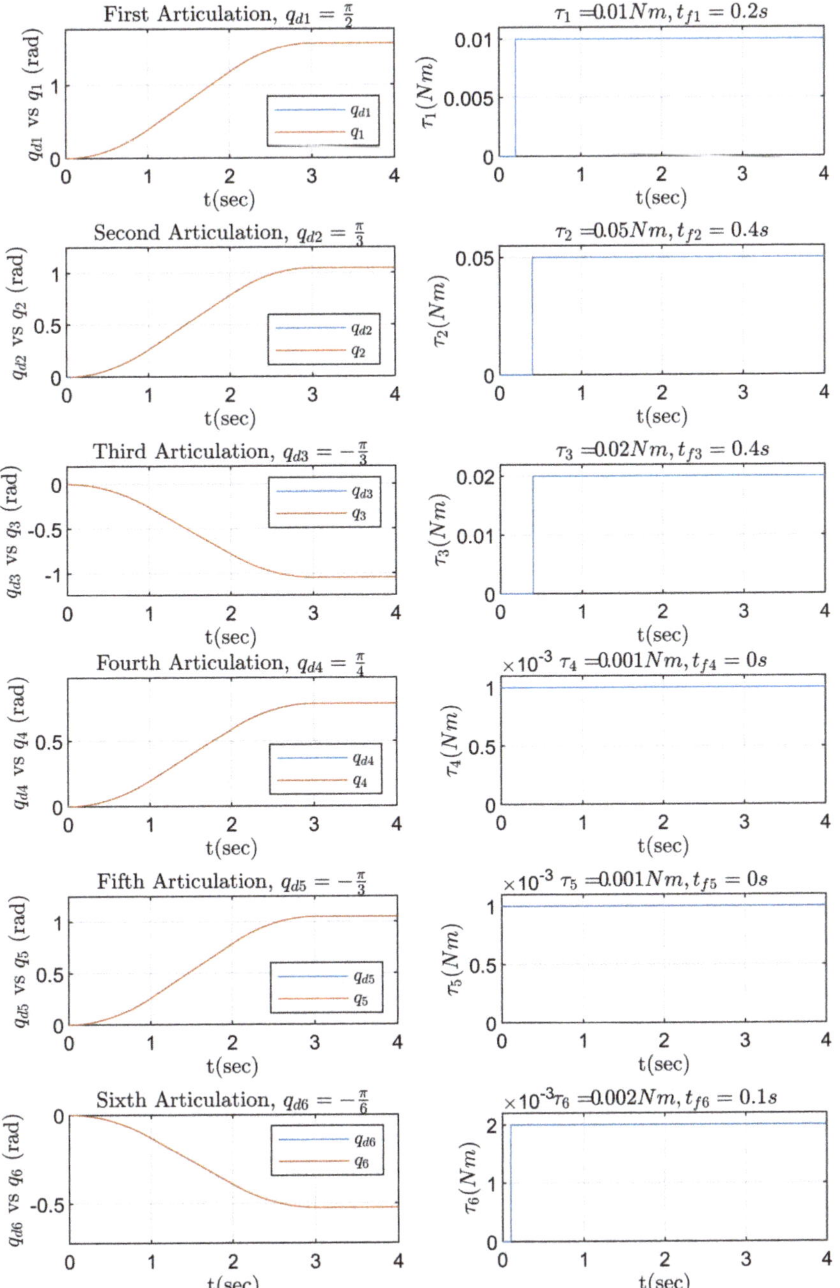

Figure 5. Simulation of the SMC with perturbations using numerical methods.

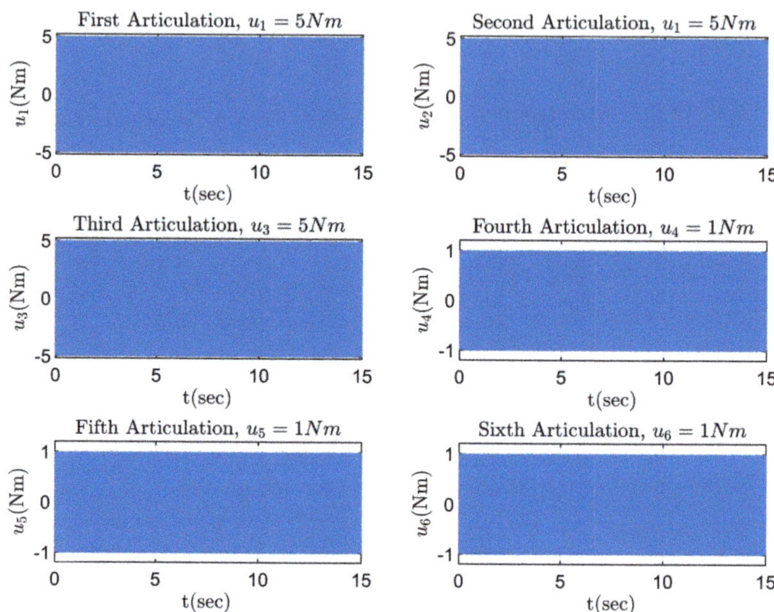

Figure 6. Simulation of the signal regulation of the SMC with perturbations using numerical methods.

Three different techniques to measure error (ITAE, IAE and ISE) in trajectory tracking were used to compare the two controllers with perturbations, which can be seen in Figure 7.

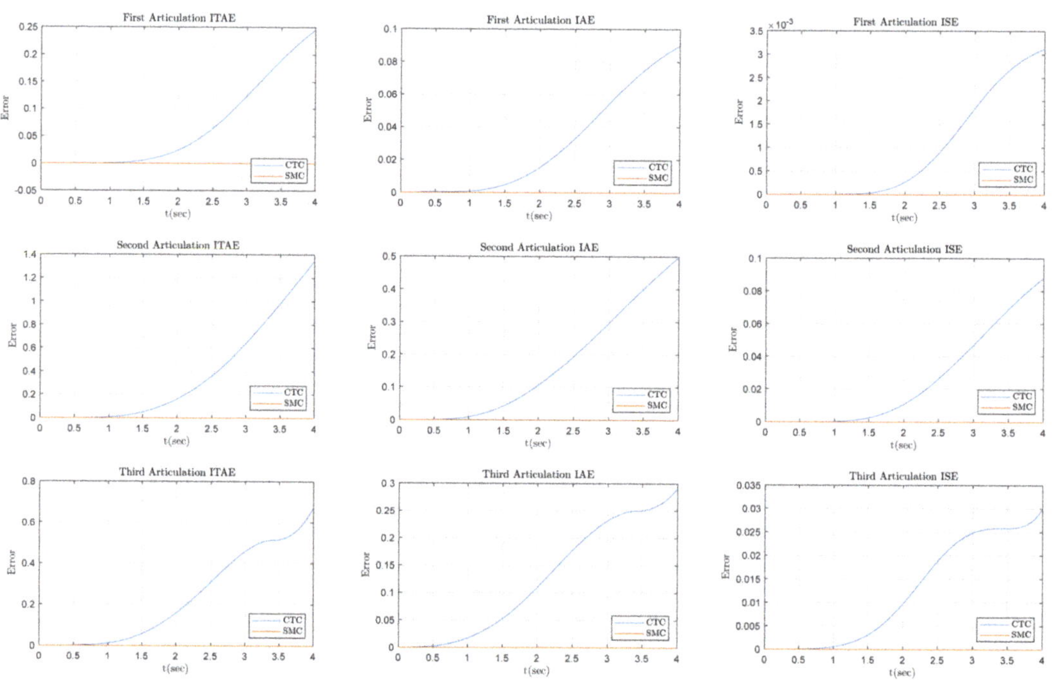

Figure 7. *Cont.*

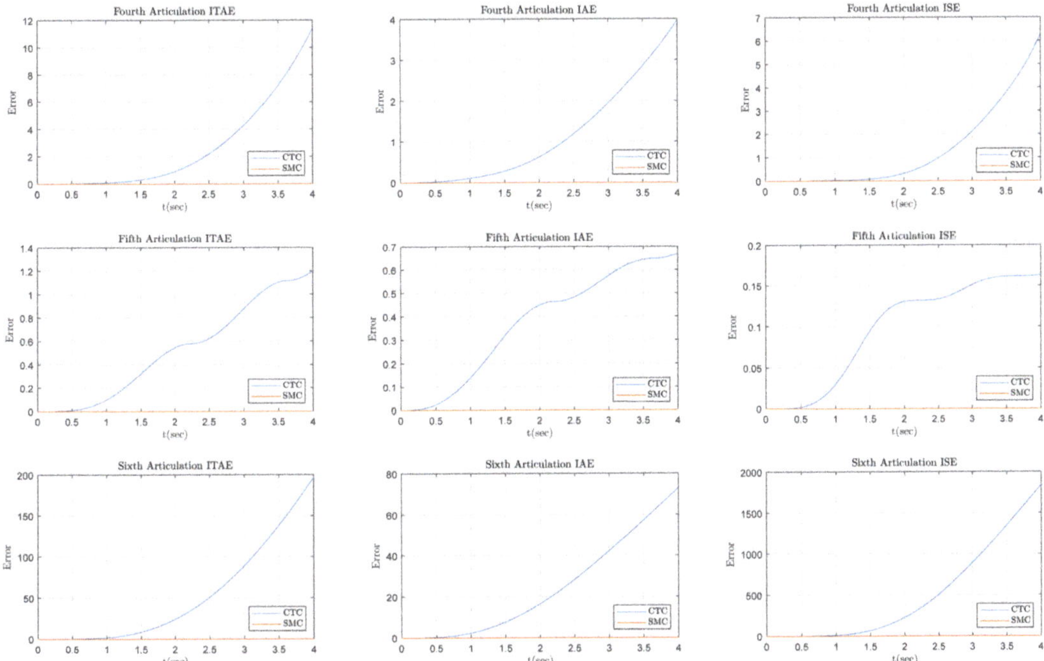

Figure 7. Error performance indexes: ITAE, IAE, and ISE.

It can be seen in Figure 5 that the manipulator followed the trajectory when we had enough gain in the SMC control, as demonstrated in Section 3, and that the time derivative of the Lyapunov function was always negative. However, it can be seen in Figure 6 that the effect of using a discontinuous controller caused the applied torque to oscillate within the manipulator.

The comparison between the SMC and CTC with perturbation can be seen in Figure 7. This comparison demonstrated the robustness of the SMC against perturbations. The CTC had a visible and high magnitude of error, but the SMC error was not visible. In the following section, an SMC was applied to a 6DOF manipulator using a self-developed electronic interface (SDEI) in MATLAB. The advantages and disadvantages of the proposed control scheme are also discussed.

5. Implementation and Experimental Results

As we showed in Section 4, SMC offered a simpler control law compared to CTC, which only needs to know the gain of the sign function. The numerical simulations showed that the SMC achieved asymptotic stability in the presence of perturbations. For these reasons, SMC was selected for the trajectory tracking of a 6DOF manipulator driven by DC motors. The manipulator was located at the "Laboratorio de Control de Electromecanismos" at the "Tecnológico de Chihuahua". The SDEI manipulator is shown in Figure 8.

To apply the SMC, we had to understand the control law in terms of the software and hardware that were being used. As the dynamic model of the manipulator with actuators had a voltage input, the sign function changed the polarity of the voltage in the DC motor. A H-bridge was in charge of this. Additionally, the voltage in the H bridge (power converter) was the gain in sliding mode. The H bridge that was selected for this study was an L298 motor drive, which could handle two DC motors simultaneously. Two systems were used for the experiment: the control system and the power system. The control system had the following main elements, each with its own communication protocols:

- SDEI (UART-SPI);
- 4 PIC18F26K20;
- FTDI FT232RL (USB-UART).

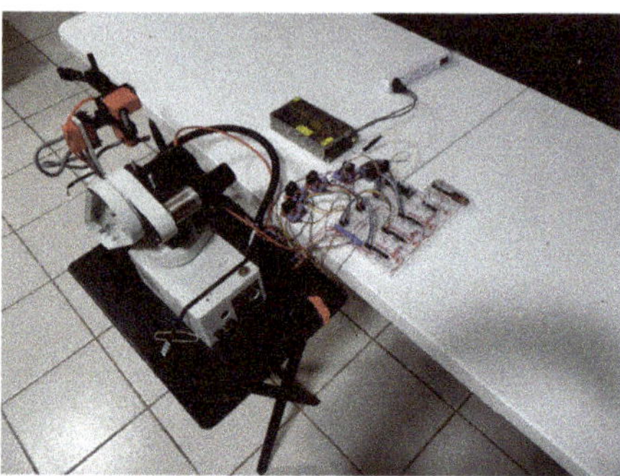

Figure 8. The 6DOF manipulator with SDEI.

The control system started in the computer, which contained the control algorithm within the MATLAB environment. MATLAB sent the information via a virtual serial port to the FTDI (USB), then the SDEI sent the information to the SEDI via the UART protocol. SEDI had four PIC18F26K20s, one of which was the master that received the data from FTDI and sent them to slaves via SPI. The slaves were in charge of generating the PWM that was needed for the power system and reading the position sensors of the serial manipulator. The slaves sent the data positions to the master and, using the same sequence as before, the data reached MATLAB. MATLAB computed the control law with the information obtained and sent it to the FTDI; then, the process was repeated again. It took 5 mili-seconds for the SDEI to complete the cycle, which was the sample time of the system. Diagrams of the experimental setup and the SDEI configuration are shown in Figure 9.

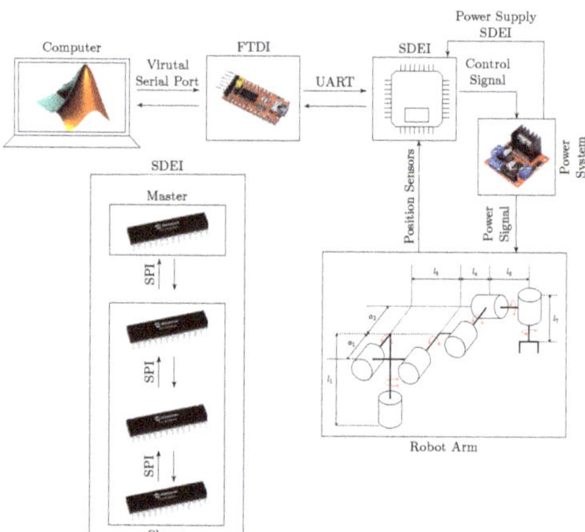

Figure 9. SDEI diagram.

The power system was responsible for supplying the voltage to the SDEI, the H bridge, and the power converters. The elements of the power system were the following:

- Power supply (12 Volts);
- Four H-bridges;
- DC–DC converter (12 Volts − 5 Volts);
- DC–DC converter (12 Volts − 3.3 Volts).

The DC–DC converters were connected to the power supply of 12 Volts. One of them reduces the voltage to 3.3 V to power the SDEI. The other reduces the voltage to 5 V to power the 2 H-bridges. The 2 H bridges with 5 V are connected to wrist motors, while the 2 other H bridges are connected to the power supply (12 V). These H-bridges are in charge of the power of the arm motors. A diagram of the power system is presented in Figure 10.

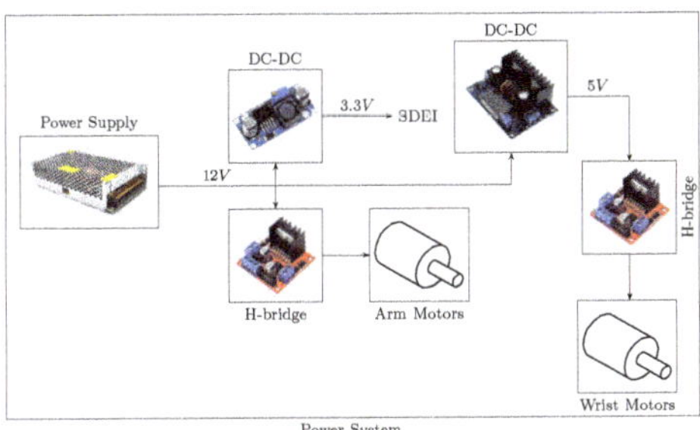

Figure 10. The power system.

With SDEI, communication was achieved between MATLAB and the manipulator. The control algorithm (SMC) and its reference, the LSPB trajectory in this case, were programmed in MATLAB. Figure 11 shows the control diagram.

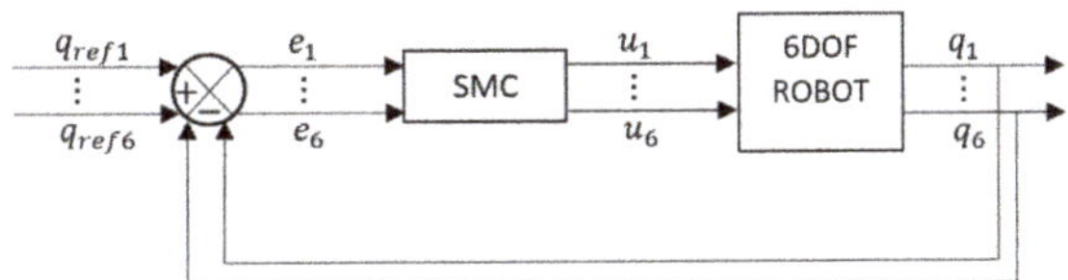

Figure 11. The control diagram.

The physical arm only communicated with the position sensor, which meant that we could only measure the position error, but we needed both the position and velocity errors for the sliding surface. Due to this, a backward Euler method was used as a differentiator to estimate the velocity error. To accomplish this, we use the following.

$$\dot{e} = \frac{de(t)}{dt} = \lim_{\Delta t \to 0} \frac{e(t) - e(t - \Delta t)}{\Delta t} \tag{35}$$

where Δt is the sample time and $e(t - \Delta t)$ is the previous error value. As the control was implemented in a digital environment, the limit of (35) could not reach zero and (35) was approximated in the following way:

$$\dot{e} \approx \frac{e(t) - e(t - \Delta t)}{\Delta t} \tag{36}$$

In other words, when the sample time was shorter, Equation (36) obtained a better approximation of Equation (35). We needed to know whether the voltage in each DC motor could follow the desired trajectory. As mentioned above, these voltages were the sliding mode gains in voltage terms. Using a numerical simulation, we graphed Equation (33) and the maximum of this graph was the torque required to reach the surface and stay there for the entire period of time. Since we did not know how much torque was needed, the simulation started with a gain of 5 nm for the arm motors and 1 nm for the wrist motors. For the physical experiment, each DOF followed an LSPB. The final times of the LSPB were $\mathbf{t}_f = \begin{bmatrix} 10 & 12 & 9 & 6 & 8 & 7 \end{bmatrix}$ and the final degrees of the LSPB are $\mathbf{q}_f = \begin{bmatrix} \frac{\pi}{3} & \frac{\pi}{2} & \frac{\pi}{4} & \frac{\pi}{3} & \pi & \frac{\pi}{2} \end{bmatrix}$. Figures 12–14 present these results.

It can be seen in Figure 12 that the manipulator followed the desired trajectory, which meant that the correct gains were used. To obtain the minimum required gain, we calculated the maximum of each torque (L_1 to L_6) using the Lyapunov analysis shown in Figure 14 and converted it into voltage using Equation (8). With these taken into account, the maximum torques τ_0, the minimum required voltages V_0, and the voltage in each motor V_m are shown in Table 3.

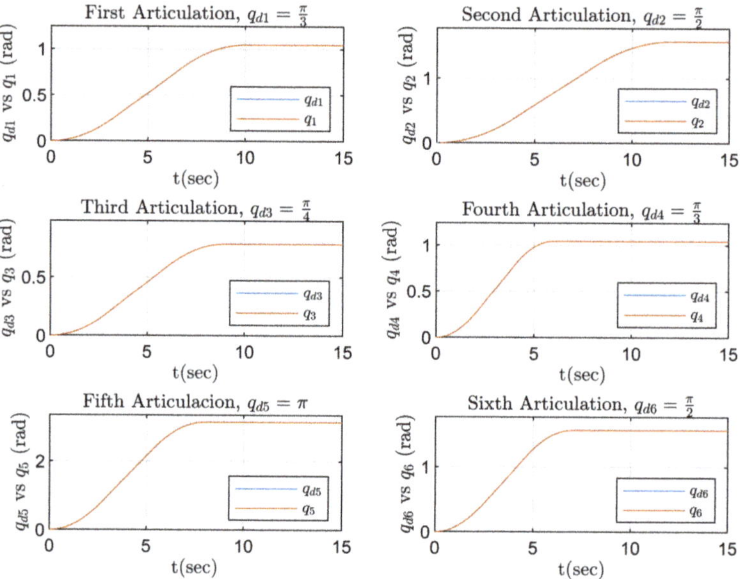

Figure 12. Trajectory tracking simulation of the applied SMC.

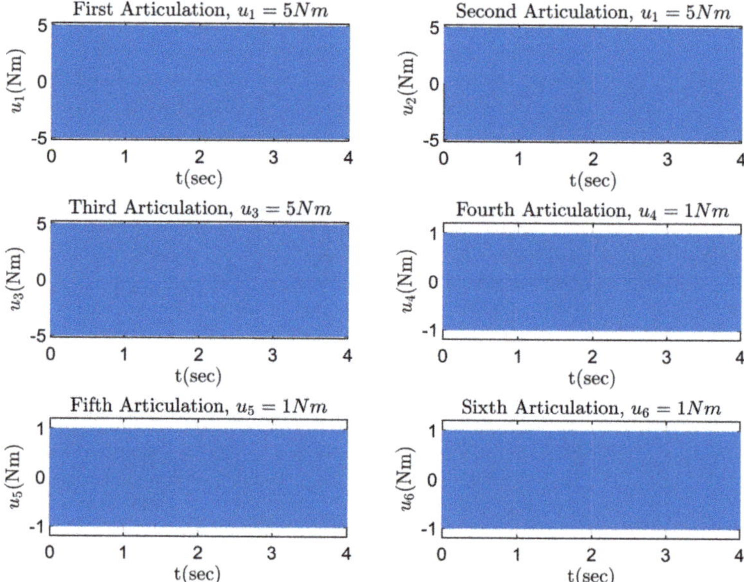

Figure 13. Input signal control.

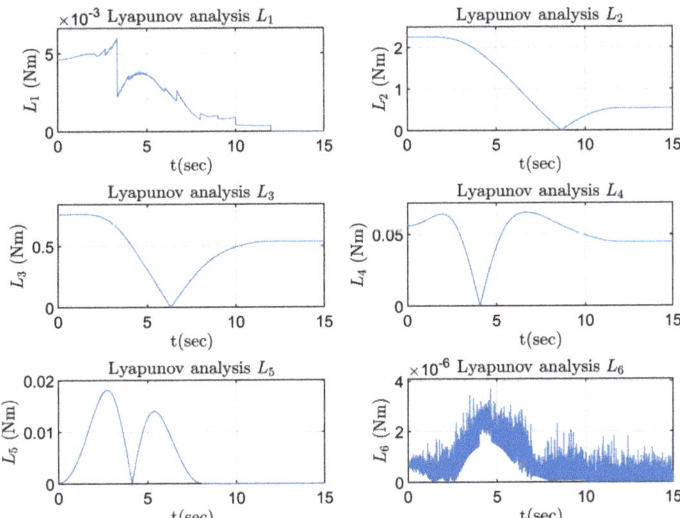

Figure 14. The signal of the Lyapunov analysis.

Table 3. Comparison between the voltage that was needed to follow the trajectory and the voltage of the electronic interface.

DOF	τ_0 (nm)	V_0 (Volts)	V_m (Volts)
1	0.0060	0.2665	12
2	2.2668	11.6858	12
3	0.7644	10.1807	12
4	0.0652	4.1253	5
5	0.0182	1.1529	5
6	3.71×10^{-6}	2.35×10^{-3}	5

When we compared the voltage in each motor V_m to the minimum required voltage (sliding-mode gain in voltage terms) V_0, $V_0 < V_m$. Therefore, the voltage in each motor was sufficient to track the LSPB trajectory. In this manner, the verification of the sliding mode was completed, and a physical application was performed. Figures 15 and 16 show the results of the physical implementation. As expected, the gain of the SMC was responsible for all tracking of the trajectories because the dynamic model of the manipulator (the known part of the ECM) was not in the control law. The consequence was a high magnitude discontinuous control law, as shown in Figure 16, which caused each DOF of the robot arm to vibrate (chatter) (see Figure 15). As can be seen in Figure 15, the physical arm followed the proposed trajectory of each DOF, as in the simulation. Thus, the voltage in each motor was sufficient. According to this result, we could validate the sliding mode gain using the Lyapunov stability analysis. This solved the problem of using iterative methods in the design of the sliding gain. A verification of this gain before this application has not been shown before in the literature. In this sense, the gain could be selected without increasing the gain in the experiments or simulations. The following advantages of the control strategy proposed in this paper were:

- The desired trajectory was perfectly followed without knowing the parameters of the manipulator;
- The SMC was a simple control strategy compared to CTC. The CTC had the dynamic model of the manipulator in its control law, which made it a complex law due to the 6DOF of the robot. For the SMC, we only had to know the sign of the surface, which did not depend on any parameter or perturbation of the manipulator. In

this way, the SMC programming was less complicated and performed better in the software environment;

- The gain validation of the SMC developed in this paper could be used as a straightforward tuning procedure. As mentioned above, the main drawback of the well-known PID controller is the tuning process. This made the SMC a better option in terms of selecting the gains of the controllers;
- There was no need to obtain the gain of the sliding mode using iterative methods in simulations or experiments.

However, even though the classic SMC had some great advantages, the following disadvantages were seen in the controller strategy when it was implemented:

- As mentioned in the literature, the chattering problem was present in the movement of the manipulator due to the discontinuous law. This could create fatigue in the actuators and the mechanical parts of the robot;
- A discontinuous law was implemented, which used all the gain in power of the electronic interface and developed an inefficient use of energy in the control law.

The most significant result of this paper was the gain validation of the SMC for its application in the experimental results. Furthermore, Lyapunov analysis was used as a simple tuning method for the gain of the SMC, which was a great advantage for this type of controller. However, the chattering phenomenon was present in robotic trajectory tracking. One of the reasons for the chattering in the manipulator was the numerical differentiator. In all of the simulations, little chattering was present because the velocity was taken directly from the model. However, in the implementation, the sample time of the SDEI induced numerical errors in the sliding surface. For better performance and a reduction in chattering, enhanced sliding mode controllers or hybrid sliding mode controllers need to be used. Similarly, when the classical SMC was applied with the numerical differentiator used in this paper, electronic interfaces with lower sample times had to be used. The approach for advanced differentiators could also reduce the chattering phenomenon.

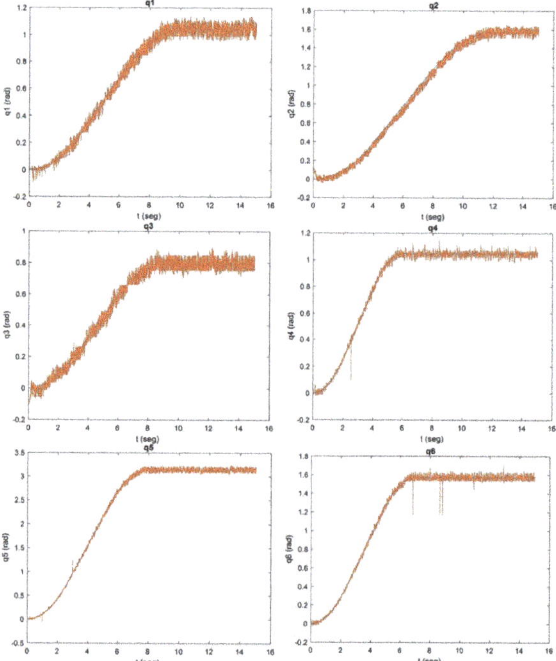

Figure 15. Tracking of the physical robot.

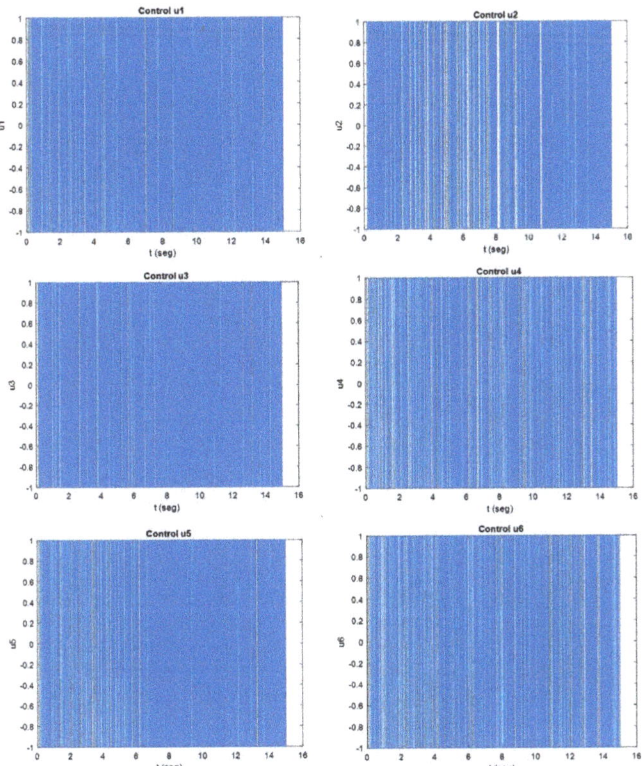

Figure 16. Signal of control applied to the physical manipulator.

6. Conclusions

In this article, the calculation of the SMC gain was performed using Lyapunov analysis and validated in a real application using a DC motor driven serial manipulator. The experimental results showed great performance in trajectory tracking with the selected gain. As we showed in Section 3, the SMC was a simple strategy compared to CTC, for which the dynamic model needed to be known a priori, and a 6DOF robot was too complex. Furthermore, in the numerical simulations, the CTC was not robust against perturbations and lost the trajectory. However, the discontinuous control law in the SMC generated chattering in the manipulator, which could cause damage to the actuators and mechanical parts. For this reason, in the future advanced SMC strategies that deal with the chattering problem must be taken into account.

Author Contributions: Conceptualization, I.R.-V., A.G.-R. and R.E.B.-A.; methodology, A.G.-R. and R.E.B.-A.; software, A.G.-R. and R.E.B.-A.; validation, A.G.-R. and R.E.B.-A.; formal analysis, A.G.-R., A.E.R.-M. and R.E.B.-A.; investigation, A.G.-R., A.E.R.-M. and R.E.B.-A.; writing—original draft preparation, A.G.-R., P.R.A.C.d.l.R., I.R.-V. and R.E.B.-A.; writing—review and editing, A.G.-R., P.R.A.C.d.l.R., I.R.-V. and R.E.B.-A.; supervision, A.G.-R., P.R.A.C.d.l.R., I.R.-V. and R.E.B.-A.; project administration, R.E.B.-A.; funding acquisition, R.E.B.-A. All authors have read and agreed to the published version of the manuscript.

Funding: This project was funded by the project thanks to the "Brazo manipulador con morfología adaptada a una plataforma móvil robotizada tipo industrial." TENM key: 6050.17-P.

Data Availability Statement: Not applicable.

Processes **2022**, *10*, 2699

Acknowledgments: The Tecnológico Nacional de México campus Instituto Tecnológico de Chihuahua is thanked for the economic funding "Brazo manipulador con morfología adaptada a una plataforma móvil robotizada tipo industrial." key: 6050.17-P. This work is grateful to CONACYT for supporting the first author with a scholarship.

Conflicts of Interest: The authors declare no conflict of interest.

Abbreviations

The following abbreviations are used in this manuscript:

CTC	Computed torque control
SMC	Sliding mode control
DOF	Degrees of freedom
ECM	Equivalent control method
NSMC	Novel sliding control method
EGWO	Extended gray wolf optimizer
SOSMC	Second-order sliding mode control
RBFN	Radial basis function network
DAM	Dual arm
MRAC	Model reference adaptive control
SPO	Sliding perturbation observer
STA	Super-twisting algorithm
ASMNN	Adaptive sliding mode neural network control
RBFNN	Radial basis function neural network
TDE	Time delay estimation
AISMC	Adaptive integral sliding mode control
6DOF	6 degrees of freedom
UAV	Unmanned aerial vehicle
ETC	Electrical time constant
MTC	Mechanical time constant
LSPB	Linear segment parabola blending
SDEI	Self-developed electronic interface

References

1. Spong, M.W.; Vidyasagar, M. *Robot Dynamics and Control*; John Wiley & Sons: Hoboken, NJ, USA, 2008.
2. Kelly, R. A tuning procedure for stable PID control of robot manipulators. *Robotica* **1995**, *13*, 141–148. [CrossRef]
3. Kelly, R.; Davila, V.S.; Perez, J.A.L. *Control of Robot Manipulators in Joint Space*; Springer Science & Business Media: Berlin, Germany, 2006.
4. Liu, M. Computed torque scheme based adaptive tracking for robot manipulators. In Proceedings of the 1995 IEEE International Conference on Robotics and Automation, Nagoya, Japan, 21–27 May 1995; Volume 1, pp. 587–590.
5. Llama, M.; Kelly, R.; Santibanez, V. Stable computed-torque control of robot manipulators via fuzzy self-tuning. *IEEE Trans. Syst. Man Cybern. Part B (Cybern.)* **2000**, *30*, 143–150. [CrossRef] [PubMed]
6. Jiang, J.; Cao, S.; Dai, Y. Research on RBF neural network model compensation and adaptive control of robot manipulators. In Proceedings of the 2016 Chinese Control and Decision Conference (CCDC), Yinchuan, China, 28–30 May 2016; pp. 516–520.
7. Shtessel, Y.; Edwards, C.; Fridman, L.; Levant, A. *Sliding Mode Control and Observation*; Springer: Berlin/Heidelberg, Germany, 2014; Volume 10.
8. Hu, Q.; Zhang, D.; Wu, Z. Trajectory planning and tracking control for 6-DOF Stanford manipulator based on adaptive sliding mode multi-stage switching control. *Int. J. Robust Nonlinear Control.* **2021**, *31*, 6602–6625. [CrossRef]
9. Utkin, V. Sliding mode control design principles and applications to electric drives. *IEEE Trans. Ind. Electron.* **1993**, *40*, 23–36. [CrossRef]
10. Haddad, S.Q.G.; Akkar, H.A. Intelligent swarm algorithms for optimizing nonlinear sliding mode controller for robot manipulator. *Int. J. Electr. Comput. Eng.* **2021**, *11*, 3943–3955. [CrossRef]
11. Pérez-San Lázaro, R.; Salgado, I.; Chairez, I. Adaptive sliding-mode controller of a lower limb mobile exoskeleton for active rehabilitation. *ISA Trans.* **2021**, *109*, 218–228. [CrossRef] [PubMed]
12. Lei, Y.l.; Wen, G.; Fu, Y.; Li, X.; Hou, B.; Geng, X. Trajectory-following of a 4WID-4WIS vehicle via feedforward–backstepping sliding-mode control. *Proc. Inst. Mech. Eng. Part D J. Automob. Eng.* **2022**, *236*, 322–333. [CrossRef]
13. Utkin, V.; Guldner, J.; Shi, J. *Sliding Mode Control in Electro-Mechanical Systems*; CRC Press: Boca Raton, FL, USA, 2017.
14. González, A.; Baray, R. Desarrollo del modelo dinámico de un brazo articulado de seis grados de libertad con comprobación. *Congr. Int. en Ing. Electrónica. Mem. ELECTRO* **2020**, *42*, 130–139.

15. Bailey, E.; Arapostathis, A. Simple sliding mode control scheme applied to robot manipulators. *Int. J. Control.* **1987**, *45*, 1197–1209. [CrossRef]
16. Chaturvedi, N.K.; Prasad, L. A comparison of computed torque control and sliding mode control for a three link robot manipulator. In Proceedings of the 2018 International Conference on Computing, Power and Communication Technologies (GUCON), Greater Noida, India, 28–29 September 2018; pp. 1019–1024.
17. Chu, X. Two degrees of freedom Cooperative Suspension Control for Maglev Wind Yaw System. *IEEE/ASME Trans. Mechatron.* **2021**. [CrossRef]
18. Gutiérrez-Oribio, D.; Mercado-Uribe, J.A.; Moreno, J.A.; Fridman, L. Robust global stabilization of a class of underactuated mechanical systems of two degrees of freedom. *Int. J. Robust Nonlinear Control.* **2021**, *31*, 3908–3928. [CrossRef]
19. Rahmani, M.; Komijani, H.; Rahman, M.H. New sliding mode control of 2-DOF robot manipulator based on extended grey wolf optimizer. *Int. J. Control. Autom. Syst.* **2020**, *18*, 1572–1580. [CrossRef]
20. Tuan, L.A.; Joo, Y.H.; Tien, L.Q.; Duong, P.X. Adaptive neural network second-order sliding mode control of dual arm robots. *Int. J. Control. Autom. Syst.* **2017**, *15*, 2883–2891. [CrossRef]
21. Tuan, L.A.; Joo, Y.H.; Duong, P.X.; Tien, L.Q. Parameter estimator integrated-sliding mode control of dual arm robots. *Int. J. Control. Autom. Syst.* **2017**, *15*, 2754–2763. [CrossRef]
22. Chen, K.Y. Robust optimal adaptive sliding mode control with the disturbance observer for a manipulator robot system. *Int. J. Control. Autom. Syst.* **2018**, *16*, 1701–1715. [CrossRef]
23. Utkin, V.I. *Sliding Modes in Control and Optimization*; Springer Science & Business Media: Berlin, Germany, 2013.
24. Kallu, K.D.; Jie, W.; Lee, M.C. Sensorless reaction force estimation of the end effector of a dual-arm robot manipulator using sliding mode control with a sliding perturbation observer. *Int. J. Control. Autom. Syst.* **2018**, *16*, 1367–1378. [CrossRef]
25. Jeong, C.S.; Kim, J.S.; Han, S.I. Tracking error constrained super-twisting sliding mode control for robotic systems. *Int. J. Control. Autom. Syst.* **2018**, *16*, 804–814. [CrossRef]
26. Yen, V.T.; Nan, W.Y.; Van Cuong, P. Robust adaptive sliding mode neural networks control for industrial robot manipulators. *Int. J. Control. Autom. Syst.* **2019**, *17*, 783–792. [CrossRef]
27. Truong, L.V.; Huang, S.D.; Yen, V.T.; Van Cuong, P. Adaptive trajectory neural network tracking control for industrial robot manipulators with deadzone robust compensator. *Int. J. Control. Autom. Syst.* **2020**, *18*, 2423–2434. [CrossRef]
28. Fei, J.; Wang, H.; Fang, Y. Novel Neural Network Fractional-Order Sliding-Mode Control With Application to Active Power Filter. *IEEE Trans. Syst. Man, Cybern. Syst.* **2021**. [CrossRef]
29. Baek, J.; Jin, M.; Han, S. A new adaptive sliding-mode control scheme for application to robot manipulators. *IEEE Trans. Ind. Electron.* **2016**, *63*, 3628–3637. [CrossRef]
30. Lee, J.; Chang, P.H.; Jin, M. Adaptive integral sliding mode control with time-delay estimation for robot manipulators. *IEEE Trans. Ind. Electron.* **2017**, *64*, 6796–6804. [CrossRef]
31. Kim, H.H.; Lee, M.C.; Kyung, J.H.; Do, H.M. Evaluation of Force Estimation Method Based on Sliding Perturbation Observer for Dual-arm Robot System. *Int. J. Control. Autom. Syst.* **2021**, *19*, 1–10. [CrossRef]
32. Dumlu, A. Design of a fractional-order adaptive integral sliding mode controller for the trajectory tracking control of robot manipulators. *Proc. Inst. Mech. Eng. Part I J. Syst. Control. Eng.* **2018**, *232*, 1212–1229. [CrossRef]
33. Mofid, O.; Mobayen, S.; Wong, W.K. Adaptive terminal sliding mode control for attitude and position tracking control of quadrotor UAVs in the existence of external disturbance. *IEEE Access* **2020**, *9*, 3428–3440. [CrossRef]
34. Liu, K.; Wang, Y.; Ji, H.; Wang, S. Adaptive saturated tracking control for spacecraft proximity operations via integral terminal sliding mode technique. *Int. J. Robust Nonlinear Control.* **2021**, *31*, 9372–9396. [CrossRef]

Article

Estimation of Chlorine Concentration in Water Distribution Systems Based on a Genetic Algorithm

Leonardo Gómez-Coronel [1], Jorge Alejandro Delgado-Aguiñaga [2],*, Ildeberto Santos-Ruiz [1] and Adrián Navarro-Díaz [3]

[1] Tecnológico Nacional de México, I. T. Tuxtla Gutiérrez, TURIX-Dynamics Diagnosis and Control Group, Carretera Panamericana S/N, Tuxtla Gutiérrez C.P. 29050, Mexico
[2] Centro de Investigación, Innovación y Desarrollo Tecnológico CIIDETEC-UVM, Universidad del Valle de México, Tlaquepaque C.P. 45601, Mexico
[3] Tecnologico de Monterrey, Escuela de Ingeniería y Ciencias, Zapopan C.P. 45138, Mexico
* Correspondence: jorge.delgado@uvmnet.edu

Abstract: This paper proposes a methodology based on a genetic algorithms (GA) to calibrate the parameters of a chlorine decay model in a water distribution system (WDS). The proposed methodology first contemplates that a GA is implemented using historical measurements of chlorine concentration at some sensed nodes to calibrate the unknown values corresponding to both the bulk and wall reaction coefficients. Once both parameters are estimated, the optimal-fit chlorine decay model is used to predict the decay of chlorine concentration in the water at each node for any concentration input at the pumping station. Then, a second GA-based algorithm is implemented to obtain the minimal chlorine concentration needed at the input to ensure that every node in the system meets the official normativity requirements for free chlorine in a WDS. The proposed methodology performed satisfactorily for a WDS simulated in EPANET with a GA implemented in MATLAB, both for the estimation of the reaction coefficients and the optimization of the required chlorine concentration at the input. Simulation results illustrate the performance of the proposed algorithm.

Keywords: model calibration; genetic algorithm; optimization; hydraulic network; water quality; chlorine

---

Citation: Gómez-Coronel, L.; Delgado-Aguiñaga, J.A.; Santos-Ruiz, I.; Navarro-Díaz, A. Estimation of Chlorine Concentration in Water Distribution Systems Based on a Genetic Algorithm. *Processes* **2022**, *11*, 676. https://doi.org/10.3390/pr11030676

Academic Editors: Jose Carlos Pinto and Iqbal M. Mujtaba

Received: 12 January 2023
Revised: 17 February 2023
Accepted: 20 February 2023
Published: 23 February 2023

1. Introduction

Water management companies focus on meeting water demand and ensuring chlorine concentration requirements. In recent decades, water pollution has unfortunately increased worldwide and as a result, continuous monitoring to maintain the integrity of drinking water has gained increasing importance. Water quality is a broad term that is related to many aspects of both the physics and the chemistry of drinking water. According to [1], the quality of the water is closely related to parameters such as turbidity, temperature, color, taste, solids in suspension, pH, and the concentration of many elements (such as nitrogen, fluoride and other heavy metals) and substances such as chlorine residual (Cl_2). From all of the abovementioned parameters, chlorine residual is typically the most addressed. Even if chlorine does not exist naturally in water, it is added to it for disinfection purposes, specially in the case of drinking and waste water. In [2] it is stated that the residual chlorine concentration maintained in the distribution system ensures a good sanitary quality of the water being distributed, which makes it one of the most important parameters to monitor to make certain the quality of the water in a distribution system.

As stated in [3], both laboratory and online measurements are necessary to account for chlorine concentration monitoring. Typically, bacteriological measurements (e.g., the number of *Escherichia coli* bacteria) are among the typical laboratory measurements in a water distribution system (WDS). However, laboratory testing cannot be performed online for obvious reasons and this delays the time in which the quality of the water being distributed is known. As a result, water agencies rely on online measurements to estimate the water quality in the WDS. In this way, adequate levels of chlorine allow immediate elimination of harmful bacteria and viruses and provide a protective residual throughout the drinking water distribution network [4]. However, because drinking water systems are complex, it is not easy to ensure that residuals are maintained throughout the distribution network under permissible levels. On the one hand, excessive chlorine dosages may consequently lead to formation of chlorine by-products, some of which are carcinogenic and may also cause birth-related problems such as low birth weight and genetic malformations [5]. On the other hand, an insufficient chlorine residual could cause the multiplication of pathogenic bacteria, which in turn could trigger public health problems with severe consequences.

Since chlorine reacts with organic matter found in water, its concentration decreases over time. Thus residual chlorine decay models can predict this decrease, including chlorine levels to comply with the normative standard that establishes the lower and upper bounds in which free chlorine can be found. The World Health Organization (WHO) indicates that to adequately treat drinking water, a minimal chlorine concentration of 0.2 mg/L should be maintained right to the point of consumer delivery, and that to avoid harmful effects to public health, it should not be found in concentrations higher than 5 mg/L [6]. In particular, this paper will address the indications given by the the Mexican norm NOM-127-SSA1-1994 [7], which indicates that free chlorine in a WDS must not be found in concentrations lower than 0.2 mg/L or higher than 1.5 mg/L.

In the literature, several approaches that analytically describe and predict the decay of free chlorine concentration in drinking WDS have been reported. In [8] a sequential steady-state approach compared to field-measured data was reported. In the same way, a mass-transfer-based model approach was presented in [9]. The authors highlighted that such an approach takes into consideration first-order reactions of chlorine that take place in both the bulk flow and in the pipe wall, demonstrating how higher decay rates of chlorine are associated with smaller pipe diameters and higher flow velocities. The effect of fluid velocity and pipe inner radius in chlorine propagation was studied in [10]. Here an EPANET model of the WDS was used, and it was shown how the same variables that affect the propagation of residual chlorine levels can potentially affect disinfection performance. Following this direction, in [11] a computational scheme was proposed to accurately and efficiently treat kinetic processes in chlorine concentration models. It was observed that both a higher accuracy and a better computational efficiency were achieved when comparing it with similar schemes. The effect of water temperature in chlorine decay was also addressed in [12]. It was evidenced how chlorine concentration experiences a faster decay rate in fresh samples than in those re-chlorinated with sodium hipochlorite. Moreover, a combined first- and second-order model was presented in [12]. This approach showed a more accurate model behavior compared with a conventional first-order model. It was pointed out that chlorine decay is mainly influenced by its initial concentration, which explains why chlorine decays faster during the initial stages of the propagation process. An inverse-model-based technique to estimate the input chlorine concentration needed to meet a specified value at a desired node in the WDS was explored in [13]. Results evidenced an excellent agreement between the inverse method and the direct simulation technique. Interval state observers were also used in [3] to model the decay of chlorine concentration, showing sufficient numerical efficiency for on-line implementations of the proposed method.

Computational models have also been used to simulate this chemical phenomena. In [14], a computer model was used to predict the spatial and temporal distribution of a residual constituent with first-order reaction rate in a pipe network under variable unsteady flow conditions. Even if a good agreement with results from a standard extended-period simulation during the initial stages was obtained, the chlorine concentration at different nodes in the network presented deviation when the flow regime became more unsteady. The main issue noted about computational models of chlorine concentration and decay was that simulation environments commonly require to know the bulk reaction coefficient and wall reaction coefficient with accuracy. In the last decades, the well-known EPANET software has been used to simulate, besides hydraulic behavior, the decay and diffusion of substances such as chlorine in a WDS. In particular, the *Multi-Species eXtension* (EPANET-MSX) allows for the consideration of multiple interacting species in the bulk flow and on the pipe walls. This capability has been incorporated into both a stand-alone executable program and a toolkit library of functions that programmers can use to build customized applications. Several contributions have been reported by using this approach, as in [15], where the performance of the proposed model as well as an nth-order decay kinetics was tested for the full-scale modeling of chlorine concentration in a WDS. It was shown how a similar level of accuracy can be achieved with the tested models, provided that there is a good calibration of the decay reaction coefficients. Following this direction, a new model of chlorine-wall reaction for simulation of the chlorine concentration in a WDS was proposed in [16], which provides the accurate chlorine modeling required to improve disinfection strategies in a WDS when implemented in EPANET-MSX (or similar simulation software). Alternative simulation software environments have also been used to model the decay of chlorine concentration, such as the work presented in [17], where the authors presented an alternative methodology based on AQUASIM to model chlorine concentration in a distribution network; however, the authors stated that further evaluation of the methodology on a larger scale WDS is needed. In [18], a methodology to efficiently calibrate chlorine decay models was proposed and applied to a part of the Barcelona drinking water distribution network. Measurements and simulated chlorine concentrations at nodes with a sensor were compared to calibrate the chlorine bulk reaction coefficient. From the artificial-intelligence (AI)-based approach, a work was presented in [19] where a genetic algorithm (GA) was used in a real WDS together with particle swarm optimization (PSO) and a hybrid GA-PSO as three different AI-based approaches to calibrate the wall reaction coefficient while minimizing the residuals for chlorine concentrations. The obtained results clearly demonstrated that the hybrid GA-PSO outperformed the other two algorithms when considering only one decision variable (the wall reaction coefficient).

Based on the abovementioned research and since water quality monitoring has become more and more important worldwide for its inherent impact on public health, the development of more efficient water quality monitoring techniques remains in challenge nowadays. For this reason the main contribution presented in this paper is an alternative method based on a GA for the estimation of both the bulk and wall reaction coefficients in a WDS by using a reduced number of chlorine meters. This approach allows the chlorine concentration along the WDS to be estimated with accuracy, which in turn aids the water management staff to adopt a better feed-chlorine strategy. To the best of our knowledge, no previous investigations have focused on the estimation of both the bulk and wall reaction coefficients simultaneously by means of a GA.

This paper is organized as follows: Sections 2 and 3 provide with the theoretical background on the genetic algorithm and the reactions within a pipe involved in the decay of chlorine concentration in water, respectively. Section 4 explains in depth the proposed methodology for the GA-driven calibration of the bulk and wall reaction coefficients and the estimation of the minimal required chlorine concentration at the input. Section 5 illustrates the obtained simulation results, and finally Section 6 discusses conclusions and future work proposals.

2. The Genetic Algorithm

This paper proposes the use of the genetic algorithm (GA) to calibrate the unknown parameters of the chlorine-decay model in EPANET for the WDS under study. The steps followed by the GA to perform the calibration of the desired parameters are:

1. Creation: An initial population is created and distributed as uniformly as possible across a pre-defined search area.

2. Selection: A selection mechanism is used to assign a probability for each individual in the population that their genes will prevail to the next generation. This selection mechanism is commonly a fitness/cost function computed for each individual, where the fittest individuals are awarded a higher probability.

3. Cross-over: Once the fittest individuals are selected, the cross-over stage takes place to create the following generation. For this purpose, the genes of two *parent* solutions are recombined to create two new *offspring* solutions.

4. Mutation: The last evolutionary stage is the random mutation of some of the genes in the newly-created population. This random mutation serves two purposes: (1) it stops the population from becoming uniform and helps maintain some level of diversity in the individuals; (2) it prevents the algorithm from failing to find locally-optimal solutions instead of the desired global optimal.

This process takes place during a pre-set number of generations or while the mean performance of the newly created population is significantly better than the previous one. Once any termination criteria are met, the GA provides the best-performing solution. The entire process is represented graphically in Figure 1, which is designed on the basis of [20,21].

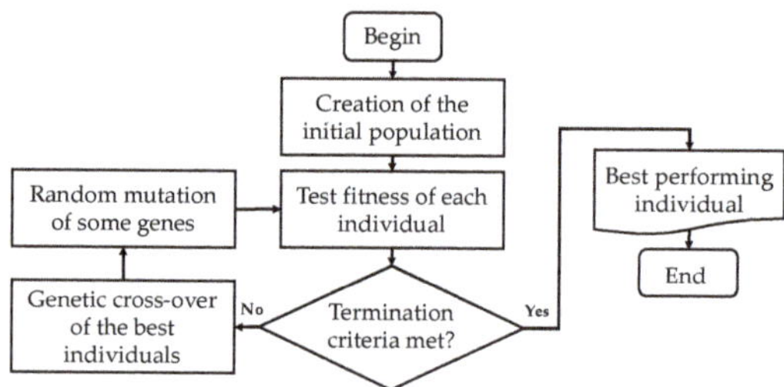

Figure 1. The genetic algorithm process as a block diagram.

3. Reaction Zones within a Pipeline

The decay of a substance by reaction as it travels through a water distribution system is an important topic of study. To model this decay, the rate at which this substance reacts and the relation between this reaction and the substance concentration must be known. According to [22], the reaction can occur in two different ways: (1) within the bulk flow; and (2) with the material along the pipe wall. The two ways in which reaction of a substance can occur are visualized in Figure 2, where free chlorine (HOCL) reacts with natural organic matter (NOM) in the bulk flow to produce the disinfectant by-product (DBP), and is also transported through the boundary layer of the pipe wall to oxidize the iron (Fe) released from through-wall corrosion.

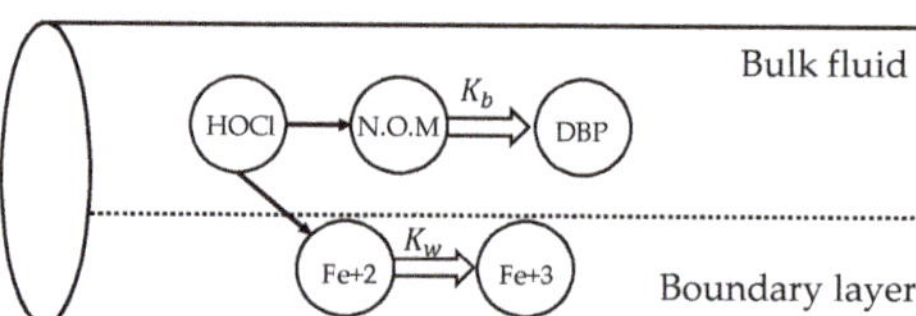

Figure 2. Reaction zones within a pipeline.

3.1. Bulk Reactions

Reaction occurring in the bulk flow is modeled with n-th order kinetics, where the instantaneous rate of reaction R (mass/volume/time) is assumed to be concentration-dependent, according to [22] as:

$$R = K_b C^n,\tag{1}$$

where K_b is known as the bulk reaction coefficient, C is the reactant concentration (mass/volume) and n is the reaction order. The units of K_b are concentration raised to the $(1 - n)$ power divided by time, and its value is positive for the case of growth reactions and negative for decay reactions. Cases where a limiting concentration exists in the ultimate growth or loss of the substance can also be considered, where Equation (1) is expressed as:

$$R = K_b(C_L - C)C^{(n-1)} : n > 0, K_b > 0, \tag{2}$$

$$R = K_b(C - C_L)C^{(n-1)} : n > 0, K_b < 0, \tag{3}$$

where C_L is the limiting concentration of the substance. Some cases of well-known kinetic models are provided in Table 1.

Table 1. Different kinetic models.

Model	Parameters
First-order decay	$C_L = 0, K_b < 0, n = 1$
First-order saturation growth	$C_L > 0, K_b > 0, n = 1$
Zero-order kinetics	$C_L = 0, K_b <> 0, n = 0$
No reaction	$C_L = 0, K_b = 0$

The diffusion of chlorine in a WDS is an example of a first-order decay reaction, and thus for first-order reactions ($n = 1$) K_b has units of 1/day. Even if an experimental estimation of K_b can be performed, the reliability of the estimated value depends on the careful measurement of the reactant concentration at each sampling time (which is prone to human mistake) as well as taking into consideration the effects of temperature in the decay of the substance. This is a strong argument to use a computer-based estimation of this parameter.

3.2. Wall Reactions

The rate at which chlorine concentration decay reactions take place near the wall of the pipe can be considered to be dependent on the concentration in the bulk flow by the expression:

$$R = (A/V)K_wC^n, \tag{4}$$

where K_w is known as the wall reaction coefficient and A/V is the surface area per unit volume within a pipe segment. This latter term converts the mass reacting per unit of wall area to a per unit volume basis. Typically, K_w values range from 0 to 1.5 m/day. The value of K_w should be adjusted to consider any mass-transfer limitation in moving reactants and products between the bulk flow and the wall. The wall reaction coefficient depends on the temperature and can also be affected by the age of the pipe, since the increase of encrustation, tuberculation or corrosion produces a higher roughness coefficient, resulting in a greater head-loss through the pipe. Some evidence has shown that the same processes that impact the roughness coefficient in a pipe also tend to increase the reactivity of the inner wall with some types of chemical species, in particular chlorine and other types of disinfectants. The value of K_w can be modeled as a function of the coefficient used to describe the roughness of the pipeline (this is, the chosen head-loss model) as shown in Table 2.

Table 2. Wall reaction formulas related to head-loss.

Head-Loss Model	Wall Reaction Coefficient
Hazen–Williams	$K_w = F/C$
Darcy–Weisbach	$K_w = -F/\log_{10}(\varepsilon/D)$
Chezy–Manning	$K_w = Fn$

In the table, C is the Hazen–Williams C-factor, ε is the absolute roughness of the pipe, D is the pipe diameter and n is the Manning roughness coefficient. The coefficient F should be developed according to site-specific field measurements and it will have different values and units depending on the head-loss equation used.

4. Proposed Methodology

4.1. Estimation of K_b and K_w

Considering the availability of sensors in practice is always a limitation, the idea is to estimate the chlorine concentration in the whole WDS (all nodes) at any time regardless of the reduced number of chlorine sensors. To do that, the proposed methodology considers a few chlorine measurements at some nodes. It is also considered that those measurements are obtained for a duration of at least two days, i.e., 48 h. The performance of the proposed methodology is tested using synthetic data generated via a simulation. To do that, initial conditions of the sourced chlorine at the pumping station and initial chlorine concentration at the nodes are assumed to be known, whereas the bulk and wall reaction coefficients are considered as unknown variables. Then, for every node, the chlorine concentration value at each sampling time is obtained by performing a simulation process through the well-known EPANET-MATLAB Toolkit [23].

This methodology relies on a WDS model with an input format (.inp) suitable for the EPANET software, for which the hydraulic parameters roughness coefficient ε and minor loss coefficient K are well-adjusted such that it is reliable in terms of hydraulic behavior (see [22]). Let us now consider that n denotes a reduced number of chlorine sensors installed along the WDS and their measurements are recorded and stored in a matrix $\alpha \in \mathbb{R}^{m \times n}$ where m denotes the number of samples. Then, the GA-driven calibration process begins as follows: a population of solution vectors $\mathbf{P} = [\mathbf{c}_1, \mathbf{c}_2, \mathbf{c}_3, \ldots, \mathbf{c}_{\mathcal{P}}]$ is created where $\mathcal{P}$ is selected as the population size and each solution vector $\mathbf{c}$ is structured as:

$$\mathbf{c}_i = \begin{bmatrix} K_{b_i} & K_{w_i} \end{bmatrix}, \tag{5}$$

where K_{b_i} is the i-th candidate bulk reaction coefficient and K_{w_i} is the i-th candidate wall reaction coefficient, such that:

$$K_{b_i} \in (lb, ub); \quad K_{w_i} \in (lb, ub),$$

with lb and ub denoting the lower and upper bounds, respectively. Then, $\forall \mathbf{c} \in \mathbf{P}$ the candidate set of K_{b_i} and K_{w_i} values are assigned into the .inp EPANET file as the bulk and wall reaction coefficients of each pipe, and thus a 24 h simulation is performed to compute the chlorine concentration along the WDS. The software-driven simulation provides the estimation of the chlorine concentration at each node in the network, where attention should be directed to those nodes with a sensor. Next, let us denote that $\widehat{\alpha} \in \mathbb{R}^{m \times n}$ is a matrix containing the estimations of chlorine concentration at nodes for which the corresponding chlorine measurements are also available such that an error matrix $e \in \mathbb{R}^{m \times n}$ can be computed as follows:

$$e = \alpha - \widehat{\alpha}, \tag{6}$$

where column 1 of e contains the estimation error for the first measured node and column 2 contains the estimation error for the second measured node, and so on. For each node the mean squared estimation error (mse) can be computed as:

$$\text{mse}_j = e_{\text{all},j}^{\mathsf{T}} e_{\text{all},j} / m, \qquad \forall j \in \{1, 2, 3, \ldots, n\}. \tag{7}$$

Finally, the global mean squared error (mse_{G}) containing the overall estimation error is computed as:

$$\text{mse}_{\text{G}} = \frac{1}{n} \sum_{j=1}^{n} \text{mse}_j, \tag{8}$$

where mse_G is set as the optimization goal for the GA, that is, a proposed **c** vector is found that contains the optimal values for K_b and K_w that minimize the value of mse_G. It is not expected that a single generation will provide the optimal values and thus the best-performing **c** vectors are selected for the cross-over and mutation stages, which are repeated iteratively until a satisfying match is found. The methodology presented in this paper is programmed with the aid of the OpenWater Analytics EPANET-MATLAB Toolkit [23], which is an open-source software used for interfacing the EPANET tools in a MATLAB programming environment. As a pseudocode, this process is summarized as follows:

Pseudocode 1: GA-driven calibration for K_b and K_w reaction coefficients

---

INPUT: Array $\alpha \in \mathbb{R}^{m \times n}$:

$$
\alpha = \begin{bmatrix}
N_{1,1}(k-m) & N_{1,2}(k-m) & N_{1,3}(k-m) & N_{1,4}(k-m) & \cdots & N_{1,n}(k-m) \\
\vdots & \vdots & \vdots & \vdots & \cdots & \vdots \\
N_{m-1,1}(k-1) & N_{m-1,2}(k-1) & N_{m-1,3}(k-1) & N_{m-1,4}(k-1) & \cdots & N_{m-1,n}(k-1) \\
N_{m,1}(k) & N_{m,2}(k) & N_{m,3}(k) & N_{m,4}(k) & \cdots & N_{m,n}(k)
\end{bmatrix} \tag{9}
$$

where columns $N_1\,N_2, N_3, \ldots, N_n$ contain the chlorine concentration at nodes with a sensor and m denotes the number of samples.

LOAD into memory the .inp file of the WDS.
SET the quality time-step in the .inp file for simulation as the sampling rate of the measurements.
CREATE the array of sensed nodes $\mathbf{N} = [N_1, N_2, N_3, \ldots, N_n]$.
DECLARE the GA parameters: population size $\mathcal{P}$, maximum generations $\mathcal{G}$, minimum tolerance $e_{\min}$, lower lb and upper ub search area bounds.
CREATE the initial population $\mathbf{P} = [\mathbf{c}_1, \mathbf{c}_2, \mathbf{c}_3, \ldots, \mathbf{c}_{\mathcal{P}}]$; $\mathbf{c}_i = \begin{bmatrix} K_{b_i} & K_{w_i} \end{bmatrix}$; $i = [1, 2, 3, \ldots, \mathcal{P}]$.
$C_{\text{gen}} = 1$
IF $C_{\text{gen}} < \mathcal{G}$
    FOR each $\mathbf{c}_i \in \mathbf{P}$
        SET K_{b_i} and K_{w_i} in the .inp file.
        COMPUTE the water-quality time series with help of the EPANET-MATLAB Toolkit
        BUILD the array $\widehat{\alpha} \in \mathbb{R}^{m \times n}$ similarly as α.
        FOR $j = [1, 2, 3, \ldots, n]$
            FOR $k = [1, 2, 3, \ldots, m]$
                $e_{k,j} = \alpha_{k,j} - \widehat{\alpha}_{k,j}$
            END FOR
        END FOR
        INITIALIZE $\text{mse} \in \mathbb{R}^{1 \times n}$
        FOR $j = [1, 2, 3, \ldots, n]$
            $\text{mse}_j = \left(e_{\text{all},j}^{\mathsf{T}} e_{\text{all},j} \right) / m$
        END FOR
        $\text{mse}_G = \frac{1}{n} \sum_{j=1}^{n} \text{mse}_j$
    END FOR
    CHOOSE the **c** values that provide the lowest mse_G.
    IF lowest $\text{mse}_G > e_{\min}$
        CREATE new population $\mathbf{P} = [\mathbf{c}_1, \mathbf{c}_2, \mathbf{c}_3, \ldots, \mathbf{c}_{\mathcal{P}}]$, $C_{\text{gen}} \leftarrow C_{\text{gen}} + 1$ and **CONTINUE**.
    ELSE
        END ALGORITHM
    END IF
ELSE
    END ALGORITHM
END IF

OUTPUT: Values for K_b and K_w that minimize mse_G.

4.2. Analysis of the Minimal Chlorine Concentration in the WDS

Once a set of K_b and K_w values that minimize the overall error have been calibrated, the minimal required input chlorine must be determined. For this, let us denote $\beta \in \mathbb{R}^{m \times \mathcal{N}}$, where $\mathcal{N}$ denotes the total number of nodes in the WDS and β is a matrix containing the m simulated values for the chlorine concentration at every node. From β, a vector γ can be built such that $\gamma = \min(\beta)$. It is now safe to state that the minimal value in γ corresponds to the overall minimal value for the experiment time. Let us denote this overall minimal value as $\min_\gamma$. To meet the official normativity requirements for the WDS, ideally, this $\min_\gamma$ must be contained between the lower and upper bounds (0.2 to 1.5 mg/L) listed in the official norm [7]. Please note that at this point it is unknown which chlorine concentration is needed at the input to ensure that a chlorine concentration of at least of 0.2 mg/L is met at each node in the WDS anytime. Since [7] also provides a maximum permitted value of 1.5 mg/L, it is not possible to simply add chlorine at the input of the WDS until the minimal value is met at the expense of exceeding the permitted upper bound. Under said conditions, the problem now turns into finding the optimal chlorine concentration between the stated lower and upper bounds at the input to ensure that the minimal normativity value is met. This analysis will also contemplate the possibility that the WDS will require a re-design if it is not possible to meet the normativity requirements under the permitted source chlorine concentration input.

The proposed optimization process is explained as follows:

1. First, the calibrated K_b and K_w values are set into an .inp file as the actual bulk and wall reaction coefficients of the pipeline.
2. Then, a GA is set to minimize the value s_Q corresponding to the chlorine concentration at the input as a restriction is also satisfied: that $\min_\gamma > \min_Q$, where $\min_Q$ is the overall minimal required chlorine concentration in the WDS.
3. By setting a proposed value s_Q to the model and performing the simulation of the chlorine decay, matrix β and vector γ are obtained as previously explained.
4. Then, the fitness of a given s_Q value is defined as follows:

    (a) If $\min_\gamma < \min_Q$ then a very low fitness is assigned to s_Q;
    (b) If $\min_\gamma \geq \min_Q$ then the fitness of s_Q is defined as the difference between the two values.

5. The search process continues until a $s_Q \in [\,0.2, 1.5\,]$ (mg/L) value is found that minimizes the difference between $\min_\gamma$ and $\min_Q$.

As a pseudocode, this process is described as follows:

Pseudocode 2: GA-driven calibration for minimal required Q value

---

INPUT: $\min_Q$

LOAD into memory the .inp file with the previously calibrated K_b and K_w values.
DEFINE the population size $\mathcal{P}$, the maximum generations $\mathcal{G}$ and the minimum tolerance e_{min}.
DEFINE the lower (lb) and upper bound (ub) for the search area as stated by the normativity.
CREATE the initial population $\mathbf{P} = \left[s_{Q_1}, s_{Q_2}, s_{Q_3}, \ldots, s_{Q_{\mathcal{P}}}\right]; i = [1, 2, 3, \ldots, \mathcal{P}]$.
$C_{gen} = 1$
IF $C_{gen} < \mathcal{G}$
    FOR each $s_{Q_i} \in \mathbf{P}$
        ASSIGN s_{Q_i} to the model.
        PERFORM the water-quality simulation with help of the EPANET-MATLAB Toolkit
        CREATE matrix $\beta \in \mathbb{R}^{m \times \mathcal{N}}$.
        $\gamma = \min(\beta)$
        $\min_\gamma = \min(\gamma)$
        IF $\min_\gamma < \min_Q$
            Error = 1000 (Very high value)
        ELSE IF $\min_\gamma \geq \min_Q$
            Error = $\min_\gamma - \min_Q$
        END IF
    END FOR
    CHOOSE the s_Q values that provide the lowest Error.
    IF lowest Error $> e_{min}$.
        CREATE new population $\mathbf{P} = \left[s_{Q_1}, s_{Q_2}, s_{Q_3}, \ldots, s_{Q_{\mathcal{P}}}\right]$, $C_{gen} \leftarrow C_{gen} + 1$ and **CONTINUE**.
    ELSE
        END ALGORITHM
ELSE
    END ALGORITHM
END IF

OUTPUT: Optimal value for s_Q that ensures $\min_\gamma \geq \min_Q$.

---

5. Simulation Results

To demonstrate the performance of the proposed methodology, data from a simulation environment will be used. In Figure 3 the system under study is presented, whereas its general topology is presented in Figure 4.

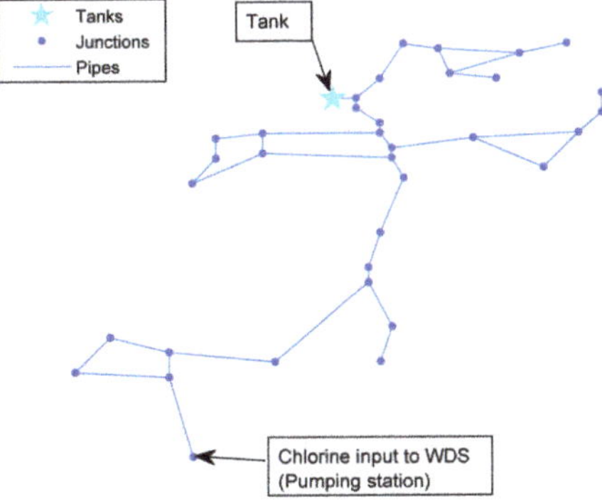

Figure 3. WDS under study.

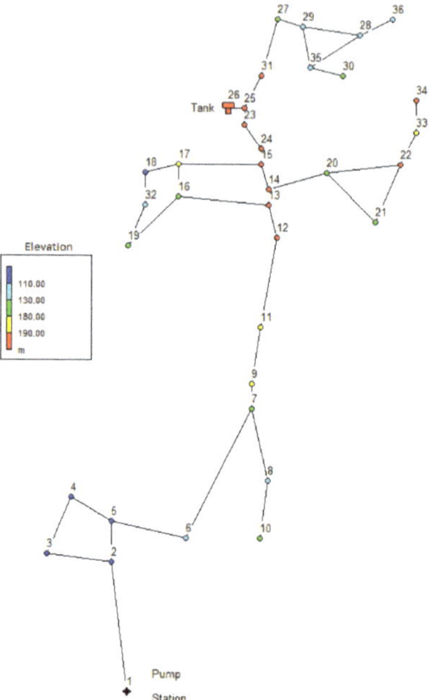

Figure 4. General topology of the WDS under study.

The system is made up of 1 pumping station (node N_1) that is also the input node for chlorine, 1 tank (node N_{36}), 34 connection nodes, and 40 pipe segments. It is assumed that only nodes N_5, N_{10}, N_{15}, N_{20}, and N_{25} have chlorine concentration meters installed; thus the EPANET-simulated chlorine concentration values at said nodes will be used as the reference value for calibration, while in experimental applications of this methodology these values can be replaced by physical measurements without major changes. At the pumping station, the input chlorine concentration is 0.8 mg/L, and also the initial chlorine concentration condition at the entire WDS is 0.5 mg/L. Note that the values corresponding to K_b and K_w are still unknown. Figure 5 shows the evolution of the chlorine concentration value as simulated by EPANET for nodes N_5, N_{10}, N_{15}, N_{20}, and N_{25} for a period of 198,000 s = 55 h.

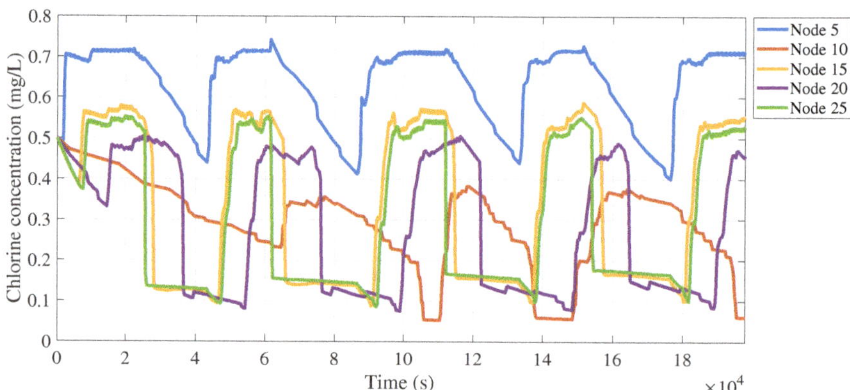

Figure 5. Historical evolution of chlorine concentration.

The chlorine concentrations illustrated in Figure 5 will be considered as the calibration objective and thus the GA-based methodology will try to adjust the values of the K_b and K_w coefficients in the .inp file of the system that when set as the actual bulk and wall reaction coefficients provide a low-error fit between the EPANET-simulated values (obtained with the real K_b and K_w coefficients) and the GA-fitted-model estimated chlorine concentrations. After the calibration process was finished, the obtained results for K_b and K_w are:

$$K_b = -0.3008\,[1/\text{day}] \qquad K_w = -0.9984\,[\text{ft}/\text{day}] = -0.3043\,[\text{m}/\text{day}]. \tag{10}$$

Figures 6–10 show the comparison between the EPANET-simulated and the GA-fitted-model chlorine concentration values in nodes 5, 10, 15, 20, and 25 respectively.

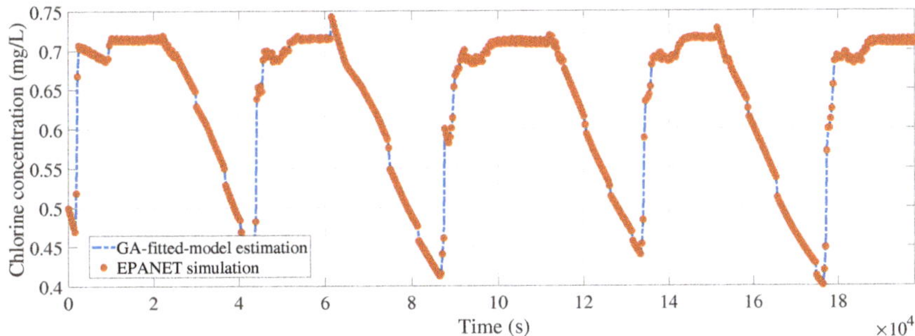

Figure 6. Calibrated model fitting for node 5 (rmse = 2.0835×10^{-4} mg/L).

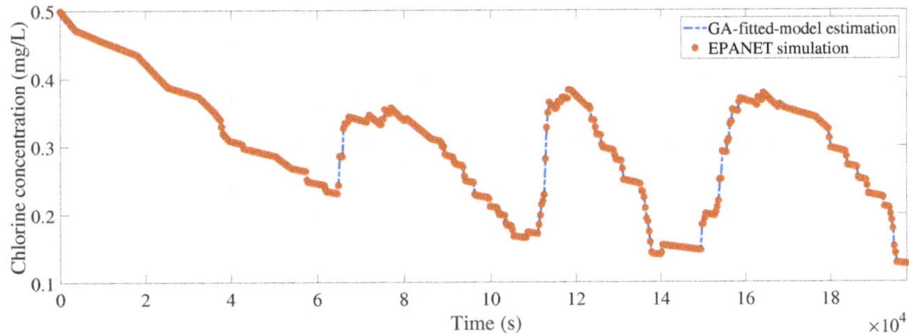

Figure 7. Calibrated model fitting for node 10 (rmse = 7.2047×10^{-5} mg/L).

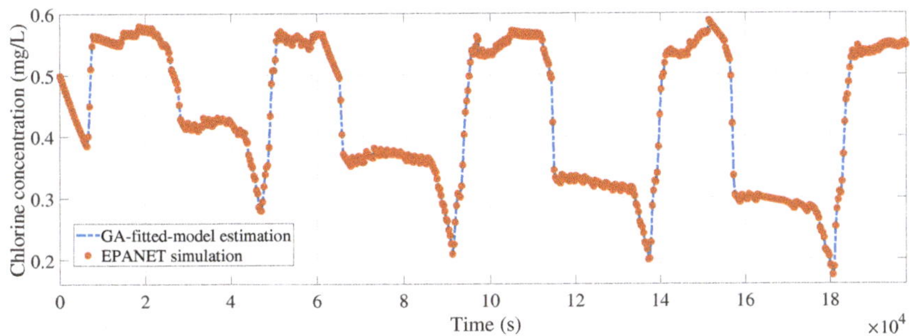

Figure 8. Calibrated model fitting for node 15 (rmse = 2.7089×10^{-4} mg/L).

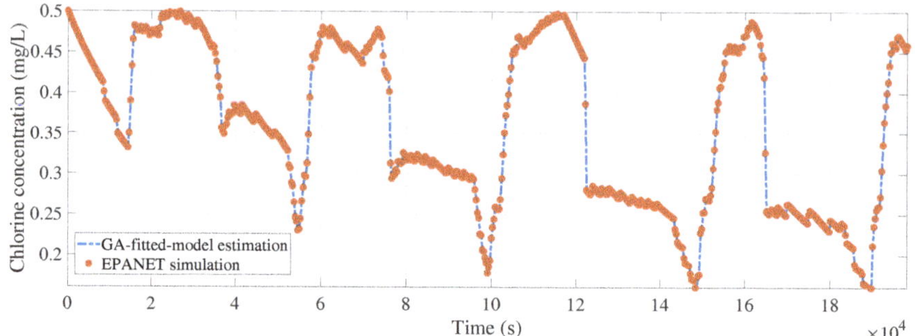

Figure 9. Calibrated model fitting for node 20 (rmse = 1.0999×10^{-4} mg/L).

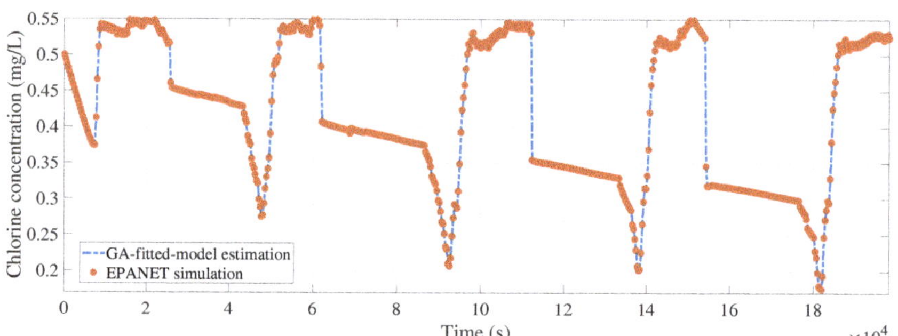

Figure 10. Calibrated model fitting for node 25 (rmse = 4.0421×10^{-4} mg/L).

Given that for each tested node the calibrated model provides an estimation that closely matches the real behavior, it is safe to ensure that the calibrated K_b and K_w values will estimate with accuracy the chlorine concentration at any other node. Even if Figures 6–10 represent the evolution of chlorine concentration for a total time of 55 h only for the tested nodes, with the new well-calibrated chlorine-decay model it is possible to determine the chlorine concentration evolution at any other node in the system. Moreover, knowing the minimal concentration in the entire WDS for the duration of the simulation is now possible. It should be noted that the evolution shown in Figures 6–10 considers 0.8 mg/L as the source chlorine concentration at the input node (s_Q) (N$_1$ in Figure 3), as well as an initial chlorine concentration (i_Q) at each node of 0.5 mg/L. Even if both initial and source chlorine concentration values are contained within the range indicated by the official norm, it is not safe to say that normativity values will be met at each node in the WDS. After performing a chlorine-decay simulation using the calibrated model, Figure 11 provides the minimal overall chlorine concentration at each node.

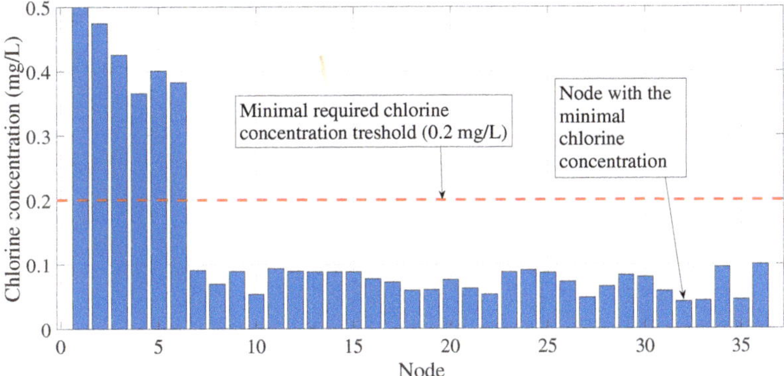

Figure 11. Minimal chlorine concentration at each node when $s_Q = 0.8$ mg/L and $i_Q = $ is 0.5 mg/L.

From Figure 11 it is clear that only nodes 1 to 6 comply with the normativity requirements with every other node dropping to a minimal chlorine concentration below 0.1 mg/L. This indicates that a higher chlorine concentration (still within the official normativity requirements) needs to be sourced to the WDS to meet the requirements. It will be assumed that chlorine is sourced to the WDS as a constant concentration at the input node (N_1). It is also theorized that to minimize the amount of sourced chlorine, the initial concentration needs to be the highest possible value. Calibration of s_Q is performed through the GA-based methodology proposed in Section 4.2, while considering a i_Q value of 1.5 mg/L (upper permitted bound according to the official norm). After applying the proposed methodology, the obtained minimal s_Q value that ensures that at all times with at least 0.2 mg/L of chlorine concentration are met for any point in the system is:

$$s_Q = 1.271229 \, \text{mg/L} \tag{11}$$

Please note that it was assumed that the initial concentration at each node in the system is already at its highest possible value of 1.5 mg/L. This condition is assumed only because in a simulated environment it is possible to freely manipulate this value, while for experimental implementations only the initial chlorine concentration at nodes with chlorine meters can be known, and thus their initial concentration values shall be included as calibration variables in the GA-based algorithm. Figure 12 shows the minimal chlorine concentration at each node when $s_Q = 1.271229$ mg/L and $i_Q = 1.5$ mg/L.

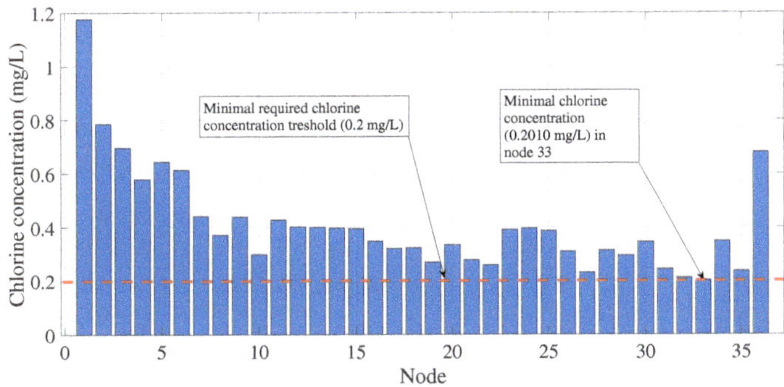

Figure 12. Minimal chlorine concentration at each node when $s_Q = 1.271229$ mg/L and $i_Q = $ is 1.5 mg/L.

In Figure 13 the evolution of the chlorine concentration at every measured node is illustrated. It is shown how during the simulated 55 h chlorine concentration remains always within the established limits of 0.2 and 1.5 mg/L.

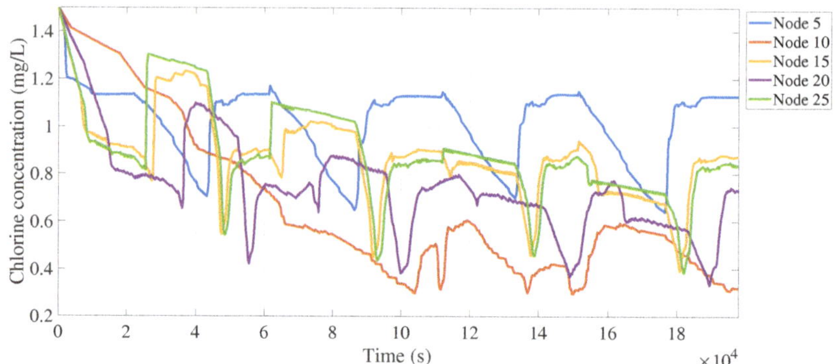

Figure 13. Evolution of *measured* chlorine concentration with $s_Q = 1.271229$ mg/L and $i_Q = $ is 1.5 mg/L.

Discussion of the Results

Application of the proposed methodology allowed for calibration of the required chlorine concentration at the input node to ensure that the minimal normativity-required value was met for every node in the network. The first calibration process to estimate the values of K_b and K_w coefficients was performed under an input chlorine concentration of $s_Q = 0.8$ mg/L and an initial chlorine concentration of $i_Q = 0.5$ mg/L. Actually, once the K_b and K_w parameters are calibrated, it is possible to perform a quality simulation at the uppermost condition for the initial chlorine concentration ($i_Q = 1.5$ mg/L), which yields that the minimal chlorine concentration threshold is not met, since many nodes are still under the required 0.2 mg/L concentration, as shown in Figure 14.

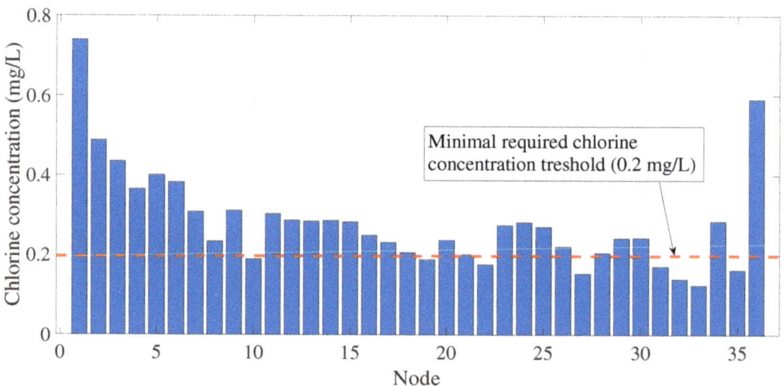

Figure 14. Minimal chlorine concentration at each node when $s_Q = 0.8$ mg/L and $i_Q = $ is 1.5 mg/L.

The importance of the calibrated model now relies on the fact that simulation of any what-if scenario is possible, and furthermore the re-design of input conditions ensures the required chlorine concentration no matter the time of day or the analyzed node. As shown previously, the calibrated chlorine-decay model is able to compute chlorine concentrations for every node at every time interval, which makes it a powerful tool in analyzing already existing networks as well as during the design and implementation stages in the creation of a new WDS. Following this line, it could be possible that even if the chlorine decay has already been finely adjusted, an s_Q value cannot be found to ensure that a chlorine concentration within the official norm is met for every node (e.g., the required $s_Q > 1.5$ mg/L for $\min_Q > 0.2$ mg/L). Models not converging to a satisfactory solution may require the adjustment of some physical parameters of the WDS as well as analyzing the possibility of sourcing chlorine at more than one node to ensure that the normative standard is met in the entire system. For example, in Figure 12 it can be inferred that for the initial chlorine concentration value for each node lower than 1.5 mg/L, the required s_Q would be slightly higher than the calibrated 1.271 mg/L, and thus a point will be reached where even if $s_Q = 1.5$ mg/L (its theoretical highest possible), it would be impossible to meet the minimal required chlorine concentration value as stated by the normativity. In other words, the re-design of the network should be performed to allow chlorine input in many different nodes since only one input node did not allow to meet the normativity requirements. For the particular WDS tested in this study, it was not possible to achieve a $0.2 < s_Q < 1.5$ [mg/L] when $i_Q = 0.8$ mg/L and thus it was decided to run the simulation at $i_Q = 1.5$ mg/L; however in experimental implementations of this work this initial condition may not be able to be achieved and further alternatives should be analyzed. It should be noted that even if the results presented in this paper considered that the permitted chlorine concentration values range according to the mexican NOM-127-SSA1-1994, the algorithm can be adjusted to meet the requirements given by a different normativity framework (e.g., the WHO recommendations).

It is important to note that the tested methodology proposes an alternative approach to most of the state-of-the-art investigations, where the major part of the related work addresses the issue of modeling of chlorine decay in a WDS while not exactly focusing on the calibration of the bulk and wall reaction coefficients, which is the main objective of the proposed methodology. Following this line, in [9], aside from the wall reaction coefficient, the initial conditions of the simulation were also considered as calibration variables (the bulk reaction coefficient was determined independently in a laboratory), while in [19] a hybrid PSO-GA demonstrated a better performance than a pure GA in finding an optimal match between observed and simulated chlorine concentration when the only calibration variable considered was the wall reaction coefficient. Moreover, both [9,19] took into consideration data from a real WDS. Even if not directly comparable, the authors stated that the methodology proposed in this paper can demonstrate an improvement from previous related studies since this work also contemplated calibration of the bulk reaction coefficient (not addressed by previous research) as well as the estimation of the minimal required source-chlorine concentration to meet normativity requirements. The performed study can be extended in future implementations by considering as well the unknown initial conditions as a calibration variable once this method is tested with non-synthetic data.

6. Conclusions

This work presented an efficient methodology on the basis of genetic algorithms for both the calibration of the bulk coefficient and wall reaction coefficient in the chlorine-decay model of a WDS. Moreover, this approach allowed to determine if the input chlorine concentration dosage ensures the compliance of the normative along the whole WDS. Simulation results illustrate a good performance in obtaining a calibrated chlorine-decay model that is able to estimate the chlorine concentration at those nodes without sensors considering the availability of a reduced number of chlorine sensors. The relevance of the presented work relies on the fact that water agencies require precise knowledge on the quality of water delivered to consumers from a healthcare perspective as well as to ensure compliance with the applicable official normativity. On the other hand, this research work expanded upon the results obtained in previous investigations by considering both the bulk and wall reaction coefficients as decision variables during the optimization process. Additionally, such a GA-based approach provided a good performance in simultaneously calibrating both variables no matter the reduced availability of chlorine sensors. It also expanded the state-of-the-art by presenting an application of the calibrated model in which the chlorine concentration rationed at the pumping station is optimized to ensure the compliance of official normativity for public health purposes. Finally, the presented methodology was shown to be a useful tool for the analysis of WDS encountered in practice as well as the re-design of a WDS that does not meet the applicable official normative. Implementation of this methodology in a real WDS will be part of future developments.

Author Contributions: Conceptualization, L.G.-C. and J.A.D.-A.; Data curation, L.G.-C.; Formal analysis, J.A.D.-A.; Methodology, L.G.-C. and J.A.D.-A.; Project administration, J.A.D.-A.; Software, L.G.-C.; Supervision, J.A.D.-A.; Validation, L.G.-C.; Visualization, I.S.-R. and A.N.-D.; Writing—original draft, L.G.-C., J.A.D.-A., I.S.-R. and A.N.-D.; Writing—review and editing, L.G.-C., J.A.D.-A., I.S.-R. and A.N.-D. All authors have read and agreed to the published version of the manuscript.

Funding: The APC was funded by Tecnológico de Monterrey.

Institutional Review Board Statement: Not applicable.

Informed Consent Statement: Not applicable.

Data Availability Statement: Not applicable.

Acknowledgments: This work was supported by Tecnológico de Monterrey through APC funding, by CONACyT through a scholarship for the first author under grant 801871, by Tecnológico Nacional de México (TecNM), and achieved during a research stay of the first author at CIIDETEC-UVM Guadalajara. Second author would like to thank Ing. Gómez for his significant achievement in this short-research stay, Santos-Ruiz and Adrián "El Gallo Claudio" Navarro-Díaz for fruitful interactions in recent collaborations.

Conflicts of Interest: The authors declare no conflict of interest.

References

1. Omer, N.H. Water quality parameters. *Water-Qual.-Sci. Assess. Policy* **2019**, *18*, 1–34.
2. Davis, M.L. *Water and Wastewater Engineering: Design Principles and Practice*; McGraw-Hill Education: New York, NY, USA, 2010.
3. Langowski, R.; Brdys, M. Monitoring of chlorine concentration in drinking water distribution systems using an interval estimator. *Int. J. Appl. Math. Comput. Sci.* **2007**, *17*, 199. [CrossRef]
4. García-Ávila, F.; Avilés-Añazco, A.; Ordoñez-Jara, J.; Guanuchi-Quezada, C.; del Pino, L.F.; Ramos-Fernández, L. Modeling of residual chlorine in a drinking water network in times of pandemic of the SARS-CoV-2 (COVID-19). *Sustain. Environ. Res.* **2021**, *31*, 1–15. [CrossRef]
5. Nono, D.; Odirile, P.T.; Basupi, I.; Parida, B.P. Assessment of probable causes of chlorine decay in water distribution systems of Gaborone city, Botswana. *Water SA* **2019**, *45*, 190–198. [CrossRef]
6. World Health Organization. Principles and Practices of Drinking-Water Chlorination: A Guide to Strengthening Chlorination Practices in Small-to Medium Sized Water Supplies. 2017. Available online: https://apps.who.int/iris/handle/10665/255145 (accessed on 11 January 2023).
7. Secretaría de Gobernación, México. NOM-127-SSA1-1994, Salud Ambiental. Agua Para Uso y Consumo Humano. Límites Permisibles de Calidad y Tratamientos a Que debe Someterse el Agua Para su Potabilización. Diario Oficial de la Federación. 2000. Available online: https://www.dof.gob.mx/nota_detalle.php?codigo=2063863&fecha=31/12/1969#gsc.tab=0 (accessed on 11 January 2023).
8. Males, R.M.; Grayman, W.M.; Clark, R.M. Modeling water quality in distribution systems. *J. Water Resour. Plan. Manag.* **1988**, *114*, 197–209. [CrossRef]
9. Rossman, L.A.; Clark, R.M.; Grayman, W.M. Modeling chlorine residuals in drinking-water distribution systems. *J. Environ. Eng.* **1994**, *120*, 803–820. [CrossRef]
10. Clark, R.M.; Rossman, L.A.; Wymer, L.J. Modeling distribution system water quality: Regulatory implications. *J. Water Resour. Plan. Manag.* **1995**, *121*, 423–428. [CrossRef]
11. Park, K.; Kuo, A.Y. A multi-step computation scheme: decoupling kinetic processes from physical transport in water quality models. *Water Res.* **1996**, *30*, 2255–2264. [CrossRef]
12. Hua, F.; West, J.; Barker, R.; Forster, C. Modelling of chlorine decay in municipal water supplies. *Water Res.* **1999**, *33*, 2735–2746. [CrossRef]
13. Islam, M.R.; Chaudhry, M.H.; Clark, R.M. Inverse modeling of chlorine concentration in pipe networks under dynamic condition. *J. Environ. Eng.* **1997**, *123*, 1033–1040. [CrossRef]
14. Islam, M.R.; Chaudhry, M.H. Modeling of constituent transport in unsteady flows in pipe networks. *J. Hydraul. Eng.* **1998**, *124*, 1115–1124. [CrossRef]
15. Monteiro, L.; Figueiredo, D.; Dias, S.; Freitas, R.; Covas, D.; Menaia, J.; Coelho, S. Modeling of chlorine decay in drinking water supply systems using EPANET MSX. *Procedia Eng.* **2014**, *70*, 1192–1200. [CrossRef]
16. Fisher, I.; Kastl, G.; Sathasivan, A. New model of chlorine-wall reaction for simulating chlorine concentration in drinking water distribution systems. *Water Res.* **2017**, *125*, 427–437. [CrossRef] [PubMed]
17. Madzivhandila, V.A.; Chirwa, E. Modeling chlorine decay in drinking water distribution systems using aquasim. *Chem. Eng. Trans.* **2017**, *57*, 1–6.
18. Nejjari, F.; Puig, V.; Pérez, R.; Quevedo, J.; Cuguero, M.; Sanz, G.; Mirats, J. Chlorine decay model calibration and comparison: Application to a real water network. *Procedia Eng.* **2014**, *70*, 1221–1230. [CrossRef]

19. Minaee, R.P.; Afsharnia, M.; Moghaddam, A.; Ebrahimi, A.A.; Askarishahi, M.; Mokhtari, M. Calibration of water quality model for distribution networks using genetic algorithm, particle swarm optimization, and hybrid methods. *MethodsX* **2019**, *6*, 540–548. [CrossRef] [PubMed]
20. Mirjalili, S.; Mirjalili, S. Genetic algorithm. In *Evolutionary Algorithms and Neural Networks: Theory and Applications*; Springer: Berlin/Heidelberg, Germany, 2019; pp. 43–55.
21. Katoch, S.; Chauhan, S.S.; Kumar, V. A review on genetic algorithm: past, present, and future. *Multimed. Tools Appl.* **2021**, *80*, 8091–8126. [CrossRef] [PubMed]
22. Rossman, L.A.; Woo, H.; Tryby, M.; Shang, F.; Janke, R.; Haxton, T. *EPANET 2.2 User Manual*; Technical Report EPA/600/R-20/133; U.S. Environmental Protection Agency: Washington, DC, USA, 2020.
23. Eliades, D.G.; Kyriakou, M.; Vrachimis, S.; Polycarpou, M.M. EPANET-MATLAB Toolkit: An Open-Source Software for Interfacing EPANET with MATLAB. In Proceedings of the 14th International Conference on Computing and Control for the Water Industry (CCWI), Amsterdam, The Netherlands, 7–9 November 2016. [CrossRef]

Article

Actuator FDI Scheme for a Wind Turbine Benchmark Using Sliding Mode Observers

Vicente Borja-Jaimes [1], Manuel Adam-Medina [1,*], Jarniel García-Morales [1], Gerardo Vicente Guerrero-Ramírez [1], Betty Yolanda López-Zapata [2] and Eduardo Mael Sánchez-Coronado [3]

[1] Electronic Engineering Department, TecNm/National Center for Research and Technological Development (CENIDET), Cuernavaca, Morelos 62490, Mexico; vicentebj@cenidet.edu.mx (V.B.-J.); jarniel.gm@cenidet.tecnm.mx (J.G.-M.); gerardo.gr@cenidet.tecnm.mx (G.V.G.-R.)

[2] Department of Mechatronic Engineering, Polytechnic University of Chiapas, Tuxtla Gutierrez, Chiapas 29082, Mexico; blopez@upchiapas.edu.mx

[3] Department of Mechatronic Engineering, Technological University of the Center of Veracruz, Cuitlahuac, Veracruz 94910, Mexico; eduardo.sanchez@utcv.edu.mx

* Correspondence: manuel.am@cenidet.tecnm.mx

Abstract: This paper proposes a fault detection and isolation (FDI) scheme for a wind turbines subject to actuator faults in both the pitch system and the drive train system. The proposed scheme addresses fault detection and isolation problems using a fault estimation approach. The proposed approach considers the use of a particular class of sliding mode observers (SMOs) designed to maintain the sliding motion even in the presence of actuator faults. The fault detection problem is solved by reconstructing the actuator faults through an appropriate analysis of the nonlinear output error injection signal, which is required to keep the SMO in a sliding motion. To ensure accurate fault reconstruction, only two conditions are required, namely that the faults are bounded and they meet the matching condition. A scheme based on a bank of SMOs is proposed to solve the fault detection and isolation problem in the pitch system. For the drive train system, a scheme using only one SMO is proposed. The performance of the proposed FDI scheme is validated by using a wind turbine benchmark model subjected to several actuator faults. Normalized root mean square error (NRMSE) analysis is performed to evaluate the accuracy of the actuator fault estimations.

Keywords: fault detection and isolation (FDI); sliding mode observer (SMO); wind turbines

Citation: Borja-Jaimes, V.; Adam-Medina, M.; García-Morales, J.; Guerrero-Ramírez, G.V.; López-Zapata, B.Y.; Sánchez-Coronado, E.M. Actuator FDI Scheme for a Wind Turbine Benchmark Using Sliding Mode Observers. *Processes* **2023**, *11*, 1690. https://doi.org/10.3390/pr11061690

Academic Editors: Francisco Ronay López-Estrada and Guillermo Valencia-Palomo

Received: 18 April 2023
Revised: 21 May 2023
Accepted: 25 May 2023
Published: 1 June 2023

1. Introduction

Nowadays, wind turbines contribute a large part of the world's electrical energy production from renewable sources. In 2022, their second-highest growth in history was registered, with 93.6 GW of installed capacity added, which is only 1.8% below the historical record reached in 2020. The addition in 2022 carries the global cumulative installed capacity of wind energy to 837 GW, which, on average, represents a growth of 12% per year [1]. The remarkable growth that wind turbines have experienced has led to a growing demand for greater efficiency and reliability. However, the size, the complexity of the components, and the stochastic behavior of the wind present significant challenges when trying to maintain an operation with a preset efficiency [2]. Furthermore, wind turbines, like any complex system, are susceptible to faults. Various factors including environmental factors, manufacturing defects, and lack of maintenance can cause wind turbine faults, leading to significant damage and operational disruptions. Additionally, wind turbines are installed in remote and isolated locations, which complicates preventive and corrective maintenance actions, whereby the incidence of a fault can trigger breakdowns or even the destruction of the turbine if it is not detected [3]. This situation has generated interest in the application of fault detection and isolation (FDI) methods in wind turbines, especially in critical wind turbine components such as the pitch and drive train systems [4].

FDI is crucial for ensuring the reliable and safe operation of wind turbines. The primary objective of an FDI scheme is to generate a warning when an unusual situation occurs in the operation of the system and then to find its source and location [5]. Overviews and surveys of FDI methods for wind turbines appear in [6–9]. Most of the works reported in the literature can be divided into two large groups: data-driven and model-based methods. Many data-driven techniques have been developed to detect faults in wind turbines based on fuzzy systems and neural networks [10], kernel methods [11], deep neural networks (DNN) and principal component analysis (PCA) [12], classifier fusion [13], and condition monitoring systems [14]. Of these data-driven approaches, condition monitoring systems are the most effective because they provide substantial information about irregularities in the system. However, one drawback of most condition monitoring methods is the slow and tedious data collection and interpretation.

Model-based FDI methods utilize mathematical models of the wind turbine system to simulate its behavior under normal and faulty conditions. These methods rely on comparing the model predictions with sensor measurements to detect and isolate faults [15]. One of the most common methods of model-based FDI schemes is the observer-based method. This method involves designing observers, also known as estimators or filters, to estimate the internal states of the wind turbine based on signals measured by sensors. By comparing the estimated states with the measured states, deviations can be detected and attributed to specific faults [16]. Many observer-based FDI schemes have been developed for wind turbines. In [17], an adaptive observer FDI scheme is proposed. The observer is used to estimate both sensor and actuator faults in the benchmark model proposed in [18]. An algorithm for detecting pitch actuator faults using interval observers is proposed in [19]. The algorithm is trained with healthy parameters of the system, and it is assumed that in the presence of faults, the values of the parameters do not stray too far from their fault-free condition, which is unrealistic. A model-based FDI scheme for the pitch system was proposed in [20]. The scheme uses an extended Kalman filter to detect faults. The Kalman filter applies multiple model-adaptive estimators to approximate the states of the pitch system. A scheme to detect both actuator and sensor faults in a wind turbine benchmark model based on an unknown input observer is proposed in [21]. For fault detection, it employs residual signals; for fault isolation, it utilizes estimating states, output signals, and control signals.

The vast majority of proposed FDI schemes for both sensor and actuator faults in wind turbines use a residual generation approach. In a residual-based approach, residuals are ideally expected to be zero during fault-free operation and non-zero in a failed operation [22]. However, since a wind turbine is subject to disturbances, parameter uncertainty, and mismatches between the mathematical model and the actual system, these discrepancies can produce non-zero residuals even in a fault-free operation, creating a false alarm problem. The false alarms can affect FDI performance to such an extent that it may become useless. To overcome these difficulties, the model-based FDI scheme must be made insensitive to modeling uncertainty but sensitive to faults, which is called a robust FDI scheme [23].

A robust FDI scheme is proposed in [24]. The scheme considers actuator and sensor faults in both the pitch and drive train systems. Robustness is achieved by decoupling the faulty dynamics from the system using coordinate transformations.

Sliding mode techniques have historically demonstrated robustness properties for a certain class of matched uncertainties [25]. In particular, sliding mode observers (SMOs) have been used for both fault detection and isolation as well as fault-tolerant control (FTC) schemes [26]. Alternatives to taking advantage of the inherent robustness of the SMO for fault detection and isolation problems have been explored in wind turbines. A model-based FDI scheme for simultaneously detecting sensor and actuator faults is presented in [27]. The proposed scheme uses an adaptive SMO to accurately estimate both system states and disturbances as part of an active FTC system. An FDI scheme using SMOs is designed in [28]. The SMOs are combined with a residual signal generator for detecting sensor faults in both the pitch and drive train systems in a wind turbine benchmark model. The proposed schemes require the designer to know the behavior of the fault to establish an adequate threshold to be able to perform fault detection. However, this is an impractical approach since fault behavior is continually changing [29]. Thus, an FDI scheme using an observer-based approach can only detect faults considered in the fault propagation analysis of the system for which a detection threshold is established. Recently, fault diagnosis methods that not only detect and isolate faults but are also capable of making a complete estimate of faults have attracted great interest. Complete fault information is crucial to an FTC system that can effectively mitigate fault effects, enabling the wind turbine to operate safely and with reduced downtime [30]. A fault estimation approach using adaptive and parameter estimation schemes for a wind turbine is proposed in [31]. The parameter estimation scheme is utilized to estimate the values of specific parameters that may be affected by faults or abnormalities in the wind turbine. A robust actuator multiplicative fault estimation method with unknown input decoupling is presented in [32]. The method employs a multiplicative fault model and an unknown input decoupling technique to mitigate the influence of unknown inputs on the fault estimation process. Although some effective fault estimation techniques for wind turbines have been presented, most of them have focused on parametric faults, leaving the application of fault estimation techniques in critical components of the wind turbine, such as the pitch and drive train systems, as an unresolved issue.

This paper addresses the problem of designing an FDI scheme for actuator faults in the pitch and drive train systems of a wind turbine using a fault estimation approach. The proposed method can detect and isolate the source of actuator faults. In addition, it provides a complete reconstruction of the faults. Therefore, the fault estimation method provides a direct estimate of the size and severity of the faults, which can be crucial for FTC schemes.

The main contributions of this paper are as follows:

- An FDI scheme for both the pitch and drive train systems that does not require explicit information from the fault. The only fault information required is the fault to be bounded.
- In contrast to the works reported in [20,23], the proposed FDI scheme employs the concept of an equivalent output error injection term. This allows a complete reconstruction of actuator faults while providing accurate estimates of states, regardless of fault occurrence.
- A simple method for actuator fault reconstruction using a low-pass filter (LPF) that analyzes the so-called nonlinear output error injection, in contrast to the works presented in [20,23], which use more complex methods.

This paper is organized as follows. The wind turbine model, the design methodology of an SMO for actuator fault reconstruction, and the proposed actuator FDI scheme for both the pitch and the drive train systems are presented in the Section 2. Simulation results of the proposed FDI scheme are presented in the Section 3. A discussion of the results is presented in the Section 4. Finally, conclusions are presented in the Section 5.

2. Materials and Methods

2.1. Mathematical Model of the Wind Turbine

A wind turbine converts wind energy into electrical energy by using the aerodynamic force of the turbine rotor blades. The input to the system is the wind speed, which causes the rotation of the turbine blades. This motion spreads through the turbine rotor. The drive train couples this motion with the generator shaft. In turn, the generator is responsible for transforming mechanical energy into electrical energy. The electrical power generated by the system is controlled by modifying the aerodynamics of the turbine, adjusting the angle of inclination of the blades, or modifying the speed of rotation of the generator shaft. In any case, the objective of the control system is to maintain the required electrical power. Figure 1 shows the interconnection of different subsystems that make up a wind turbine and their interaction with the control used to regulate electrical power. This model addresses the wind turbine at a system level and provides mathematical models with simplicity and sufficient accuracy for all subsystems. The variables associated with Figure 1 are defined as follows: v_ω is the wind speed, τ_r is the rotor torque, τ_g is the generator torque, ω_r is the rotor speed, ω_g is the generator speed, E_g is the electrical power, $\tau_{g,ref}$ is the reference torque of the generator, β is the pitch angle, and β_{ref} is the reference angle of the pitch.

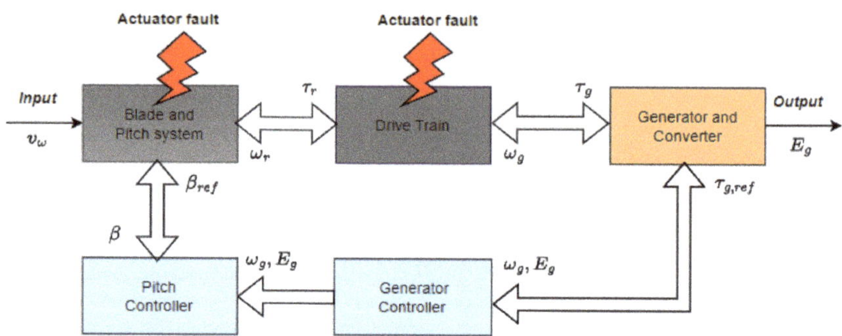

Figure 1. Wind turbine benchmark model.

For a more detailed description of the different subsystems of the wind turbine benchmark model and its internal connection, see [18]. The subsystems models are briefly described below.

2.1.1. Blade and Pitch Model

This system combines both aerodynamic and hydraulic pitch models. The wind turbine's nonlinear aerodynamic is modeled as a torque acting on the blades. The torque is given as

$$\tau_r(t) = \frac{1}{2}\pi\rho r^3 v_\omega^2(t)S(\lambda, \beta) \tag{1}$$

where $S(\lambda, \beta)$ represents a mapping of the coefficients, which depends on the tip speed ratio λ and the pitch angle β. r is the rotor ratio, ρ is the density of the air, and v_ω is the wind speed. Since the pitch system is composed of three identical subsystems its aerodynamics are stated as follows:

$$\tau_r(t) = \sum_{1 < i \leq 3} \frac{\pi\rho r^3 v_{\omega,i}^2(t)S(\lambda, \beta_i)}{6} \tag{2}$$

It is important to highlight that this model is valid for small differences between the pitch angles.

The hydraulic pitch system is modeled as a closed-loop second-order transfer function described as

$$\frac{\beta(s)}{\beta_{ref}(s)} = \frac{\psi^2}{s^2 + 2\xi\psi s + \psi^2} \tag{3}$$

where β is the measurement pitch position angle and β_{ref} is the reference input provided by the wind turbine controller. ξ is the damping factor and ψ is the natural frequency.

2.1.2. Drive Train Model

The drive train system uses a simple two-mass model. Therefore, the nominal dynamics of the drive train system can be represented by

$$j_r\dot{\omega}_r(t) = \tau_r(t) - k_{dt}\theta(t) - (h_{dt} + b_r)\omega_r(t) + \frac{h_{dt}}{N_g}\omega_g(t) \tag{4}$$

$$j_g\dot{\omega}_g(t) = \frac{\eta_{dt}k_{dt}}{N_g}\theta(t) + \frac{\eta_{dt}h_{dt}}{N_g}\omega_r(t) - \left(\frac{\eta_{dt}h_{dt}}{N_g^2} + b_g\right)\omega_g(t) - \tau_g(t) \tag{5}$$

$$\dot{\theta}(t) = \omega_r(t) - \frac{1}{N_g}\omega_g(t) \tag{6}$$

where $\omega_r(t)$ and $\omega_g(t)$ are the rotor and generator speed and $\theta(t)$ denotes the torsion angle of the drive train system. The inputs $\tau_g(t)$ and $\tau_r(t)$, represent the torque of the rotor and generator, respectively. J_r and J_g are the moment of inertia of the low- and high-speed shafts, respectively. b_r and b_g are the viscous frictions of the low- and high-speed shafts, respectively. k_{dt} is the torsion stiffness of the drive train system, h_{dt} represents the torsion damping coefficient of the drive train, N_g is the gear ratio, and η_{dt} is the efficiency percentage of the drive train system.

2.1.3. Generator and Converter Model

The dynamics of the generator and converter can be modeled as a first-order transfer function because the frequency range used in the benchmark model is much smaller than the speed of the electrical subsystem and its controllers. The joint dynamics of the converter and generator are given by

$$\frac{\tau_g(s)}{\tau_{g,ref}(s)} = \frac{\gamma}{s + \gamma} \tag{7}$$

where $\tau_{g,ref}$ is the reference value and γ is the cutoff frequency. The electrical power produced by the generator is given by

$$E_g(t) = \eta_g \omega_g(t) \tau_g(t) \tag{8}$$

where E_g is the power produced by the generator and η_g is the efficiency of the generator.

2.1.4. Faults in Wind Turbines

In general, a fault refers to an error or defect in something that causes it to not work properly or function as intended [33]. In the context of wind turbines, a fault typically refers to a malfunction or failure in the wind turbine's components, systems, or operations that can lead to reduced performance, downtime, or even safety risks. The faults can occur in various parts of the wind turbine, including the pitch system, the drive train system, the generator, the electrical system, and the control system [34]. In addition, wind turbine faults can occur in both the actuators and sensors of any of the systems that make up the wind turbine. However, in this paper, we only focus on actuator faults in the pitch and the drive train systems.

According to a report published by the National Renewable Energy Laboratory (NREL) in the US, the most common faults in wind turbines include electrical system faults, blade damage, and mechanical faults [35]. Faults in wind turbines can have several consequences affecting the performance and safety of the wind turbine. Two critical components in the wind turbine are the pitch and the drive train systems. If a fault in any of these components is not detected on time, this can cause severe consequences for the wind turbine, such as reduced energy generation, increased downtime, higher maintenance and repair costs, and safety risks. Early detection and isolation in the pitch system as well as the drive train system are crucial for ensuring the optimal performance, reliability, and safety of the wind turbine [36]. Therefore, in this paper, we focus on the problem of fault detection and isolation in the pitch and drive train systems. It is worth noting that simultaneous faults in wind turbines can occur, although they are relatively rare. Accordingly, the case of simultaneous faults is not considered.

The severity of actuator faults in wind turbines can vary depending on the specific fault and its impact on the wind turbine's operation. Based on their severity, actuator faults are classified as abrupt, incipient, and intermittent faults. These actuator faults can be modeled using signal models that include the effect of the fault on the output of the actuator.

In normal wind turbine operation, the output of the actuator increases smoothly in response to the input. However, during an abrupt actuator fault, the output of the actuator may experience sudden changes or fluctuations, leading to a non-smooth response. Hence, to model an abrupt actuator fault, a square signal model is used. In contrast, an incipient actuator fault in a wind turbine is a fault that is just beginning to develop, and its behavior is characterized by subtle changes in the actuator's output. Over time, the behavior of an incipient fault may become more pronounced, with the output of the actuator becoming increasingly irregular or unstable. Then, an incipient fault can be modeled using a time-varying signal model that allows for the characterization of the actuator's output over time [37]. Therefore, to simulate an incipient actuator fault, both sinusoidal and sawtooth signal models are proposed.

2.2. Design of Sliding Mode Observers for Actuator Fault Reconstruction

In recent years, SMOs have been successfully applied to solve the problem of estimating the states in dynamical systems. One of their most interesting properties, which has been used most in the literature, is robustness. This allows an SMO to successfully solve the estimation problem despite the presence of disturbances or uncertainties bounded within the system. In essence, an SMO is a mathematical replica of the system fitted by means of feedback of the output estimation error through a nonlinear function, which provides an interesting solution to this issue. The SMO can force the output estimation error to zero in finite time by determining the bound on the magnitude of the disturbance acting on the system. Consequently, the observer states converge asymptotically to the system states. Under this condition, it is said that a sliding motion takes place [38].

During the sliding motion, the nonlinear output estimation error contains information about unknown signals affecting the system. By suitably filtering the nonlinear output estimation error, unknown signals can be obtained. Therefore, by modeling faults as unknown signals, it is possible to use an SMO to reconstruct and, thus, detect and isolate these faults. The SMO design methodology is described below.

Consider the linear dynamical system subject to actuator faults described by

$$\dot{x}(t) = Fx(t) + Gu(t) + Mf_a(t, x, u) \tag{9}$$

$$y(t) = Hx(t) \tag{10}$$

where $F \in R^{n \times n}$, $G \in R^{n \times m}$, $H \in R^{p \times n}$, and $M \in R^{n \times q}$ with $p \geq q$. Assume that the matrices G, H, and M are the full rank, and the function $f_a : R^+ \times R^n \times R^m \to R^q$ is deemed to represent an actuator fault that is assumed to be bounded so that $\|f_a(t, x, u)\| \leq \alpha$ with $\alpha \in R^+$.

In addition, consider that the dynamical system given in Equations (9) and (10) satisfies the following two conditions:

- $rank(HM) = q$;
- the invariant zeros of the system represented by the triple (F, G, H) must lie in $\mathbb{C}^-$.
- Under these conditions, there is a linear change of coordinates T, such that the new coordinate the system can be written as

$$\dot{x}_1(t) = F_{11}x_1(t) + F_{12}x_2(t) + G_1u(t) \tag{11}$$

$$\dot{x}_2(t) = F_{21}x_1(t) + F_{22}x_2(t) + G_2u(t) + M_2f_a(t, x, u) \tag{12}$$

$$y(t) = Cx(t) \tag{13}$$

where $x_1 \in \mathbb{R}^{n-p}$, $x_2 \in \mathbb{R}^p$, and F_{11} has stable eigenvalues. The coordinate system described above is used as a platform for the design of an SMO. The SMO structure that will be considered can be written in the form

$$\dot{\hat{x}}_1(t) = F_{11}\hat{x}_1(t) + F_{12}\hat{x}_2(t) + G_1u(t) - F_{12}e_y(t) \tag{14}$$

$$\dot{\hat{x}}_2(t) = F_{21}\hat{x}_1(t) + F_{22}\hat{x}_2(t) + G_2 u(t) - (F_{22} - F_{22}^s)e_y(t) + w(t) \tag{15}$$

where F_{22}^s is a stable design matrix, and the discontinuous function $w(t)$ is defined as

$$w(t) = \begin{cases} -\sigma\|M_2\|\dfrac{P_2 e_y(t)}{\|P_2 e_y(t)\|} & \text{if } e_y(t) \neq 0 \\ 0 & \text{otherwise} \end{cases} \tag{16}$$

where $P_2 \in R^{p \times p}$ is a symmetric positive definite Lyapunov matrix for F_{22}^s. If the state estimation error and the output estimation errors are defined as $e_1(t) = \hat{x}_1(t) - x_1(t)$ and $e_y(t) = \hat{x}_2(t) - x_2(t)$, respectively, it is straightforward to show that the dynamical of the error is given by

$$\dot{e}_1(t) = F_{11}e_1(t) \tag{17}$$

$$\dot{e}_y(t) = F_{21}e_1(t) + F_{22}^s e_y(t) + w(t) - M_2 f_a(t, x, u) \tag{18}$$

It is shown by Edwards and Spurgeon in [38] that the dynamical of the error in Equations (17) and (18) is quadratically stable, and a sliding motion is achieved in finite time, forcing both $e_y(t)$ and $\dot{e}_y(t)$ to zero. Therefore, the dynamical system in Equations (14) and (15) can be considered as an SMO for the dynamical system in Equations (11)–(13). However, it is more convenient to express it in terms of the original coordinates system as

$$\dot{\hat{x}}(t) = F x(t) + G u(t) - K_l e_y(t) + K_{nl} w(t) \tag{19}$$

where K_l and K_{nl} are the linear and nonlinear gain matrices given by

$$K_l = T^{-1} \begin{bmatrix} F_{12} \\ F_{22} - F_{22}^s \end{bmatrix}, \tag{20}$$

and

$$K_{nl} = \|M_2\| T^{-1} \begin{bmatrix} 0 \\ I_p \end{bmatrix} \tag{21}$$

The output estimation error injection term, which is a nonlinear discontinuous signal, is defined as

$$w(t) = \begin{cases} -\sigma\|M_2\|\dfrac{P_2 H e_x(t)}{\|P_2 H e_x(t)\|} & \text{if } H e_x(t) \neq 0 \\ 0 & \text{otherwise} \end{cases} \tag{22}$$

where the state estimation error is defined as $e_x(t) = \hat{x}(t) - x(t)$ and the scalar σ is chosen so that $\sigma > \|f_a(t, x, u)\|$. It is important to note that even when the model considered for the design of the SMO has actuator fault signals, the dynamics of the observer do not depend on the faults. Therefore, the convergence of the observer is achieved even in the presence of actuator faults. The SMO design methodology described in this part is applied in the next subsection to develop an actuator FDI scheme for a wind turbine benchmark model.

2.3. Actuator FDI Scheme Based on Sliding Mode Observers

The traditional approach to FDI schemes using SMOs ensures that the sliding mode is interrupted when a fault occurs in the system, and the fault information is obtained by analyzing the residual signals. Under nominal fault-free conditions, the residuals are expected to be zero or nonzero when a fault occurs. In addition, most approaches that use residual generation can detect and isolate faults, but they do not provide information about the faults. In contrast to the traditional FDI methods, this paper addresses the fault detection and isolation problem by using a fault estimation approach based on the analysis of the nonlinear output error injection signal. One of the benefits of this approach compared to other SMOs based on FDI schemes is that the sliding motion is not broken, even when a fault appears. This allows the use of the SMO as a state estimator and, most importantly, as an estimator of faults in the system.

Figure 2 shows the proposed architecture for actuator fault reconstruction using an SMO. The scheme is based on a mathematical model of the system, which describes the dynamics of the plant, actuators, and sensors. The model includes the fault description in the actuator, which is modeled as an additive signal. Then, an SMO is designed to achieve asymptotic convergence of the estimation error to zero. The SMO uses the plant outputs, inputs, and the difference between the system's outputs and the observer's outputs, named output estimation error. The output estimation error signal is fed back into the observer via a nonlinear function, which provides two advantages. Firstly, the SMO can force the output estimation error to zero in finite time, even in the presence of actuator faults acting in the system. This is possible because the nonlinear output error injection signal compensates for the effect of actuator faults throughout the system to maintain the sliding motion. Therefore, the observer states converge asymptotically to the system states.

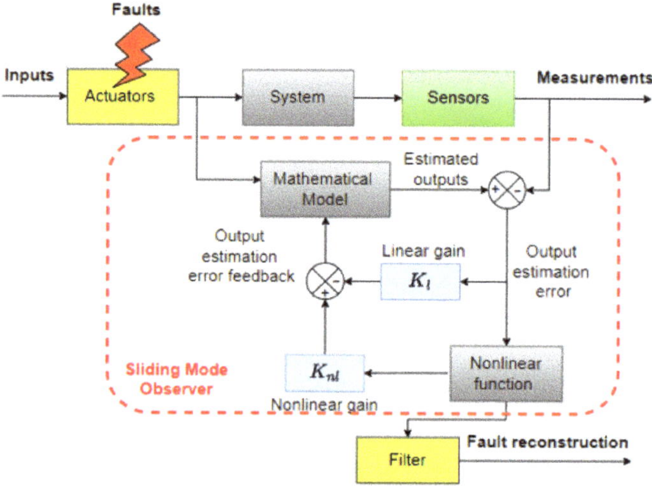

Figure 2. Actuator FDI scheme based on a sliding mode observer.

Secondly, once the output estimation error is forced to zero, it is said that a sliding motion takes place. During the sliding movement, the nonlinear output estimation signal injected into the SMO contains information about the faults affecting the system actuator. Then, by applying a suitable filtering process to the nonlinear output estimation error, the actuator fault can be reconstructed.

The idea behind this approach is to take advantage of the properties of the sliding motion and use the nonlinear output estimation error injection signal to reconstruct the actuator faults. Once a sliding motion is attained, $e_y(t) = 0$ and $\dot{e}_y(t) = 0$ are forced to zero in finite time, so the dynamics of the output estimation error in Equations (17) and (18) can be written as

$$0 = F_{21}e_1(t) + w(t) - M_2 f_a(t, x, u) \tag{23}$$

Given that F_{11} is stable, it follows that $e_1(t) \to 0$ and, therefore, $w(t) = M_2 f_a(t, x, u)$. That is, the fault information is contained in the nonlinear injection signal $w(t)$. Since the signal $w(t)$ switches at a very high frequency, one way to estimate the actuator fault is using a low-pass filter (LPF). The key point here is that to achieve a sliding motion in the presence of faults, it is only required that the magnitude of the fault is bounded and that it meets the matching condition, namely that the fault lies within the range space of the input distribution matrix of the system. However, the sliding motion cannot be maintained if the condition $\sigma > \|f_a(t, x, u)\|$ is not satisfied for $t_0 > 0$, where t_0 represents the time at which the sliding motion is achieved.

2.3.1. Actuator FDI Scheme for the Pitch System

The pitch actuator is an important component of a wind turbine and is responsible for controlling the position and orientation of the turbine blades to optimize energy capture and maintain safe operation. If a fault in the pitch actuator is not detected on time, it can have serious consequences for the safe and efficient operation of the wind turbine.

For example, a pitch actuator fault can cause the turbine blades to become misaligned. This can cause stress and damage to other components of the turbine, such as the drive train and the generator. Over time, this can lead to more serious damage and the need for costly repairs or replacements.

There are several potential causes for pitch actuator faults, which can be classified into external (environmental factors) and internal causes. Environmental factors, such as exposure to wind, rain, and saltwater, can cause corrosion, erosion, or damage to actuator components. On the other hand, the internal sources of pitch actuator faults are the hydraulic system and the electrical motor in charge of turbine blade movement. For example, wiring issues or power surges can cause damage to the electrical components of the actuator, while hydraulic fluid leaks or clogs can impair the function of hydraulic actuators [39].

Figure 3 shows the proposed architecture for actuator fault detection and isolation in the pitch system. The proposed scheme consists of a bank of three SMOs to estimate the outputs of each of the three pitch subsystems. As each pitch subsystem is independent of the others, one observer receives only the input u_{pb} and output y_{pb} signals from a single pitch subsystem, that is, the observer is sensitive only to the faults of the monitored pitch subsystem, which ensures a solution to the fault isolation problem once a fault is detected.

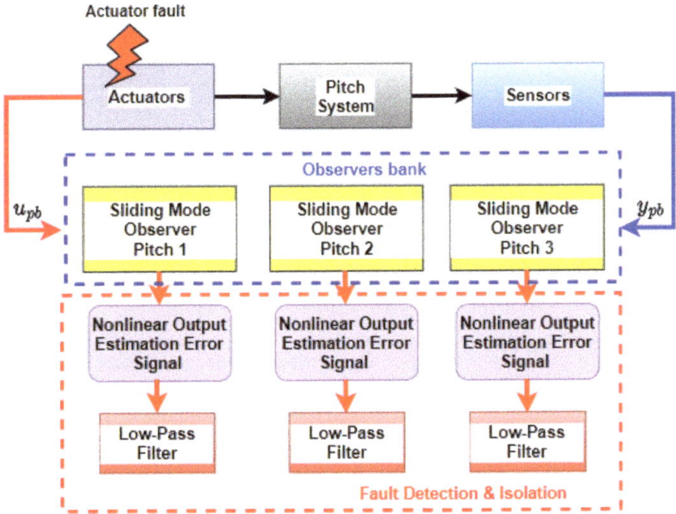

Figure 3. Actuator FDI scheme for the pitch system.

Since the nonlinear output estimation error injection signal compensates for the effect of faults in the SMO in order to maintain the sliding motion, fault detection is performed through the reconstruction of fault signals by filtering the nonlinear output error injection signal of each pitch subsystem.

As the pitch system is composed of three independent and identical subsystems, the SMO design presented in this section is applied identically to the other two subsystems.

The design of the SMO for one of the pitch subsystems considers the state space model given by

$$\dot{x}_{pb}(t) = F_{pb}x_p(t) + G_{pb}u_{pb}(t) + M_{pb}f_a(t) \tag{24}$$

$$y_{pb}(t) = H_{pb}x_{pb}(t) \tag{25}$$

where $x_{pb} \in \mathbb{R}^2$ is the state vector, $y_{pb} \in \mathbb{R}^2$ is the output vector, $u_{pb} \in \mathbb{R}$ is the input vector, $f_a(t)$ is any actuator fault signal, in this case of one dimension, and the matrices F_{pb}, G_{pb} and H_{pb} are defined as follows:

$$F_{pb} = \begin{bmatrix} 0 & 1 \\ -\psi^2 & -2\xi\psi \end{bmatrix}, \tag{26}$$

$$G_{pb} = \begin{bmatrix} 0 \\ \psi^2 \end{bmatrix}, \tag{27}$$

and

$$H_{pb} = \begin{bmatrix} 1 & 0 \\ 0 & 1 \end{bmatrix}. \tag{28}$$

From Equation (19), an SMO for each pitch subsystem can be written as

$$\dot{\hat{x}}_{pb}(t) = F_{pb}\hat{x}_{pb}(t) + G_{pb}u_{pb}(t) - K_l e_y(t) + K_{nl}w_{pb}(t) \tag{29}$$

$$\hat{y}_{pb}(t) = H_{pb}\hat{x}_{pb}(t) \tag{30}$$

where $\hat{x}_{pb}(t)$ and $\hat{y}_{pb}(t)$ denote the estimated states and outputs, respectively. K_l and K_{nl} are fixed appropriate gain matrices selected to ensure the stability and convergence of the error dynamics. $w_{pb}(t)$ represents a discontinuous switched output estimation error injected into the SMO to induce a sliding motion and is given by

$$w_{pb}(t) = \begin{cases} -\sigma\|M_2\| \dfrac{P_2 H_{pb}e_x(t)}{\|P_2 H_{pb}e_x(t)\|} & \text{if } H_{pb}e_x(t) \neq 0 \\ 0 & \text{otherwise} \end{cases} \tag{31}$$

To verify the convergence of the SMO, the simulation diagram shown in Figure 2 is used. The parameters used to simulate the pitch system model are listed in Table 1.

Table 1. Parameters of the wind turbine model.

Symbol	Description	Value
b_r	viscous friction of the low-speed shaft	7.11 Nms/rad
b_g	viscous friction of the high-speed shaft	45.6 Nms/rad
ξ	damping factor	0.6
ψ	natural frequency	11.11 rad/s
h_{dt}	torsion damping coefficient	775.49 Nms/rad
k_{dt}	torsion stiffness	2.7×10^9 Nm/rad
J_r	moment of inertia of the low-speed shaft	55×10^6 Kg.m^2
J_g	moment of inertia of the high-speed shaft	390 Kg.m^2
N_g	gear ratio	95
η_{dt}	efficiency of the drive train	0.97

For simulation purposes, the input to the pitch system is set as a step signal of magnitude 10 and it is considered fault-free, that is, $f_a(t) = 0$, and the fault coupling matrix is set as $M_{pb} = [0, 0]^T$. The SMO initial conditions are set as zero, while the initial conditions of the states of the pitch system are set as 5 and -10, respectively. The linear and nonlinear gain matrices of the SMO are given as:

The linear and nonlinear gain matrices of the SMO are given as

$$K_l = \begin{bmatrix} 5 & 1 \\ -123.43 & -9.332 \end{bmatrix} \tag{32}$$

and

$$K_{nl} = \begin{bmatrix} 1 & 0 \\ 0 & 1 \end{bmatrix}. \tag{33}$$

The Lyapunov design matrix P_2 for the switched signal is defined as

$$P_2 = \begin{bmatrix} 0.1 & 0 \\ 0 & 0.125 \end{bmatrix} \tag{34}$$

the design matrix $M_2 = 1$, and the scalar $\sigma = 6$.

The results of the SMO convergence are shown in Figures 4 and 5. Figure 4 shows both the pitch angle estimation and the pitch angle estimation error. It can be seen that the observer estimate converges asymptotically to the system state even if their initial conditions are different; therefore, the convergence of the SMO is assured. It can be observed that after 0.5 s, a sliding motion takes place, i.e., the estimation error is forced to reach and subsequently remain on the sliding surface and, thus, a sliding mode motion is said to take place.

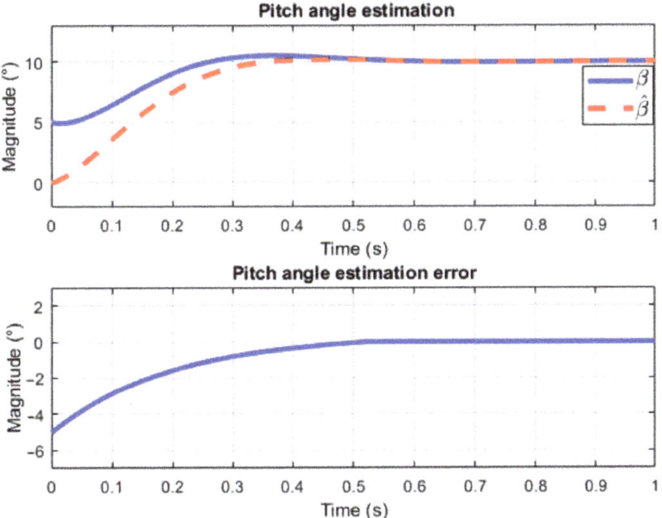

Figure 4. Convergence of the pitch angle estimation.

Figure 5 shows both the pitch speed estimation and the pitch speed estimation error. It can be seen that the estimated state converges asymptotically to the system state after approximately 0.6 s. This is indicative that a sliding motion is taking place on the sliding surface. It is important to mention that the difference in the speed of convergence in the estimation of each state variable is due to the fact that the initial conditions of the state variables are different.

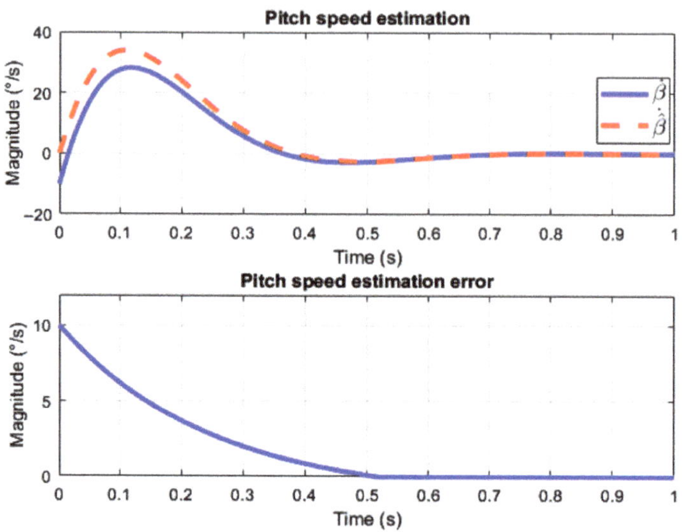

Figure 5. Convergence of the pitch speed estimation.

2.3.2. Actuator FDI Scheme for the Drive Train System

Actuator faults in the drive train system of a wind turbine can have significant consequences for the safety and operation of the wind turbine. Actuator faults in the drive train can be caused by various factors, including mechanical wear and tear, electrical faults, and environmental factors such as temperature and moisture. One common actuator fault in the drive train is caused by gear tooth damage, which inevitably results in gear ratio changes. The friction coefficient in the drive train changes slowly with time and can evolve over months or years. This change can also result in changes in the dynamics of the drive train system. The drive train system bearings are also likely to be faulty. Fatigue and wear due to heavy loads are inevitable. Moreover, pitting and impending cracks are also causes of faults. Generally, faults in the drive train system require time-consuming and costly maintenance. In order to reduce the potential impact of these faults in the drive train system, an effective fault diagnosis scheme is needed.

The drive train system connects the low-speed shaft to the high-speed shaft of the wind turbine. From the mathematical representation of the drive train shown in Equations (4)–(6), it can be observed that it has two input channels, namely, the rotor and the generator torque. Actuator faults are considered to occur in both the rotor and generator torques. However, since the case of simultaneous failures rarely occurs, we consider that only one actuator fault occurs at a time; that is, the case of simultaneous faults is not considered. Actuator fault signals are modeled as additive bounded signals acting in each input channel of the drive train system. It is important to note that the rotor torque cannot be measured; therefore, different approaches have been proposed to estimate it. Because the focus of this study is on the design of an actuator fault reconstruction scheme and not on the actuator itself, it is assumed that the rotor torque is known.

Figure 6 shows the proposed scheme for performing actuator fault detection and isolation in the drive train system. It considers the design of a single SMO to estimate both states and outputs. The difference between the measured drive train variables and the observer outputs is injected into the SMO using a nonlinear function. This term forces the output estimation error to zero in finite time and compensates for the effects of actuator faults in the SMO to maintain a sliding motion. Because the drive train system has two measured outputs, the nonlinear output error injection term is a two-dimensional vector. Each component of this vector is injected into only one input channel of the drive train system in such a way that it compensates only the actuator fault acting on its corresponding input channel. By independently filtering each component of the nonlinear output error injection vector, both fault reconstruction and fault isolation in the drive train system are achieved.

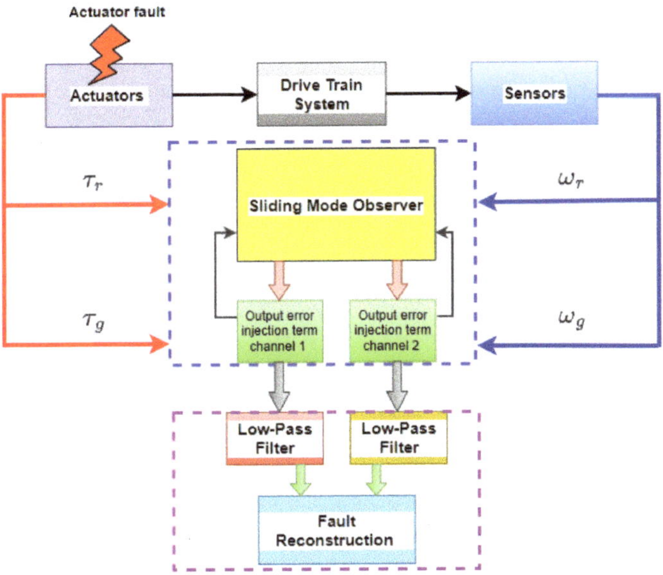

Figure 6. Actuator FDI scheme for the drive train system.

The SMO methodology presented in previous sections is applied below to the drive train system model. For the design of the SMO, it is considered that the drive train model in state space is written as

$$\dot{x}_{dt}(t) = F_{dt}x_{dt}(t) + G_{dt}u_{dt}(t) + M_{dt}f_a(t) \tag{35}$$

$$y_{dt}(t) = H_{dt}x_{dt}(t) \tag{36}$$

where $x_{dt}(t) \in \mathbb{R}^3$ is the state vector, $u \in \mathbb{R}^2$ is the input vector, $y_{dt}(t) \in \mathbb{R}^2$, $f_a(t)$ is any actuator fault signal, M_{dt} is a fault distribution matrix, and F_{dt}, G_{dt}, and H_{dt} are matrices defined as

$$F_{dt} = \begin{bmatrix} -\dfrac{h_{dt}+b_r}{J_r} & \dfrac{h_{dt}}{N_g J_r} & \dfrac{-k_{dt}}{J_r} \\ \dfrac{\eta_{dt}h_{dt}}{N_g J_g} & \dfrac{-\dfrac{\eta_{dt}h_{dt}}{N_g^2}-b_g}{J_g} & \dfrac{\eta_{dt}k_{dt}}{N_g J_g} \\ 1 & \dfrac{-1}{N_g} & 0 \end{bmatrix}, \tag{37}$$

$$G_{dt} = \begin{bmatrix} \dfrac{1}{J_r} & 0 \\ 0 & -\dfrac{1}{J_g} \\ 0 & 0 \end{bmatrix}, \tag{38}$$

and

$$H_{dt} = \begin{bmatrix} 1 & 0 & 0 \\ 0 & 1 & 0 \end{bmatrix} \tag{39}$$

From Equation (19), the sliding mode observer for the drive train system model can be written as

$$\dot{\hat{x}}_{dt}(t) = F_{dt}\hat{x}_{dt}(t) + G_{dt}u_{dt}(t) - K_l e_y(t) + K_{nl}w_{dt}(t) \tag{40}$$

$$\hat{y}_{dt}(t) = H_{dt}\hat{x}_{dt}(t) \tag{41}$$

where $\hat{x}_{pb}(t)$ and $\hat{y}_{pb}(t)$ denote the estimate of state and output, respectively. K_l and K_{nl} are appropriate gain matrices selected to ensure the stability and convergence of the dynamics of error, and $w_{dt}(t)$ represents a discontinuous switched component to induce a sliding motion in the SMO. To verify the convergence of the SMO, the simulation diagram shown in Figure 2 was implemented. The parameters used to simulate the drive train system model are shown in Table 1. The linear and nonlinear gain matrices of the observer are given as

$$K_l = \begin{bmatrix} 10 & -0.0035 \\ 0.0203 & 19.8829 \\ 1 & -0.0095 \end{bmatrix} \tag{42}$$

$$K_{nl} = \begin{bmatrix} 1 & 0 \\ 0 & 1 \\ 0 & 0.0001 \end{bmatrix}, \tag{43}$$

and the nonlinear output error injection signal $w_{dt}(t)$ is given as

$$w_{dt}(t) = \begin{cases} -\sigma\|M_2\| \dfrac{P_2 H_{dt}e_x(t)}{\|P_2 H_{dt}e_x(t)\|} & \text{if } H_{dt}e_x(\text{t}) \neq 0 \\ 0 & \text{otherwise} \end{cases} \tag{44}$$

where the scalar $\sigma = 6$, the design matrix $P_2 = [0.1, 0; 0, 0.125]$, the matrix $M_2 = 1$, and the estate estimation error is defined as $e_x(t) = \hat{x}_{dt}(t) - x_{dt}(t)$. For simulation purposes, the two inputs of the drive train system are set as step-type signals of magnitude 10 and 100, respectively. For the time being, the system is considered fault-free, that is, $f_a(t) = 0$, and the fault coupling matrix $M_{dt} = [0, 0, 0]^T$. The observer initial conditions are considered to be zero, while the initial conditions for the drive train system are chosen as $x_{dt} = [1, -1, 1]^T$. The results of both observer outputs and output estimation errors are shown in Figures 7 and 8. Figure 7 shows that after approximately 1.2 s, a perfect tracking between system output and observer output is achieved and, therefore, a sliding motion takes place, i.e., the output estimation error is forced to reach and subsequently remain on the sliding surface. It can be seen that once the sliding motion is achieved, the state estimation errors evolve according to a first-order decay.

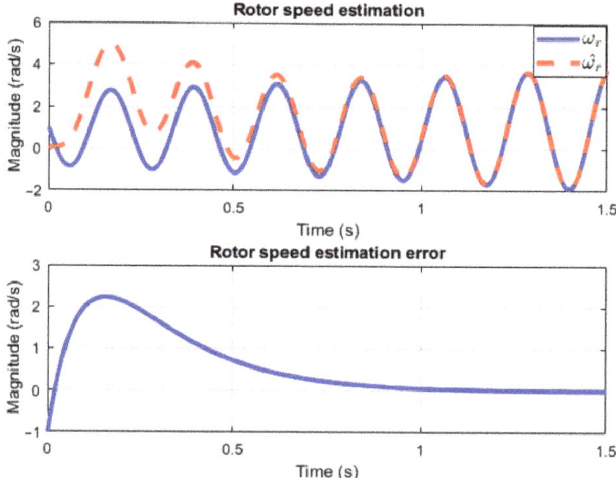

Figure 7. Convergence of the rotor speed estimation.

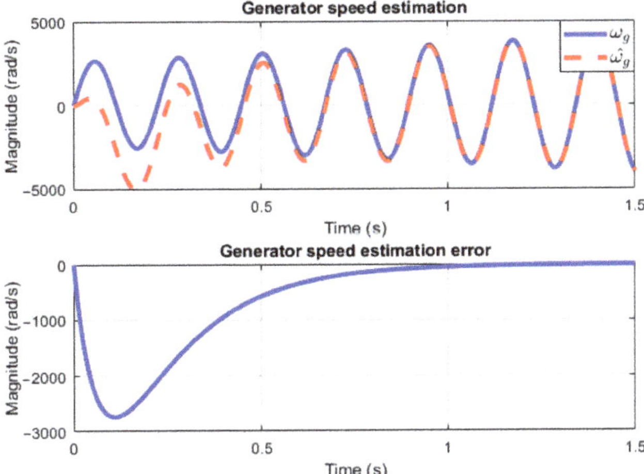

Figure 8. Convergence of the generator speed estimation.

3. Results

In this section, the results of the proposed SMO-based FDI scheme are presented. To evaluate the effectiveness of the proposed FDI scheme, a set of different actuator faults on a wind turbine benchmark model are employed. The test only considers actuator faults in the pitch and the drive train systems. Bounded actuator faults modeled as additive signals are considered. The actuator faults are applied to the input channels of both the pitch and the drive systems. The faults are set as follows:

- Fault 1: A sinusoidal fault signal presented in pitch system 1 given by $f_{a1} = 5 \times in(0.5t)$ in the time period from 2.5133 to 47.7522 s.
- Fault 2: A sinusoidal fault signal presented in pitch system 2 given by $f_{a2}(t) = 5 \times sin(0.5t)$ in the time period from 0 to 50 s.
- Fault 3: A sinusoidal fault signal presented in pitch system 3 given by $f_{a3}(t) = 10 \times sin(0.5t)$ in the time period from 0 to 50 s.
- Fault 4: A square fault signal presented in the rotor torque of the drive train system given by $f_{a4}(t) = 5 \times \text{square}(0.5t)$ in the time period from 6.2832 to 40.2100 s.
- Fault 5: A sinusoidal fault signal presented in the generator torque of the drive train system given by $f_{a5}(t) = 5 \times sin(0.5t)$ in the time period from 1.2566 to 37.6991 s.
- Fault 6: A sawtooth fault signal presented in the generator torque of the drive system given by $f_{a6}(t) = 5 \times sawtoot(0.5t)$ in the time period from 10.0531 to 43.9800 s.

3.1. Results of the FDI Scheme in the Pitch System

The problem of fault detection and isolation in the pitch system is solved by using the architecture shown in Figure 3. Since the output error injection signal injected into the SMO compensates for the dynamics added by the faults in the pitch system, the sliding motion is not interrupted, even in the presence of actuator faults. The dynamic of these faults is recovered using an LPF implemented through a first-order differential equation. By choosing the time constant of the filter to be tiny, but larger than the sampling time used on the computer to implement the LPF, the dynamics of the faults in the pitch system can be reconstructed accurately.

To measure how much better the method is at reproducing the actuator fault in the system, the normalized root mean square error (NRMSE) criterion is used, which proportions a fit percentage value.

The simulation of the proposed FDI scheme in the pitch system is obtained considering the initial conditions of the SMO to be zero, while the initial conditions of the pitch system states are set as $x_{pb}(t) = [2 - 1]^T$. The system input has been arbitrarily set as a step function of magnitude 10. The fault lock matrix for the pitch actuator is chosen as $M_{pb} = [0, 1]^T$ and the scaling parameter of the nonlinear output error injection term is chosen as $\sigma = 6$.

The results of the reconstruction of Fault 1 in pitch system 1 are shown in Figure 9. In order to obtain the highest accuracy in the actuator fault reconstruction, a heuristic process is performed to find the best choice of the LPF time constant. In Table 2, three different filter time constants are presented. From the results obtained, it can be concluded that the best fit is achieved by using a time constant of 0.06 s, with which a fit percentage of 82.9103% is obtained.

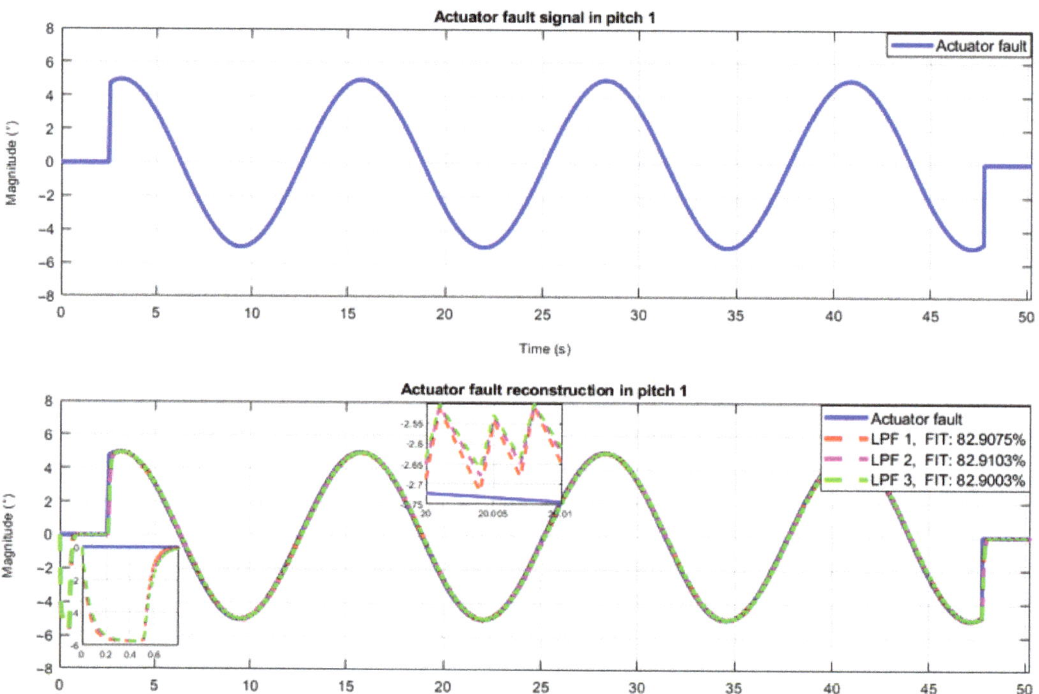

Figure 9. Reconstruction of actuator Fault 1 in pitch system 1.

Table 2. Results of the actuator fault reconstruction in the pitch system.

Faults	Time Constant LPF 1 [s]	Time Constant LPF 2 [s]	Time Constant LPF 3 [s]	Best FIT [%]
Fault1	0.050	0.06	0.065	82.9103
Fault2	0.085	0.09	0.095	82.8394
Fault3	0.020	0.03	0.040	67.1775

Figure 10 shows the results of the actuator fault reconstruction in pitch system 2. It can be seen that the best-fit percentage is obtained with LPF 2, which is implemented with a time constant of 0.09 s (see Table 2).

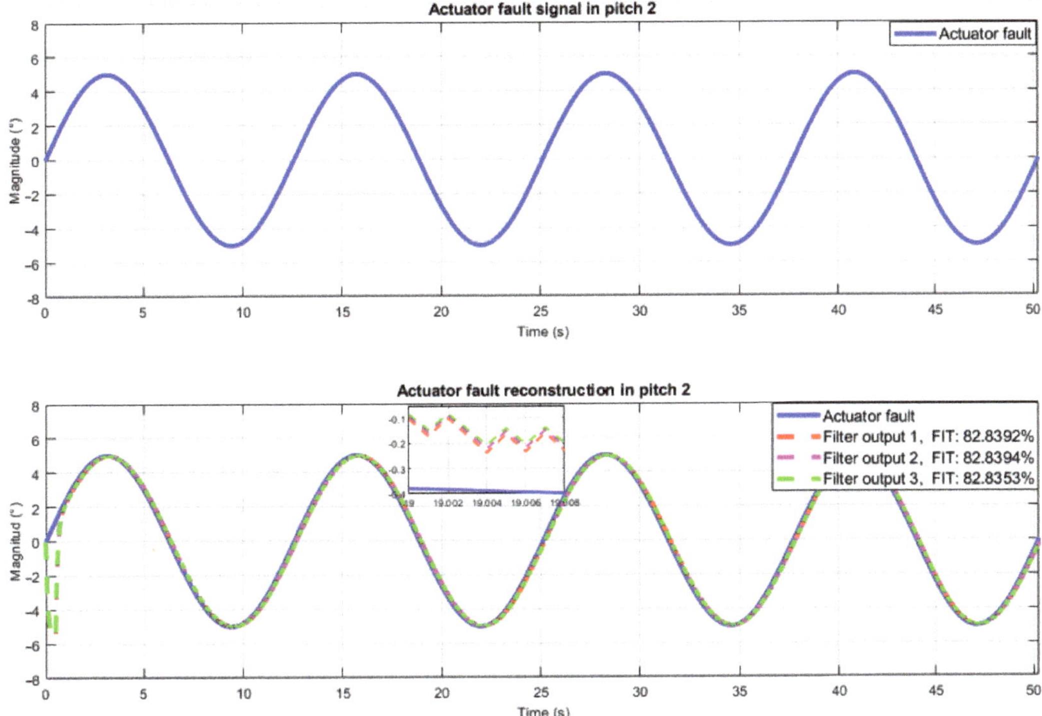

Figure 10. Reconstruction of actuator Fault 2 in the pitch system 2.

It is important to underline that for a sliding motion to take place and, subsequently, for it to be maintained in the presence of actuator faults, it is necessary to select the scalar σ so that the inequality $\sigma > \|f_a(t)\|$ is satisfied for all $t_0 > 0$. Since Faults 1 and 2 satisfy this condition, the sliding motion is maintained despite the dynamics introduced by the actuator faults. This ensures that an accurate reconstruction of the faults can be performed, as shown in Figures 9 and 10.

In the time intervals where inequality $\sigma > \|f_a(t)\|$ is not fulfilled, two issues occur; the first is that the sliding motion cannot be maintained and the second is that the matched disturbance rejection property is lost and, therefore, it is not possible to perform the actuator fault reconstruction.

Figure 11 shows the results of the actuator FDI scheme for Fault 3. It can be observed that since the boundary of Fault 3 is greater than the scalar σ, the inequality $\sigma > \|f_{a3}(t)\|$ cannot be verified for all $t_0 > 0$. This impedes the achievement of an adequate reconstruction of Fault 3 in the intervals where the amplitude fault is greater than the scalar σ, as shown in Figure 11.

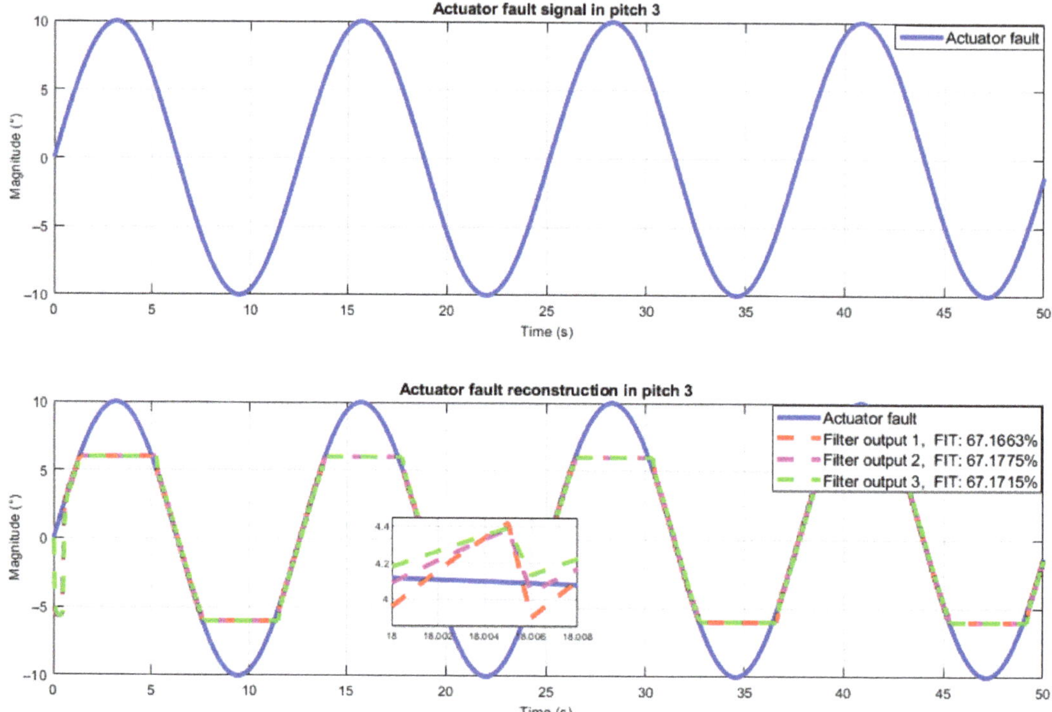

Figure 11. Reconstruction of actuator Fault 3 in the pitch system 3.

3.2. Results of the FDI Scheme in the Drive Train System

To solve the fault isolation problem in the drive train system, the proposed scheme shown in Figure 6 is implemented. Two different LPFs are used to isolate the actuator faults, acting on each one of the input channels of the drive train system. When an actuator fault occurs in channel 1 of the system, the output of LPF 1 is expected to reconstruct the fault while the output of LPF 2 remains at zero because the second input channel of the system is free of fault.

The fault detection problem is solved using the estimation approach described earlier (see Section 2). The main objective of the proposed approach is to maintain the sliding motion even in the presence of actuator faults. The second objective is to recover the fault dynamic through a filtering process of the output error injection signal, which is used to keep the dynamic in a sliding motion. The filtering process is performed using an LPF, which is implemented by a first-order differential equation. To evaluate the effectiveness of the reconstruction, NRMSE analysis, which proportioned a fit percentage number, is used.

The simulation of the proposed FDI scheme for the drive train system has been obtained considering the initial conditions of the SMO to be $\hat{x}(t) = [0,0,0]^T$, while the initial conditions of the drive train system are established as $x_{dt}(t) = [1,-1,1]^T$.

The fault lock matrix for rotor torque fault is set as $M_{dt} = [1,0,0]^T$, while that for generator torque fault is set as $M_{dt} = [0,1,0]^T$. The scalar σ is chosen as 6. System inputs and the SMO gains are selected as stated in Section 2.

Figure 12 shows Fault 4 acting on the first input channel of the system, which affects the signal of the rotor torque. Since the sliding motion is maintained despite the presence of a fault, by passing the nonlinear output error injection term of channel 1 through LPF 1, the fault reconstruction is achieved. From a heuristic process, the optimal filter time constant is 0.0150 s, with a fit of 90.6006%.

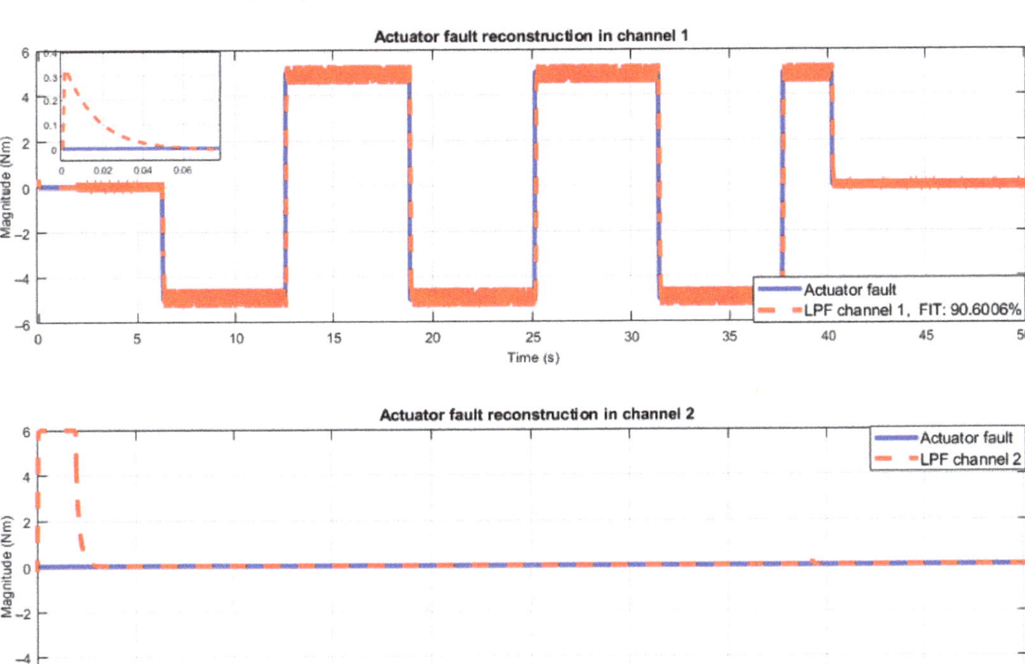

Figure 12. Reconstruction of actuator Fault 4 in the drive train system.

As LPF 2 filters the output error injection term injected into channel 2 of the SMO and it is free of fault, it can be seen that its output remains at zero. It is important to note that the zero deviation in the output of LPF 2 at the start of the simulation is due to the delay introduced by the filter. This delay is also present in the reconstruction of the fault. It can be observed that after the delay introduced by the filtering process, fault isolation and reconstruction are carried out precisely.

Figure 13 shows the fault reconstruction when Fault 5 is acting in the second input channel of the drive train. Since the inequality $\sigma > \|f_{a5}(t)\|$ is satisfied for all $t_0 > 0$, by applying an LPF with a time constant of 0.04 s, a fit of 90.1299% is achieved in the fault reconstruction. It can be observed that during the time the fault remains active, the output of LPF of channel 1 remains at zero, which indicates that the fault is present in channel 2 of the system. Hence, it can be established that the fault is properly isolated.

Figure 14 shows the results of the reconstruction of sawtooth Fault 6. Since Fault 6 is applied in the second input channel of the drive train system, by filtering the output error injection term injected into channel 2 with an LPF whose time constant is 0.024 s, a fit of 87.2779% is obtained.

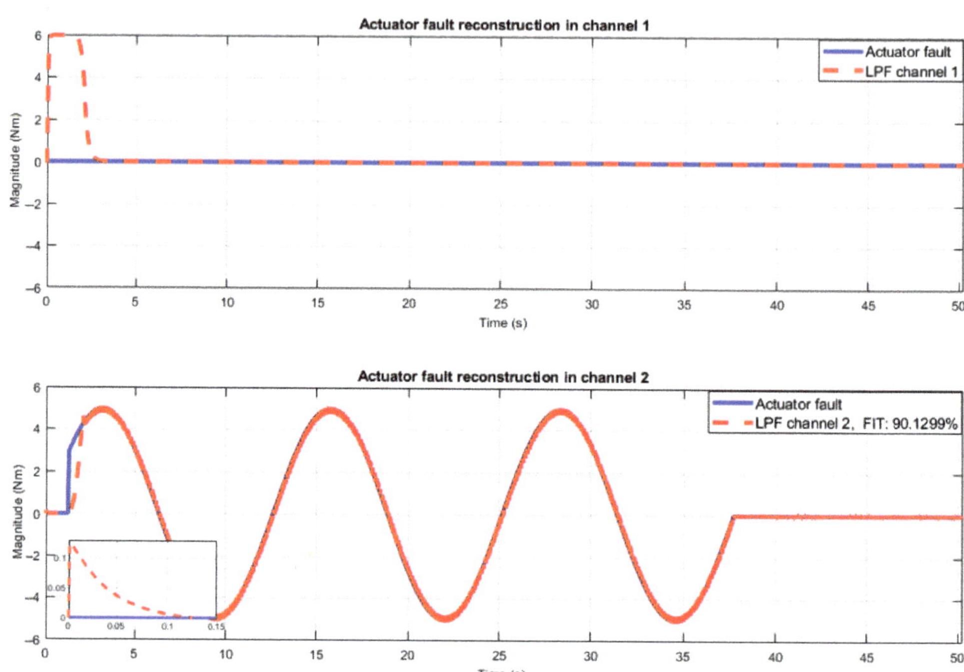

Figure 13. Reconstruction of actuator Fault 5 in the drive train system.

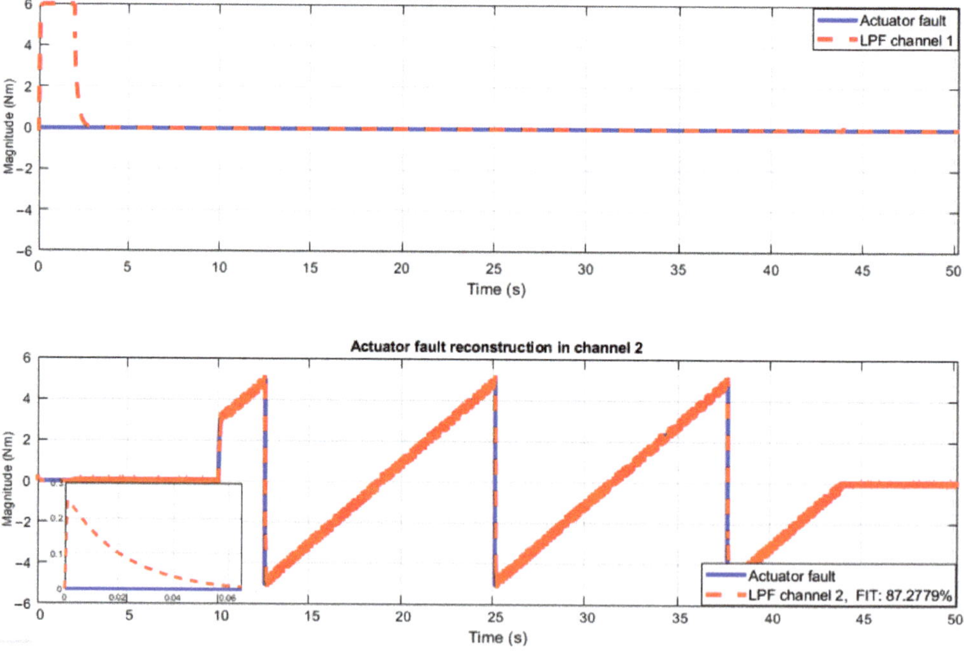

Figure 14. Reconstruction of actuator Fault 6 in the drive train system.

From the output of the LPF applied to the output error injection term injected into channel 1, it can be seen that it remains at zero during the time that the fault is active. Therefore, it can be concluded that the fault occurred in channel 2 of the system. Table 3 presents the time constants used in the LPF for each of the faults in the drive train system, as well as the best fit in percentage obtained when performing the fault reconstruction.

Table 3. Results of the actuator fault reconstruction in the drive train system.

Faults	Time Constant LPF [s]	Best FIT [%]
Fault 4	0.0150	90.6006
Fault 5	0.0400	90.1299
Fault 6	0.0200	87.2779

3.3. Factors Affecting the Fault Reconstruction Performance

Several factors can affect the performance of fault reconstruction using an SMO in wind turbines. Here are some key factors to consider:

1. The characteristics of the faults themselves can impact the performance of fault reconstruction. From the results presented in Section 3, it is noted that factors such as fault magnitude, fault duration, and fault dynamics (e.g., abrupt or gradual changes) affect the accuracy of the fault reconstruction. Different types of faults require specific adaptations in both the SMO design (the σ-parameter) and the LPF time constant to ensure effective fault estimation.
2. The design of the SMO, including its parameter selection and initial conditions, affects the fault reconstruction performance. In order to implement the SMO-based FDI scheme in a wind turbine, the initial condition of the observer must be assigned prior to the beginning of the wind turbine operation. However, the issue of how best to choose those initial condition values has apparently been completely ignored in control engineering textbooks. As a consequence, observer initial conditions in industrial applications are usually set to zero by default. A study of the effect of the SMO initial conditions on the performance of the fault reconstruction was realized. Table 4 shows Fault 5 reconstruction performance using different SMO initial conditions, keeping the initial condition of the drive train system fixed. From the results, it can be noted that for SMO initial conditions close to system conditions, reconstruction performance improves, while for SMO initial conditions far from the initial conditions of the system, the performance worsens. This is an expected result since the choice of initial conditions impacts the transient behavior and convergence speed of the SMO.

Table 4. Effects of the SMO initial conditions in Fault 5 reconstruction.

Initial Conditions	Time Constant LPF [s]	Best FIT [%]
$\hat{x} = \begin{bmatrix} 0 & 0 & 0 \end{bmatrix}^T$	0.0400	90.1299
$\hat{x} = \begin{bmatrix} 0 & 0 & 1 \end{bmatrix}^T$	0.0400	92.2562
$\hat{x} = \begin{bmatrix} 3 & 2 & 3 \end{bmatrix}^T$	0.0350	88.5118
$\hat{x} = \begin{bmatrix} 5 & 5 & 4 \end{bmatrix}^T$	0.0350	87.5168
$\hat{x} = \begin{bmatrix} 100 & 90 & 100 \end{bmatrix}^T$	0.0350	79.2138

The observer's parameters should be appropriately chosen to achieve fast and accurate estimation of both system states and fault signals. Improper observer design can result in slow convergence, estimation errors, or undesirable oscillations in the reconstructed fault signals. A study of the effect of the SMO σ-parameter on the performance of the fault reconstruction was realized. Table 5 shows Fault 5 reconstruction performance using different values of the σ-parameter. From the results, it is noted that for σ values less than the magnitude of the fault, a substantial degradation in the reconstruction of Fault 5 is observed. However, increasing the σ value does not substantially improve the performance of the fault reconstruction. Consequently, an appropriate selection of the σ value must satisfy the condition $\sigma > \|f_a(t)\|$.

Table 5. Effects of σ in Fault 5 reconstruction.

Parameter σ	Time Constant LPF [s]	Best FIT [%]
$\sigma = 6$	0.0400	90.1299
$\sigma = 5$	0.0350	89.8022
$\sigma = 3$	0.0350	66.5234
$\sigma = 8$	0.0400	90.5583
$\sigma = 10$	0.0450	90.7927
$\sigma = 20$	0.0800	90.9546

3. From the point of view of the implementation of the proposed scheme, factors such as sampling rate and computational resources can affect the fault reconstruction performance. The sampling rate at which sensor data are collected and processed affects the temporal resolution of fault reconstruction. A higher sampling rate allows for finer detection and tracking of fault dynamics, but it also increases computational requirements. The sampling rate should be chosen carefully to balance the need for accuracy with practical implementation considerations. Moreover, the computational resources available for implementing the SMO influence its performance. A higher computational capability enables faster calculations and can facilitate real-time fault reconstruction. Insufficient computational resources may result in slower estimation or limitations in the complexity of the observer algorithm.

4. The selection and quality of sensors used to measure the system outputs and inputs also impact the performance of fault reconstruction. High-quality sensors with low noise and suitable measurement range contribute to accurate state estimation.

4. Discussion

An observer-based FDI scheme that uses a particular class of SMO to identify and isolate actuator faults using an estimation approach was presented. Fault detection and fault isolation problems are solved by analyzing the nonlinear output error injection signal required to keep the SMO in a sliding motion. The proposed actuator FDI scheme is evaluated with six different actuator faults, and the accuracy of the proposed approach is validated using the NRMSE method. In contrast to the works presented in [20,23] a simple architecture using an LPF is shown be sufficient to perform an accuracy fault reconstruction, with no false alarm or missed detection.

The proposed FDI scheme does not require explicit information from the fault. The only fault information required is the fault to be bounded. Thus, the actuator fault reconstruction signal is obtained online and without prior knowledge of the fault, which is of vital importance for the FDI scheme to be implemented in real applications.

It is important to highlight, however, that the dynamics of faults, the filter time constant, the observer initial conditions, and the σ value of the SMO affect fault reconstruction accuracy. To find the optimal filter time constant, a heuristic process is used. However, a method to automatically find the optimal filter time constant could be used. The initial conditions of the SMO were shown to affect the fault reconstruction performance. A better fault reconstruction performance is achieved with SMO initial conditions close to the initial condition of the system.

The σ value was shown to have an impact on the fault reconstruction accuracy. If the σ value is smaller than the upper bound of the fault, an accurate fault reconstruction is not possible. However, values of σ much larger than the upper bound of the fault were not shown to lead to a significant increase in the fault reconstruction precision. Furthermore, high values of σ were shown to have a prejudicial effect on the accuracy of state estimation due to the undesirable phenomenon known as chattering.

5. Conclusions

In this paper, an actuator observer-based FDI scheme for the pitch and drive train systems of a wind turbine is proposed. The proposed scheme employs a special class of SMOs that have the capacity to drive the output estimation error to zero in finite time, even in the presence of actuator faults. The approach adopted here uses the nonlinear output estimation error signal to perform actuator fault reconstruction. This error signal is used to drive the observer dynamics and converges the estimated output to the actual output of the system. Thus, it compensates for fault dynamics in the SMO by switching at high frequency as a function of the output estimation error. To reconstruct actuator faults in both the pitch and drive systems, an approach using an LPF on the nonlinear output estimation error signal is adopted. The actuator fault reconstruction signal is obtained online and without prior knowledge of the fault and, thus, it is easy to implement in real time. Factors such as the initial conditions and SMO settings were demonstrated to influence the fault reconstruction performance. The simulation results and NRMSE analysis have demonstrated the effectiveness of the proposed scheme for performing actuator fault reconstruction. In future work, the SMO-based FDI scheme will be integrated with an FTC system.

Author Contributions: Conceptualization, V.B.-J. and M.A.-M.; methodology, V.B.-J. and M.A.-M.; software, V.B.-J. and J.G.-M.; validation, V.B.-J. and M.A.-M.; formal analysis, G.V.G.-R. and V.B.-J.; investigation, V.B.-J., M.A.-M. and B.Y.L.-Z.; resources, G.V.G.-R., B.Y.L.-Z., E.M.S.-C. and J.G.-M.; writing—original draft preparation, V.B.-J., B.Y.L.-Z., J.G.-M. and E.M.S.-C.; writing—review and editing, M.A.-M., E.M.S.-C., V.B.-J., B.Y.L.-Z., G.V.G.-R. and J.G.-M.; supervision, V.B.-J. All authors have read and agreed to the published version of the manuscript.

Funding: This research received no external funding.

Data Availability Statement: Not applicable.

Acknowledgments: The development of this project is the product of the support provided by the National Council of Science and Technology (CONACYT) and the National Center for Technological Research and Development(CENIDET).

Conflicts of Interest: The authors declare no conflict of interest.

Abbreviations

The following abbreviations are used in this manuscript:

SMO	Sliding mode observer
FDI	Fault detection and isolation
NEMSE	Normalized root mean square error
DNN	Deep neural networks
PCA	Principal component analysis
LPF	Low-pass filter
NREL	National Renewable Energy Laboratory
FTC	Fault-tolerant control

Nomenclature

Matrix

F	System matrix
G	Input matrix
H	Output matrix
K	Observer gain matrix
M	Fault distribution matrix
P	Symmetric positive define matrix
T	Change of coordinates matrix

Vectors

u	System input vector
e	Estimation error vector
x	System state vector

$\hat{x}$	Estimate of states vector
y	System output vector
$\hat{y}$	Estimate of output vector
w	Nonlinear discontinuous function
Constants	
b	Viscous friction constant
h	Torsion damping constant
j	Moment of inertia constant
k	Torsion stiffness constant
N	Gear ratio constant
r	Rotor ratio constant
n	Real number
p	Real number
m	Real number
q	Real number
Signals	
f	Fault signal
v	Wind speed signal
E	Power electrical signal
Greek letters	
α	Scalar
β	Pitch angle
ζ	Damping coefficient
η	Efficiency
θ	Torsion angle
σ	Scalar
ψ	Natural frequency
γ	Cutoff frequency
λ	Tip speed ratio
ρ	Wind density
Subscripts	
a	Actuator
dt	Drive train
g	Generator
l	Linear
nl	Nonlinear
pb	Pitch blade
r	Rotor
ω	Wind
x	State
y	Output
ref	Reference

References

1. Global Wind Energy Council. *Global Wind Report 2022*; Global Wind Energy Council: Bonn, Germany, 2022.
2. Anaya-Lara, O.; Jenkins, N.; Ekanayake, J.B.; Cartwright, P.; Hughes, M. *Wind Energy Generation: Modelling and Control*; John Wiley & Sons: Chichester, UK, 2009.
3. Wymore, M.L.; Van Dam, J.E.; Ceylan, H.; Qiao, D. A survey of health monitoring systems for wind turbines. *Renew. Sustain. Energy Rev.* **2015**, *52*, 976–990. [CrossRef]
4. Luo, N.; Vidal, Y.; Acho, L. *Wind Turbine Control and Monitoring*; Springer: Cham, Switzerland, 2014; Available online: http://www.springer.com/series/1412 (accessed on 4 January 2023).
5. Chang, C.C.W.; Ding, T.J.; Ariannejad, M.; Chao, K.C.; Samdin, S.B. Fault detection and anti-icing technologies in wind energy conversion systems: A review. *Energy Rep.* **2022**, *8*, 28–33. [CrossRef]
6. Thirumarimurugan, M.; Bagyalakshmi, N.; Paarkavi, P. Comparison of fault detection and isolation methods: A review. In Proceedings of the 2016 10th International Conference on Intelligent Systems and Control (ISCO), Coimbatore, India, 7–8 January 2016; pp. 1–6. [CrossRef]
7. Liu, Z.; Zhang, L. A review of failure modes, condition monitoring and fault diagnosis methods for large-scale wind turbine bearings. *Measurement* **2020**, *149*, 107002. [CrossRef]

8. Fekih, A.; Habibi, H.; Simani, S. Fault Diagnosis and Fault Tolerant Control of Wind Turbines: An Overview. *Energies* **2022**, *15*, 7186. [CrossRef]
9. Gao, Z.; Liu, X. An Overview on Fault Diagnosis, Prognosis and Resilient Control for Wind Turbine Systems. *Processes* **2021**, *9*, 300. [CrossRef]
10. Simani, S.; Castaldi, P. Intelligent Fault Diagnosis Techniques Applied to an Offshore Wind Turbine System. *Appl. Sci.* **2019**, *9*, 783. [CrossRef]
11. Liu, W. Intelligent fault diagnosis of wind turbines using multi-dimensional kernel domain spectrum technique. *Measurement* **2018**, *133*, 303–309. [CrossRef]
12. Wen, X.; Xu, Z. Wind turbine fault diagnosis based on ReliefF-PCA and DNN. *Expert Syst. Appl.* **2021**, *178*, 115016. [CrossRef]
13. Pashazadeh, V.; Salmasi, F.R.; Araabi, B.N. Data driven sensor and actuator fault detection and isolation in wind turbine using classifier fusion. *Renew. Energy* **2018**, *116*, 99–106. [CrossRef]
14. Liu, W.; Tang, B.; Han, J.; Lu, X.; Hu, N.; He, Z. The structure healthy condition monitoring and fault diagnosis methods in wind turbines: A review. *Renew. Sustain. Energy Rev.* **2015**, *44*, 466–472. [CrossRef]
15. Lan, J.; Patton, R.J. *Robust Integration of Model-Based Fault Estimation and Fault-Tolerant Control*; Springer: Berlin/Heidelberg, Germany, 2021. [CrossRef]
16. Yang, Y.; Ding, S.X.; Li, L. On observer-based fault detection for nonlinear systems. *Syst. Control Lett.* **2015**, *82*, 18–25. [CrossRef]
17. Teng, J.; Li, C.; Feng, Y.; Yang, T.; Zhou, R.; Sheng, Q.Z. Adaptive Observer Based Fault Tolerant Control for Sensor and Actuator Faults in Wind Turbines. *Sensors* **2021**, *21*, 8170. [CrossRef] [PubMed]
18. Odgaard, P.F.; Stoustrup, J.; Kinnaert, M. Fault-Tolerant Control of Wind Turbines: A Benchmark Model. *IEEE Trans. Control Syst. Technol.* **2013**, *21*, 1168–1182. [CrossRef]
19. Pujol-Vazquez, G.; Acho, L.; Gibergans-Báguena, J. Fault Detection Algorithm for Wind Turbines' Pitch Actuator Systems. *Energies* **2020**, *13*, 2861. [CrossRef]
20. Zhu, J.; Ma, K.; Hajizadeh, A.; Soltani, M.; Chen, Z. Fault detection and isolation for wind turbine electric pitch system. In Proceedings of the 2017 IEEE 12th International Conference on Power Electronics and Drive Systems (PEDS), Honolulu, HI, USA, 12–15 December 2017; pp. 618–623. [CrossRef]
21. Odgaard, P.F.; Stoustrup, J. Unknown input observer based detection of sensor faults in a wind turbine. In Proceedings of the 2010 IEEE International Conference on Control Applications, Yokohama, Japan, 8–10 September 2010; pp. 310–315. [CrossRef]
22. Hwang, I.; Kim, S.; Kim, Y.; Seah, C.E. A Survey of Fault Detection, Isolation, and Reconfiguration Methods. *IEEE Trans. Control Syst. Technol.* **2009**, *18*, 636–653. [CrossRef]
23. Chen, J.; Patton, R.J. Robust Model-Based Fault Diagnosis for Dynamic Systems. In *The International Series on Asian Studies in Computer and Information Science*; Springer: Boston, MA, USA, 1999; Volume 3. [CrossRef]
24. Ziyabari, S.H.S.; Shoorehdeli, M.A.; Karimirad, M. Robust fault estimation of a blade pitch and drivetrain system in wind turbine model. *J. Vib. Control* **2020**, *27*, 277–294. [CrossRef]
25. Edwards, C.; Spurgeon, S.K.; Patton, R.J. Sliding mode observers for fault detection and isolation. *Automatica* **2000**, *36*, 541–553. [CrossRef]
26. Alwi, H.; Edwards, C.; Tan, C.P. Fault Detection and Fault-Tolerant Control Using Sliding Modes. In *Advances in Industrial Control*; Springer: London, UK, 2011. [CrossRef]
27. Habibi, H.; Howard, I.; Simani, S.; Fekih, A. Decoupling Adaptive Sliding Mode Observer Design for Wind Turbines Subject to Simultaneous Faults in Sensors and Actuators. *IEEE/CAA J. Autom. Sin.* **2021**, *8*, 837–847. [CrossRef]
28. Borja-Jaimes, V.; Adam-Medina, M.; López-Zapata, B.Y.; Valdés, L.G.V.; Pachecano, L.C.; Coronado, E.M.S. Sliding Mode Observer-Based Fault Detection and Isolation Approach for a Wind Turbine Benchmark. *Processes* **2021**, *10*, 54. [CrossRef]
29. Chen, W.; Ding, S.; Haghani, A.; Naik, A.; Khan, A.; Yin, S. Observer-based FDI Schemes for Wind Turbine Benchmark. *IFAC Proc. Vol.* **2011**, *44*, 7073–7078. [CrossRef]
30. Habibi, H.; Howard, I.; Simani, S. Reliability improvement of wind turbine power generation using model-based fault detection and fault tolerant control: A review. *Renew. Energy* **2018**, *135*, 877–896. [CrossRef]
31. Witczak, M.; Rotondo, D.; Puig, V.; Nejjari, F.; Pazera, M. Fault estimation of wind turbines using combined adaptive and parameter estimation schemes. *Int. J. Adapt. Control Signal Process.* **2017**, *32*, 549–567. [CrossRef]
32. Sun, X.; Patton, R.J. Robust actuator multiplicative fault estimation with unknown input decoupling for a wind turbine system. In Proceedings of the Conference on Control and Fault-Tolerant Systems (SysTol), Nice, France, 9–11 October 2013; pp. 263–268. [CrossRef]
33. Isermann, R. *Fault-Diagnosis Systems: An Introduction from Fault Detection to Fault Tolerance*; Springer: Berlin/Heidelberg, Germany, 2006; pp. 1–475.
34. Garcia Márquez, F.P.; Tobias, A.M.; Pérez, J.M.P.; Papaelias, M. Condition monitoring of wind turbines: Techniques and methods. *Renew. Energy* **2012**, *46*, 169–178. [CrossRef]
35. Sheng, S. NREL's Wind Turbine Drivetrain Condition Monitoring and Wind Plant Operation and Maintenance Research during the 2010s: A US Land-Based Perspective. *Acoust. Aust.* **2021**, *49*, 239–249. [CrossRef]
36. Lin, Y.; Tu, L.; Liu, H.; Li, W. Fault analysis of wind turbines in China. *Renew. Sustain. Energy Rev.* **2016**, *55*, 482–490. [CrossRef]
37. Abbaspour, A.; Mokhtari, S.; Sargolzaei, A.; Yen, K.K. A Survey on Active Fault-Tolerant Control Systems. *Electronics* **2020**, *9*, 1513. [CrossRef]

38. Edwards, C.; Spurgeon, S. *Sliding Mode Control: Theory and Applications*; CRC Press: London, UK, 1998. [CrossRef]
39. Olabi, A.G.; Wilberforce, T.; Elsaid, K.; Sayed, E.T.; Salameh, T.; Abdelkareem, M.A.; Baroutaji, A. A Review on Failure Modes of Wind Turbine Components. *Energies* **2021**, *14*, 5241. [CrossRef]

Article

Generalized Functional Observer for Descriptor Nonlinear Systems—A Takagi-Sugeno Approach

C. Ríos-Ruiz [1,2], G.-L. Osorio-Gordillo [1,*], C.-M. Astorga-Zaragoza [1], M. Darouach [2], H. Souley-Ali [2] and J. Reyes-Reyes [1]

[1] Tecnológico Nacional de México/CENIDET, Interior Internado Palmira S/N, Cuernavaca 62490, Mexico; crios@cenidet.edu.mx (C.R.-R.); carlos.az@cenidet.tecnm.mx (C.-M.A.-Z.); juan.rr@cenidet.tecnm.mx (J.R.-R.)

[2] CRAN-CNRS (UMR 7039), Université de Lorraine, IUT Longwy, 186, Rue de Lorraine, 54400 Cosnes et Romain, France; mohamed.darouach@univ-lorraine.fr (M.D.); harouna.souley@univ-lorraine.fr (H.S.-A.)

* Correspondence: gloria.og@cenidet.tecnm.mx

Abstract: This paper concerns the design of a generalized functional observer for Takagi–Sugeno descriptor systems. Furthermore, a generalized structure is herein introduced for purposes of estimating linear functions of the states of descriptor nonlinear systems represented into a Takagi–Sugeno descriptor form. The originality of the functional generalized observer structure is that it provides additional degrees of freedom in the observer design, which allows for improvements in the estimation against parametric uncertainties. The effectiveness of the developed design is illustrated by a nonlinear model of a single link robotic arm with a flexible link. A comparison between the functional generalized observer and the functional proportional observer is given to demonstrate the observer performances.

Keywords: functional observers; Takagi–Sugeno systems; descriptor systems

Citation: Ríos-Ruiz, C.; Osorio-Gordillo, G.-L.; Astorga-Zaragoza, C.-M.; Darouach, M.; Souley-Ali, H.; Reyes-Reyes, J. Generalized Functional Observer for Descriptor Nonlinear Systems—A Takagi-Sugeno Approach. *Processes* 2023, 11, 1707. https://doi.org/10.3390/pr11061707

Academic Editor: Wen-Jer Chang

Received: 25 April 2023
Revised: 27 May 2023
Accepted: 27 May 2023
Published: 2 June 2023

1. Introduction

A functional observer is a dynamical system that is designed to estimate one or more functions of the system states. They can be viewed as a generalization of classical state observers. For instance, suppose that there exist n functions $z_1(x), \ldots, z_n(x)$ to be estimated for an n-order system by using a functional observer, and each function is a different state $x_i, i = 1, \ldots, n$, of an n-order system, i.e., $z_1(x) = x_1, \ldots, z_n(x) = x_n$, then a functional observer coincides with a full-order state observer. However, the main advantage of a functional observer is that when a specific function needs to be estimated, it may be easier than the standard approach. It may even be the case that the system does not need to be observable but only be functional observable [1,2]. A complete study about the existence and design of functional observers can be further explored in [3–5].

Numerous works in the literature can be found about applications of functional observers. For instance, in [6] a functional observer for fault-detection of linear time-invariant systems which is designed to converge in a finite-time is proposed. In [7], a real-time implementation of a functional observer-based feedback controller is performed to control the position of a ball on a balancing table. The authors demonstrate that this task can be accomplished with only a minimum order functional observer. Their work clearly reflects the advantages to implementing functional observers compared to classical full-order state observers. Another interesting work where functional observers are used as a method to cope with unknown inputs (which can represent process uncertainties, sensor faults, communication problems or cyber-attacks) is presented in [8], where a functional observer is used to perform load frequency control for a complex power system.

The case of functional observers for descriptor systems has been discussed by several authors. For instance, in [9], the authors propose a functional observer for switched descriptor systems with Lipschitz nonlinearities. The observer is used for fault estimation purposes. Functional observers of various orders for continuous-time and discrete-time systems are treated in [10]. In [11], a finite-time functional observer for descriptor systems is presented and used for fault-tolerant control purposes. Recently, the authors in [12] presented a functional unknown input observer for linear descriptor systems with arbitrarily prescribed convergence time. The case of functional interval observers for switched descriptor systems is treated in [13].

Functional observers for Takagi–Sugeno systems are presented in [14], where the authors designed a functional observer for time-delayed systems. The effectiveness of the observer is evaluated for fault estimation in a truck-trailer system and fault detection of a permanent magnet DC motor. The case of stabilizing time-delay Takagi–Sugeno systems is presented by the same authors in a precedent paper [15]. Researchers in [16,17] present a parallel distributed compensation (PDC) feedback controller for Takagi–Sugeno systems. The approach is based on the capability of functional observers to estimate the unmeasured premise variables and/or the control law. A braking control of a suspended floater, the regulation of a power electronics converter and stabilization of a synchronous generator are used as case studies. Observer-based control for Wind Energy Centers (WEC) have been studied in [18,19].

The design of functional observers for Takagi–Sugeno descriptor systems represents the subject of this work. There exist some recent works about observer synthesis for Takagi–Sugeno descriptor systems. For instance, the authors in [20] present the design of a robust observer for Takagi–Sugeno descriptor systems with unmeasurable premise variables. The observer is used for sensor fault detection and isolation. Recently in [21], a Takagi–Sugeno observer for discrete nonlinear descriptor systems is designed. An extended Luenberger-like structure is used for the observer synthesis. Unmeasured premise variables of the Takagi–Sugeno representation and unknown inputs are considered. In the work presented in [22], the authors reveal the design of a reduced-order observer for state estimation for Takagi–Sugeno descriptor systems with time delay. The observer is then used in an observer-based sliding control scheme in order to stabilize the system.

In this paper, a functional observer for Takagi–Sugeno descriptor systems is presented. This class of functional observers is interesting because, on one hand, systems represented in a Takagi–Sugeno descriptor form take advantage of the features of descriptor systems which cover a wide range of systems represented by dynamic and static equations such as hydraulic systems [23], mechanical systems [24], biomechanical systems [25], hydrogen generation [26] and electronic circuits [27]. On the other hand, the Takagi–Sugeno approach can be used to represent a wide class of nonlinear systems such as alcoholic fermentation reactors [28], steam generators of thermal power plants [29], nonlinear vehicle dynamics [30], glucose–insulin system [31] and so on. The ability to merge both classes of systems offers enormous potential for applications across a wide variety of fields.

The main contribution of this work is the presentation of a generalized functional observer for Takagi–Sugeno descriptor systems. All the sufficient and necessary conditions for its design are given. The main feature of this observer is to exploit the generalized observer structure to estimate linear functions of the states of descriptor nonlinear systems represented into a Takagi–Sugeno descriptor form. Another significant advantage of the proposed functional observer is its capability to directly estimate a linear stabilizing control law, instead of two subsystems that are used in a classical observer-based stabilization control: (1) the observer and (2) the controller. This can be particularly advantageous in systems where a stabilizing control law is required but where an estimation of the state is not. In these cases, a functional observer can be designed to directly estimate the control law. Another case is a system with a large number of states and a small number of inputs. Here, a functional minimum order observer can stabilize the system.

A non-linear model of a single link robotic arm with a flexible link [32] is used to illustrate the design procedure of the proposed observer. The performance capabilities of the generalized functional observer (GFO) are compared with a simple proportional observer (PO).

2. Preliminaries

In this section, the notations used in this paper are introduced. A^+ denotes the generalized inverse of A and verifies $AA^+A = A$. The notation $E^\perp$ denotes a maximal row rank matrix such that $E^\perp E = 0$. When E is a full row rank matrix, $E^\perp = 0$ by convention.

The number of local models A_i, and B_i depends on the scheduling variables, $\kappa = 2^s$, where s is the number of scheduling variables. For systems with a high number of scheduling variables, the value of κ will rapidly increase, this can be a disadvantage for the calculus of the observers. Thus, this approach is suitable for nonlinear systems with a convenient-for-design number of nonlinearities. There exist different T-S models that can be obtained from a nonlinear system, the way to select the appropriate T-S model is to take into account the variable states of the final transformed system, and those that are needed to be estimated by the observer.

Consider a Takagi–Sugeno descriptor system of the form:

$$E\dot{x}(t) = \sum_{i=1}^{\kappa} w_i(t)(A_i x(t) + B_i u(t)) \tag{1}$$

$$y(t) = Cx(t) \tag{2}$$

$$z(t) = Lx(t) \tag{3}$$

where $x(t) \in \mathbb{R}^n$ is the semi-state of the system, $u(t) \in \mathbb{R}^m$ is the input vector and $y(t) \in \mathbb{R}^p$ is the measured output of the system. $z(t) \in \mathbb{R}^q$ is a linear function of the state to be estimated.

$w_i(t)$ are membership functions formed with $\rho \in \mathbb{R}^s$ scheduled variables, which can depend on states, inputs, measured variables or other exogenous variables of the system. The membership functions have the following properties:

$$\sum_{i=1}^{\kappa} w_i(t) = 1, \quad w_i(t) \geq 0 \tag{4}$$

for $i = 1, \dots, \kappa = 2^s$. Matrices $E \in \mathbb{R}^{n \times n}$, $A_i \in \mathbb{R}^{n \times n}$, $B_i \in \mathbb{R}^{n \times m}$, $C \in \mathbb{R}^{p \times n}$ and $L \in \mathbb{R}^{q \times n}$ are assumed known.

Assumption 1 ([4]). *The triplet (C, E, A_i) is partially impulse observable with respect to matrix L if the following equivalent statements hold*

$$i. \quad \text{rank} \begin{bmatrix} E & A_i \\ 0 & C \\ 0 & E \\ 0 & L \end{bmatrix} = \text{rank} \begin{bmatrix} E & A_i \\ 0 & C \\ 0 & E \end{bmatrix}$$

$$ii. \quad \text{rank} \begin{bmatrix} L \\ E \\ E^\perp A_i \\ C \end{bmatrix} = \text{rank} \begin{bmatrix} E \\ E^\perp A_i \\ C \end{bmatrix}, \quad \forall i = 1, \dots, \kappa$$

Partial impulse observability allows us to estimate the function $z(t)$ by using only the available output. It is important to note that the observer must correctly estimate the function of the state, even in the presence of impulsive behavior of the descriptor system, thus the importance of this lemma.

The following Lemma will be used later in this paper.

Lemma 1 ([33]). *Let matrices $\mathcal{B}$ and $\mathcal{Q}$ be given. The following statements are equivalent:*

1.   *There exists a matrix $\mathcal{X}$ satisfying*

$$\mathcal{B}\mathcal{X} + (\mathcal{B}\mathcal{X})^T + \mathcal{Q} < 0$$

2.   *The following condition holds*

$$\mathcal{B}^{\perp}\mathcal{Q}\mathcal{B}^{\perp T} < 0$$

Suppose the above statements hold and assume that $\mathcal{B}^{\perp}\mathcal{B} > 0$. Then matrix $\mathcal{X}$ in statement 1 is given by

$$\mathcal{X} = -\gamma\mathcal{B}^T + \sqrt{\gamma}\mathcal{L}\Gamma^{1/2}$$

where $\mathcal{L}$ is any matrix such that $||\mathcal{L}|| < 1$ and $\gamma > 0$ is any scalar such that

$$\Gamma = \gamma\mathcal{B}\mathcal{B}^T - \mathcal{Q} > 0$$

3. Problem Statement

Consider the following generalized functional observer (GFO) of the form

$$\dot{\zeta}(t) = \sum_{i=1}^{\kappa} w_i(t)(N_i\zeta(t) + J_i v(t) + F_i y(t) + H_i u(t)) \tag{5}$$

$$\dot{v}(t) = \sum_{i=1}^{\kappa} w_i(t)(S_i\zeta(t) + G_i v(t) + M_i y(t)) \tag{6}$$

$$\hat{z}(t) = \sum_{i=1}^{\kappa} w_i(t)(P_i\zeta(t) + Q_i y(t)) \tag{7}$$

where $\zeta(t) \in \mathbb{R}^{q_0}$ is the state of the observer, $v(t) \in \mathbb{R}^{q_1}$ is an auxiliary vector and $\hat{z}(t) \in \mathbb{R}^{q}$ is the estimate of $z(t)$. $N_i \in \mathbb{R}^{q_0 \times q_0}$, $J_i \in \mathbb{R}^{q_0 \times q_1}$, $F_i \in \mathbb{R}^{q_0 \times p}$, $H_i \in \mathbb{R}^{q_0 \times m}$, $S_i \in \mathbb{R}^{q_1 \times q_0}$, $G_i \in \mathbb{R}^{q_1 \times q_1}$, $M_i \in \mathbb{R}^{q_1 \times p}$, $P_i \in \mathbb{R}^{q \times q_0}$ and $Q_i \in \mathbb{R}^{q \times p}$ are constant matrices of appropriate dimensions to be determined such that $\lim_{t \to \infty}(\hat{z}(t) - z(t)) = 0$. Equation (5) is the generalized form of the observer, Equation (6) is an auxiliary vector that is used to give more degrees of liberty, finally, Equation (7) makes it possible to design an observer with $q_0 \neq q$.

Consider a parameter matrix $T \in \mathbb{R}^{q_0 \times n}$ and define the transformed error vector

$$\varepsilon(t) = \zeta(t) - TEx(t) \tag{8}$$

whose derivative is given by

$$\dot{\varepsilon}(t) = \sum_{i=1}^{\kappa} w_i(t)\Big(N_i\zeta(t) + J_i v(t) + (F_iC - TA_i)x(t) + (H_i - TB_i)u(t)\Big) \tag{9}$$

replacing $\zeta(t)$ from Equation (8) in Equation (9).

$$\dot{\varepsilon}(t) = \sum_{i=1}^{\kappa} w_i(t)\Big(N_i\varepsilon(t) + J_i v(t) + (N_iTE + F_iC - TA_i)x(t) + (H_i - TB_i)u(t)\Big) \tag{10}$$

By using the definition of $\zeta(t)$, Equations (6) and (7) can be rewritten as

$$\dot{v}(t) = \sum_{i=1}^{\kappa} w_i(t)\Big(S_i\varepsilon(t) + G_i v(t) + (S_iTE + M_iC)x(t)\Big) \tag{11}$$

and also

$$\hat{z}(t) = \sum_{i=1}^{\kappa} w_i(t)\Big(P_i\varepsilon(t) + (P_iTE + Q_iC)x(t)\Big) \tag{12}$$

By considering that the following conditions

$$N_iTE + F_iC - TA_i = 0 \tag{13}$$
$$H_i = TB_i \tag{14}$$
$$S_iTE + M_iC - 0 \tag{15}$$
$$P_iTE + Q_iC = L \tag{16}$$

$\forall i = 1,\ldots,\kappa$ are verified, then Equations (10) and (11) become

$$\dot{\varepsilon}(t) = \sum_{i=1}^{\kappa} w_i(t)\Big(N_i\varepsilon(t) + J_iv(t)\Big) \tag{17}$$

$$\dot{v}(t) = \sum_{i=1}^{\kappa} w_i(t)\Big(S_i\varepsilon(t) + G_iv(t)\Big) \tag{18}$$

and, from Equation (12) we obtain

$$\hat{z}(t) - z(t) = e_z(t) = \sum_{i=1}^{\kappa} w_i(t)\Big(P_i\varepsilon(t)\Big) \tag{19}$$

By defining an augmented state vector $\sigma(t) = \begin{bmatrix} \varepsilon(t) \\ v(t) \end{bmatrix} \in \mathbb{R}^{q_0+q_1}$, Equations (17) and (18) can be rewritten as:

$$\dot{\sigma}(t) = \sum_{i=1}^{\kappa} w_i(t)\Big(\mathbb{A}_i\sigma(t)\Big) \tag{20}$$

where $\mathbb{A}_i = \begin{bmatrix} N_i & J_i \\ S_i & G_i \end{bmatrix} \in \mathbb{R}^{(q_0+q_1)\times(q_0+q_1)}$. It can be seen that, if matrix $\mathbb{A}_i$ is Hurwitz, then $\lim_{t\to\infty} \varepsilon(t) = 0$ and $\lim_{t\to\infty} e_z(t) = 0$.

4. Observer Parameterization

In this section, a specific structure for the observer matrices is determined.

Define matrices $\Gamma = \begin{bmatrix} E \\ C \end{bmatrix} \in \mathbb{R}^{(n+p)\times n}$ and $R \in \mathbb{R}^{q_0\times n}$, which is a full row rank matrix, such that $\text{rank}\begin{bmatrix} R \\ \Gamma \end{bmatrix} = \text{rank}(\Gamma)$. In this case there always exists two matrices $T \in \mathbb{R}^{q_0\times n}$ and $K \in \mathbb{R}^{q_0\times p}$ such that,

$$TE + KC = R \tag{21}$$

which can be rewritten as

$$\begin{bmatrix} T & K \end{bmatrix}\Gamma = R \tag{22}$$

the general solution for Equation (22) is

$$\begin{bmatrix} T & K \end{bmatrix} = R\Gamma^+ - Z\big(I_{n+p} - \Gamma\Gamma^+\big) \tag{23}$$

which can be decomposed as

$$T = T_1 - Z_1T_2 \tag{24}$$
$$K = K_1 - Z_1K_2 \tag{25}$$

where $Z_1 \in \mathbb{R}^{q_0 \times (n+p)}$ is a matrix with arbitrary elements, and $T_1 = R\Gamma^+ \begin{bmatrix} I_n \\ 0 \end{bmatrix}$, $T_2 = \left(I_{n+p} - \Gamma\Gamma^+\right) \begin{bmatrix} I_n \\ 0 \end{bmatrix}$, $K_1 = R\Gamma^+ \begin{bmatrix} 0 \\ I_p \end{bmatrix}$, and $K_2 = \left(I_{n+p} - \Gamma\Gamma^+\right) \begin{bmatrix} 0 \\ I_p \end{bmatrix}$.

Now, we define matrix $\Sigma = \begin{bmatrix} R \\ C \end{bmatrix} \in \mathbb{R}^{(q_0+p) \times n}$. Replacing TE from Equation (21), in condition (13) we obtain

$$N_i(R - KC) + F_iC = TA_i \tag{26}$$

which can be written as

$$\begin{bmatrix} N_i & \tilde{K}_i \end{bmatrix} \Sigma = TA_i \tag{27}$$

where $\tilde{K}_i = F_i - N_iK \in \mathbb{R}^{q_0 \times p}$. The necessary and sufficient condition for the existence of a solution to Equation (27) is

$$\mathrm{rank} \begin{bmatrix} \Sigma \\ TA_i \end{bmatrix} = \mathrm{rank}(\Sigma)$$

the general solution to Equation (27) is

$$\begin{bmatrix} N_i & \tilde{K}_i \end{bmatrix} = TA_i\Sigma^+ - Y_{i1}\left(I_{q_0+p} - \Sigma\Sigma^+\right) \tag{28}$$

if we replace Equation (24) in Equation (28), we obtain

$$N_i = N_{i1} - Z_1 N_{i2} - Y_{i1} N_3 \tag{29}$$
$$\tilde{K}_i = \tilde{K}_{i1} - Z_1 \tilde{K}_{i2} - Y_{i1} \tilde{K}_3 \tag{30}$$

where $Y_{i1} \in \mathbb{R}^{q_0 \times (q_0+p)}$ is a matrix with arbitrary elements, and $N_{i1} = T_1 A_i \Sigma^+ \begin{bmatrix} I_{q_0} \\ 0 \end{bmatrix}$, $N_{i2} = T_2 A_i \Sigma \begin{bmatrix} I_{q_0} \\ 0 \end{bmatrix}$, $N_3 = \left(I_{q_0+p} - \Sigma\Sigma^+\right) \begin{bmatrix} I_{q_0} \\ 0 \end{bmatrix}$, $\tilde{K}_{i1} = T_1 A_i \Sigma^+ \begin{bmatrix} 0 \\ I_p \end{bmatrix}$, $\tilde{K}_{i2} = T_2 A_i \Sigma^+ \begin{bmatrix} 0 \\ I_p \end{bmatrix}$, $\tilde{K}_3 = \left(I_{q_0+p} - \Sigma\Sigma^+\right) \begin{bmatrix} 0 \\ I_p \end{bmatrix}$.

From the definition of $\tilde{K}_i$ we can deduce matrix F_i as

$$\begin{aligned} F_i &= \tilde{K}_i + N_iK \\ &= \tilde{K}_{i1} + N_{i1}K - Z_1\left(\tilde{K}_{i2} - N_{i2}K\right) - Y_{i1}\left(\tilde{K}_3 - N_3K\right) \\ &= F_{i1} - Z_1 F_{i2} - Y_{i1} F_3 \end{aligned} \tag{31}$$

where $F_{i1} = T_1 A_i \Sigma^+ \begin{bmatrix} K \\ I_p \end{bmatrix}$, $F_{i2} = T_2 A_i \Sigma^+ \begin{bmatrix} K \\ I_p \end{bmatrix}$, $F_3 = \left(I_{q_0+p} - \Sigma\Sigma^+\right) \begin{bmatrix} K \\ I_p \end{bmatrix}$.

From Equation (22), we have

$$\begin{bmatrix} TE \\ C \end{bmatrix} = \begin{bmatrix} I_{q_0} & -K \\ 0 & I_p \end{bmatrix} \Sigma \tag{32}$$

Conditions (15) and (16) can be written as

$$\begin{bmatrix} S_i & M_i \\ P_i & Q_i \end{bmatrix} \begin{bmatrix} TE \\ C \end{bmatrix} = \begin{bmatrix} 0 \\ L \end{bmatrix} \tag{33}$$

replacing Equation (32) into (33), we obtain

$$\begin{bmatrix} S_i & M_i \\ P_i & Q_i \end{bmatrix} \begin{bmatrix} I_{q_0} & -K \\ 0 & I_p \end{bmatrix} \Sigma = \begin{bmatrix} 0 \\ L \end{bmatrix} \tag{34}$$

since $\begin{bmatrix} I_{q_0} & -K \\ 0 & I_p \end{bmatrix}^{-1} = \begin{bmatrix} I_{q_0} & K \\ 0 & I_p \end{bmatrix}$ the general solution of Equation (34) is given by

$$\begin{bmatrix} S_i & M_i \\ P_i & Q_i \end{bmatrix} = \left(\begin{bmatrix} 0 \\ L \end{bmatrix} \Sigma^+ - Y_i(I_{q_0+p} - \Sigma\Sigma^+) \right) \begin{bmatrix} I_{q_0} & K \\ 0 & I_p \end{bmatrix} \tag{35}$$

where $Y_i = \begin{bmatrix} Y_{i2} \\ Y_{i3} \end{bmatrix} \in \mathbb{R}^{(q_1+q) \times (q_0+p)}$ is a matrix with arbitrary elements. The solutions for S_i, M_i, P_i and Q_i are given by

$$S_i = -Y_{i2}N_3 \tag{36}$$
$$M_i = -Y_{i2}F_3 \tag{37}$$
$$P_i = P_1 - Y_{i3}N_3 \tag{38}$$
$$Q_i = Q_1 - Y_{i3}F_3 \tag{39}$$

where $Y_{i2} \in \mathbb{R}^{q_1 \times (q_0+p)}$, $Y_{i3} \in \mathbb{R}^{q \times (q_0+p)}$, $P_1 = L\Sigma^+ \begin{bmatrix} I_{q_0} \\ 0 \end{bmatrix}$, $Q_1 = L\Sigma^+ \begin{bmatrix} K \\ I_p \end{bmatrix}$.

The estimation error (19) shows that $e_z(t) \to 0$ while $\varepsilon(t) \to 0$, so then, the function estimation error $e_z(t)$ does not depend on the choice of matrix P. Then, for simplicity we can assume that $Y_{i3} = 0$, so $P_i = P_1$ and $Q_i = Q_1$, being now, constant matrices P and Q.

Now, by using Equations (29) and (36), the error dynamics (20) can be rewritten as

$$\dot{\sigma}(t) = \sum_{i=1}^{\kappa} w_i(t) \left(\mathbb{A}_{i1} - \mathbb{Y}_i \mathbb{A}_2 \right) \sigma(t) \tag{40}$$

where $\mathbb{A}_{1i} = \begin{bmatrix} N_{i1} - Z_1 N_{i2} & 0 \\ 0 & 0 \end{bmatrix} \in \mathbb{R}^{(q_0+q_1) \times (q_0+q_1)}$, $\mathbb{Y}_i = \begin{bmatrix} Y_{i1} & J_i \\ Y_{i2} & G_i \end{bmatrix} \in \mathbb{R}^{(q_0+q_1) \times (q_0+q_1+p)}$ and

$\mathbb{A}_2 = \begin{bmatrix} N_3 & 0 \\ 0 & -I_{q_1} \end{bmatrix} \in \mathbb{R}^{(q_0+q_1+p) \times (q_0+q_1)}$.

The problem of the design is to find matrices $\mathbb{Y}_i$ and Z_1 such that matrix $\mathbb{A}_i = \mathbb{A}_{i1} - \mathbb{Y}_i \mathbb{A}_2$ is Hurwitz. This can be reached by using the linear matrix inequality (LMI) approach.

5. Functional Observer Design

Theorem 1. *There exist parameter matrices $\mathbb{Y}_i$ and Z_1 such that error dynamic system (40) is asymptotically stable if there exists a matrix*

$$X = X^T = \begin{bmatrix} X_1 & X_1 \\ X_1 & X_2 \end{bmatrix} > 0$$

such that the following LMI is satisfied:

$$N_3^{T\perp} \left(N_{i1}^T X_1 + X_1 N_{i1} - N_{i2}^T W^T - W N_{i2} \right) N_3^{T\perp T} < 0 \tag{41}$$

where $Z_1 = X_1^{-1}W$ and matrix $\mathbb{Y}_i$ can be determined as follows

$$\mathbb{Y}_i = -X^{-1}(-\gamma B^T + \sqrt{\gamma}\mathcal{L}\Omega_i^{1/2})^T \tag{42}$$

where

$$\Omega_i = \gamma BB^T - \mathcal{Q}_i > 0 \tag{43}$$

with

$$\mathcal{Q}_i = \begin{bmatrix} X_1(N_{i1} - Z_1 N_{i2}) + (N_{i1} - ZN_{i2})^T X_1 & (N_{i1} - Z_1 N_{i2})^T X_1 \\ X_1(N_{i1} - Z_1 N_{i2}) & 0 \end{bmatrix}$$

$$\mathcal{B} = \begin{bmatrix} N_3^T & 0 \\ 0 & -I \end{bmatrix}$$

and matrix $\mathcal{L}$ *is any matrix such that* $||\mathcal{L}|| < 1$ *and* $\gamma > 0$ *is any scalar such that* $\Omega_i > 0$.

Proof. Consider the following Lyapunov function candidate

$$V(\sigma(t)) = \sigma(t)^T X \sigma(t) \tag{44}$$

which derivative is given by

$$\dot{V}(\sigma(t)) = \sigma(t)^T \left[(\mathbb{A}_{i1} - \mathbb{Y}_i \mathbb{A}_2)^T X + X(\mathbb{A}_{i1} - \mathbb{Y}_i \mathbb{A}_2) \right] \sigma(t) \tag{45}$$

The asymptotic stability of Equation (40) is guaranteed only if $\dot{V}(\sigma(t)) < 0$, this leads to the following LMI

$$\mathbb{A}_{i1}^T X - \mathbb{A}_2^T \mathbb{Y}_i^T X + X \mathbb{A}_{i1} - X \mathbb{Y}_i \mathbb{A}_2 < 0 \tag{46}$$

which can be rewritten as

$$\mathcal{Q}_i + \mathcal{B}\mathcal{X}_i + (\mathcal{B}\mathcal{X}_i)^T < 0 \tag{47}$$

where $\mathcal{X}_i = -\mathbb{Y}_i^T X$, $\mathcal{Q}_i = X\mathbb{A}_{i1} + \mathbb{A}_{i1}^T X$ and $\mathcal{B} = \mathbb{A}_2^T$. According to Lemma 1, there exists a matrix $\mathcal{X}_i$ satisfying Equation (47) if and only if the following condition holds.

$$\mathcal{B}^{\perp} \mathcal{Q}_i \mathcal{B}^{\perp T} < 0 \tag{48}$$

with $\mathcal{B}^{\perp} = \begin{bmatrix} N_3^{T\perp} & 0 \end{bmatrix}$. By using the definitions of X, $\mathcal{Q}_i$ and $W = X_1 Z_1$, we obtain Equation (41). Matrix $\mathbb{Y}_i$ is obtained from Equation (42). $\square$

The following Algorithm 1 summarizes the observer design to obtain the corresponding matrices.

Algorithm 1: Methodology of the observer design

---

1.   Choose a matrix $R \in \mathbb{R}^{q_0 \times n}$ such that $\text{rank}\begin{bmatrix} R \\ \Gamma \end{bmatrix} = \text{rank}(\Gamma)$.
2.   Compute matrices $N_{i1}, N_{i2}, N_3, T_1, T_2, K_1, K_2, P_1$ and Q_1.
3.   Solve the LMI (41) to find X and Z_1.
4.   Choose a matrix $\mathcal{L}$ such that $||\mathcal{L}|| < 1$, and a scalar $\gamma > 0$ such that $\Omega_i > 0$, then determinate matrix $\mathbb{Y}_i$ as in (42).
5.   Compute all the matrices gains of the observer (5) by using (29) to determinate N_i, (42) to determinate J_i and G_i, (36)–(39) to find S_i, M_i, P_i and Q_i taking matrix $Y_{i3} = 0$. F_i is given by (31) and matrix H_i could be determined with Equation (14).

---

6. Proportional Functional Observer Case

In order to obtain a Proportional Functional Observer (PFO) from the GFO, it corresponds to the parameter matrices $S_i = 0$, $J_i = 0$, $M_i = 0$ and $G_i = 0$ which generates the following observer [34]:

$$\dot{\zeta}(t) = \sum_{i=1}^{\kappa} w_i(t) \left(N_i \zeta(t) + F_i y(t) + H_i u(t) \right)$$

$$\hat{z}(t) = \sum_{i=1}^{\kappa} w_i(t) \left(P_i \zeta(t) + Q_i y(t) \right) \tag{49}$$

and the error dynamics (20) becomes

$$\dot{\varepsilon}(t) = \sum_{i=1}^{\kappa} w_i(t)\left(\tilde{A}_{i1} - \tilde{Y}_i\tilde{A}_2\right)\varepsilon(t) \tag{50}$$

where $\tilde{A}_{i1} = N_{i1} - Z_1 N_{i2}$, $\tilde{Y}_i = Y_{i1}$ and $\tilde{A}_2 = N_3$. Matrices $\mathcal{Q}_i$, $\mathcal{B}$ and $\mathcal{X}_i$ of Theorem 1 become:

$$\mathcal{Q}_i = X(N_{i1} - Z_1 N_{i2}) + (N_{i1} - Z_1 N_{i2})^T X,$$
$$\mathcal{B} = N_3^T,$$
$$\mathcal{X}_i = -Y_{i1}^T X$$

The observer matrices can be obtained following Algorithm 1.

7. Mathematical Model

The mathematical model chosen to show the performance of the generalized observer is a linear-rotational vibration system with an uncertainty in one of the spring rigidity values (see Figure 1).

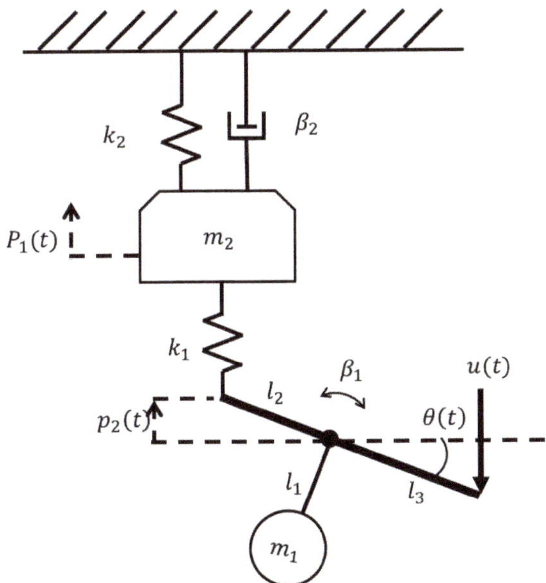

Figure 1. Single link robot arm.

The robot has the following nonlinear model:

$$m_2\ddot{p}_1(t) = -k_2 p_1(t) - \beta_2\dot{p}_1(t) - k_1(p_2(t) - p_1(t))$$
$$J_m\ddot{\theta}(t) = -m_1 g l_1 \sin(\theta(t)) + l_3 k_1(p_1(t) - \beta_1\dot{\theta}(t) - p_2(t))\cos(\theta(t)) - l_2\cos(\theta(t))u(t) \tag{51}$$
$$p_2(t) = l_3\sin(\theta(t))$$

The measurable states are

$$y_1(t) = p_1(t)$$
$$y_2(t) = \theta(t) \tag{52}$$

The parameters considered are given in Table 1.

Table 1. Parameters of model.

J_m	Moment of inertia of the pendulum weight	0.0081 Kg m^2
m_1	Mass of the pendulum weight	0.25 Kg
m_2	Linear movement mass	0.15 Kg
l_i	Distances of the system	$0.18, 0.15, 0.1$ m
k_i	Spring rigidity coefficients	$0.3, 0.25$ Nm/rad
β_i	Viscous friction coefficients	0.1 Nms/rad

Taking as the states the position of m_1 $x_1(t) = p_1(t)$, the linear speed of m_1 $x_2(t) = \dot{p}_1(t)$, the angle of the lever $x_3(t) = \theta(t)$, the angular speed of the lever $x_4(t) = \dot{\theta}(t)$, and the position of the lever $x_5(t) = p_2(t)$, we can represent the nonlinear model (51) as follows:

$$E\dot{x}(t) = A(x(t))x(t) + B(x(t))u(t) \tag{53}$$

$$y(t) = Cx(t) \tag{54}$$

where $x(t)$ is the semi-state vector $x(t) = \begin{bmatrix} x_1(t) & x_2(t) & x_3(t) & x_4(t) & x_5(t) \end{bmatrix}$, $u(t)$ is the force input, and $A(x(t))$, $B(x(t))$ are matrices containing nonlinear terms depending on the state variables. E is a singular constant matrix given as:

$$E = \begin{bmatrix} 1 & 0 & 0 & 0 & 0 \\ 0 & 1 & 0 & 0 & 0 \\ 0 & 0 & 1 & 0 & 0 \\ 0 & 0 & 0 & 1 & 0 \\ 0 & 0 & 0 & 0 & 0 \end{bmatrix} \tag{55}$$

and C is a constant matrix

$$C = \begin{bmatrix} 1 & 0 & 0 & 0 & 0 \\ 0 & 0 & 1 & 0 & 0 \end{bmatrix} \tag{56}$$

The variable terms presented in matrices $A(x(t))$, $B(x(t))$ from the nonlinear model (51) are highlighted in boxes in Equations (57) and (58)

$$A(x(t)) =$$

$$\begin{bmatrix} 0 & 1 & 0 & 0 & 0 \\ \dfrac{k_2-k_1}{m_2} & -\dfrac{\beta_2}{m_2} & 0 & 0 & -\dfrac{k_2}{m_2} \\ 0 & 0 & 0 & 1 & 0 \\ \dfrac{l_3 k_2}{J_m}\boxed{\cos(x_3(t))} & 0 & -\dfrac{m_1 g l_1}{J_m}\boxed{\dfrac{\sin(x_3(t))}{x_3(t)}} - \dfrac{\beta_1}{J_m} & -\dfrac{l_3 k_2}{J_m}\boxed{\cos(x_3(t))} \\ 0 & 0 & l_3\boxed{\dfrac{\sin(x_3(t))}{x_3(t)}} & 0 & -1 \end{bmatrix} \tag{57}$$

$$B(x(t)) = \begin{bmatrix} 0 \\ 0 \\ 0 \\ -\dfrac{l_2}{J_m}\boxed{\cos(x_3(t))} \\ 0 \end{bmatrix} \tag{58}$$

Considering the $s = 2$ scheduling variables as

$$\rho_1(t) = \frac{\sin(x_3(t))}{x_3(t)} \tag{59}$$

$$\rho_2(t) = \cos(x_3(t)) \tag{60}$$

each variable has behavior limits depending on the variation of the input and the states.

Since there are two scheduling variables, then four weighting functions are obtained:

$$\eta_0^1(t) = \frac{\overline{\rho_1} - \rho_1(t)}{\overline{\rho_1} - \underline{\rho_1}} \tag{61}$$

$$\eta_1^1(t) = 1 - \eta_0^1(t) \tag{62}$$

$$\eta_0^2(t) = \frac{\overline{\rho_2} - \rho_2(t)}{\overline{\rho_2} - \underline{\rho_2}} \tag{63}$$

$$\eta_1^2(t) = 1 - \eta_0^2(t) \tag{64}$$

where $\overline{\rho_j}$ and $\underline{\rho_j}$ are the upper and lower limit of variation of $\rho_j(t)$, respectively, for all $j = 1, 2$.

In this case $s = 2$, therefore there are $\kappa = 2^2 = 4$ membership functions, as:

$$w_1(t) = \eta_0^1 \eta_0^2 \tag{65}$$

$$w_2(t) = \eta_1^1 \eta_0^2 \tag{66}$$

$$w_3(t) = \eta_0^1 \eta_1^2 \tag{67}$$

$$w_4(t) = \eta_1^1 \eta_1^2 \tag{68}$$

Once the fuzzy sets are defined, the Takagi–Sugeno model has four rules. For each rule, there is a linear local model of the form:

$$E\dot{x}(t) = A_i x(t) + B_i u(t), \quad \forall i = 1, \ldots, \kappa = 2^s \tag{69}$$

For example, the first rule (65) corresponds to $\rho_1 = \overline{\rho_1}$ and $\rho_2 = \overline{\rho_2}$ then, lower limits are directly replaced by the nonlinear terms in matrices $A(x(t))$ and $B(x(t))$, resulting in the following local model:

$$E\dot{x}(t) = A_1 x(t) + B_1 u(t) \tag{70}$$

where $A_1 = \begin{bmatrix} 0 & 1 & 0 & 0 & 0 \\ \frac{k_2 - k_1}{m_2} & -\frac{\beta_2}{m_2} & 0 & 0 & -\frac{k_2}{m_2} \\ 0 & 0 & 0 & 1 & 0 \\ \frac{l_3 k_2}{J_m}\underline{\rho_2} & 0 & -\frac{m_1 g l_1}{J_m}\underline{\rho_1} & -\frac{\beta_1}{J_m} & -\frac{l_3 k_2}{J_m}\underline{\rho_2} \\ 0 & 0 & l_3\underline{\rho_1} & 0 & -1 \end{bmatrix}$, $B_1 = \begin{bmatrix} 0 \\ 0 \\ 0 \\ -\frac{l_2}{J_m}\underline{\rho_2} \\ 0 \end{bmatrix}$ and matrix E is defined

Eq.

in (55).

The Takagi–Sugeno model that reproduces the dynamics of the nonlinear model is given by:

$$E\dot{x}(t) = (w_1(t)A_1 + w_2(t)A_2 + w_3(t)A_3 + w_4(t)A_4)x(t) +$$
$$(w_1(t)B_1 + w_2(t)B_2 + w_3(t)B_3 + w_4(t)B_4)u(t)$$

where

$$A_1 = A(\underline{\rho_1}, \underline{\rho_2}), \quad B_1 = B_2 = B(\underline{\rho_2})$$
$$A_2 = A(\overline{\rho_1}, \underline{\rho_2}), \quad B_3 = B_4 = B(\overline{\rho_2})$$
$$A_3 = A(\underline{\rho_1}, \overline{\rho_2}), \quad A_4 = A(\overline{\rho_1}, \overline{\rho_2})$$

The Takagi–Sugeno model of the single link robot arm is

$$E\dot{x}(t) = \sum_{i=1}^{\kappa} w_i(t)(A_i x(t) + B_i u(t)) \tag{71}$$

$$y(t) = Cx(t) \tag{72}$$

Considering a constant input as $u(1) = 1N$, we can obtain the variation level of $x_3(t)$ that allows us to determine the maximum and minimum of variation of $\rho_1(t)$ and $\rho_2(t)$, to finally obtain the local matrices for the T-S model as:

$$A_1 = \begin{bmatrix} 0 & 1 & 0 & 0 & 0 \\ -3.67 & -0.67 & 0 & 0 & 1.67 \\ 0 & 0 & 0 & 1 & 0 \\ 0 & 0 & 11.57 & -12.35 & 0 \\ 0 & 0 & -0.02 & 0 & -1 \end{bmatrix}, \quad A_2 = \begin{bmatrix} 0 & 1 & 0 & 0 & 0 \\ -3.67 & -0.67 & 0 & 0 & 1.67 \\ 0 & 0 & 0 & 1 & 0 \\ 0 & 0 & -54.22 & -12.35 & 0 \\ 0 & 0 & 0.1 & 0 & -1 \end{bmatrix}$$

$$A_3 = \begin{bmatrix} 0 & 1 & 0 & 0 & 0 \\ -3.67 & -0.67 & 0 & 0 & 1.67 \\ 0 & 0 & 0 & 1 & 0 \\ 3.04 & 0 & 11.57 & -12.35 & -3.04 \\ 0 & 0 & -0.02 & 0 & -1 \end{bmatrix} \quad A_4 = \begin{bmatrix} 0 & 1 & 0 & 0 & 0 \\ -3.67 & -0.67 & 0 & 0 & 1.67 \\ 0 & 0 & 0 & 1 & 0 \\ 3.04 & 0 & -54.22 & -12.35 & -3.04 \\ 0 & 0 & 0.1 & 0 & -1 \end{bmatrix}$$

$$B_1 = B_2 = \begin{bmatrix} 0 \\ 0 \\ 0 \\ -18.24 \\ 0 \end{bmatrix}, \quad B_3 = B_4 = \begin{bmatrix} 0 \\ 0 \\ 0 \\ 0 \\ 0 \end{bmatrix}$$

and matrix C is given in (56).

8. Results

By following Algorithm 1, a Generalized Functional Observer can be designed. The matrix L was chosen in order to estimate the non-measurable state of the system.

$$L = \begin{bmatrix} 0 & 1 & 0 & 0 & 0 \\ 0 & 0 & 0 & 1 & 0 \\ 0 & 0 & 0 & 0 & 1 \end{bmatrix}$$

Considering matrix $R = \begin{bmatrix} 0 & 1 & 0 & 0 & 0 \\ 0 & 1 & 0 & 1 & 0 \\ 1 & 0 & 0 & 0 & 1 \end{bmatrix}$, the matrix gains for the GFO are:

$$N_1 = N_2 = \begin{bmatrix} -25.62 & 7.53 & 5.7 & 7.33 \\ -11.11 & -51.58 & -1.32 & -2.54 \\ -7.75 & -0.18 & -50.92 & -0.12 \\ -9.27 & 0.88 & -0.88 & -51.08 \end{bmatrix}, \quad N_3 = N_4 = \begin{bmatrix} -25.62 & 7.3 & 5.61 & 8.09 \\ -11.11 & -51.8 & -1.38 & -4.57 \\ -9.27 & -0.44 & -51.02 & -0.79 \\ -9.27 & 2.95 & -0.04 & -50.81 \end{bmatrix},$$

$$J_1 = J_2 = \begin{bmatrix} 8.79 & 8.79 & 8.79 \\ 1.78 & 1.78 & 1.78 \\ 1.56 & 1.56 & 1.56 \\ 1.63 & 1.63 & 1.63 \end{bmatrix}, \quad J_3 = J_4 = \begin{bmatrix} 8.54 & 8.54 & 8.54 \\ 1.83 & 1.83 & 1.83 \\ 1.51 & 1.51 & 1.51 \\ 1.52 & 1.52 & 1.52 \end{bmatrix}, \quad F_1 = \begin{bmatrix} 570.55 & 381.86 \\ -2544.23 & -57.2 \\ 42.8 & -1974.86 \\ 159.49 & -20.02 \end{bmatrix}, \quad F_2 = \begin{bmatrix} 570.55 & 382.74 \\ -2544.23 & -57.71 \\ 42.8 & -1908.72 \\ 159.49 & -26.19 \end{bmatrix},$$

$$F_3 = \begin{bmatrix} 539.06 & 370.16 \\ -2558.75 & -62.38 \\ 25.45 & -1975.59 \\ 368.69 & 58.01 \end{bmatrix}, \quad F_4 = \begin{bmatrix} 539.06 & 371.14 \\ -2558.75 & -63.13 \\ 25.45 & -1909.9 \\ 368.69 & 51.88 \end{bmatrix}, \quad S_1 = S_2 = S_3 = S_4 = \begin{bmatrix} 9.91 & 0 & 0 & 0 \\ 9.91 & 0 & 0 & 0 \\ 9.91 & 0 & 0 & 0 \end{bmatrix}, \quad H_1 = H_2 = \begin{bmatrix} 0 \\ -18.23 \\ 0 \\ 0 \end{bmatrix},$$

$$H_3 = H_4 = \begin{bmatrix} 0 \\ 0 \\ 0 \\ 0 \end{bmatrix}, \quad G_1 = G_2 = \begin{bmatrix} -29.2 & 3.5 & 3.5 \\ 3.5 & -29.2 & 3.5 \\ 3.5 & 3.5 & -29.2 \end{bmatrix}, \quad G_3 = G_4 = \begin{bmatrix} -29.15 & 3.55 & 3.55 \\ 3.55 & -29.15 & 3.55 \\ 3.55 & 3.55 & -29.15 \end{bmatrix},$$

$$M_1 = M_2 = \begin{bmatrix} -74.55 & -56.51 \\ -74.55 & -56.51 \\ -74.55 & -56.51 \end{bmatrix}, \quad M_3 = M_4 = \begin{bmatrix} -72.34 & -55.56 \\ -72.34 & -55.56 \\ -72.34 & -55.56 \end{bmatrix}, \quad P = \begin{bmatrix} 0 & 1 & 0 & 0 \\ 0 & 0 & 1 & 0 \\ 0 & 0 & 0 & 1 \end{bmatrix}, \quad Q = \begin{bmatrix} 51.14 & 1.38 \\ 0.44 & 38.67 \\ -2.95 & 0.04 \end{bmatrix}.$$

In order to compare the performance of the GFO, a PFO is also designed as a particular case. Considering matrix $R = \begin{bmatrix} 0 & 1 & 0 & 0 & 0 \\ 0 & 1 & 0 & 1 & 0 \\ 1 & 0 & 0 & 0 & 1 \end{bmatrix}$, the matrix gains for the PFO are:

$$N_1 = N_2 = \begin{bmatrix} 0 & 0.5 & 0 & 0 \\ -1.83 & -0.5 & 0 & 0.56 \\ 1.52 & 0 & -0.5 & -1.01 \\ 0 & -0.56 & 1.01 & -0.5 \end{bmatrix}, \quad N_3 = N_4 = \begin{bmatrix} 0 & 0.5 & 0 & 0 \\ -1.83 & -0.5 & 0 & 0.56 \\ 0 & 0 & -0.5 & 0 \\ 0 & -0.56 & 0 & -0.5 \end{bmatrix},$$

$$F_1 = \begin{bmatrix} -0.08 & 0 \\ -2.36 & -0.45 \\ 1.72 & -47.48 \\ -0.19 & -11.45 \end{bmatrix}, \quad F_2 = \begin{bmatrix} -0.08 & 0 \\ -2.36 & -0.59 \\ 1.72 & 18.56 \\ -0.19 & -11.51 \end{bmatrix}, \quad F_3 = \begin{bmatrix} -0.08 & 0 \\ -2.36 & 0.11 \\ 0 & -48.3 \\ -0.19 & 0.05 \end{bmatrix}, \quad F_4 = \begin{bmatrix} -0.08 & 0 \\ -2.36 & -0.02 \\ 0 & 17.49 \\ -0.19 & -0.01 \end{bmatrix},$$

$$H_1 = H_2 = \begin{bmatrix} 0 \\ 0 \\ -18.24 \\ 0 \end{bmatrix}, \quad H_3 = H_4 = \begin{bmatrix} 0 \\ 0 \\ 0 \\ 0 \end{bmatrix}, \quad P = \begin{bmatrix} 1 & 0 & 0 \\ 0 & 1 & 0 \\ 0 & 0 & 1 \end{bmatrix}, \quad Q = \begin{bmatrix} -0.17 & 0 \\ 0 & -11.85 \\ 0.56 & -1.01 \end{bmatrix}.$$

Considering a constant unit step input $u(t) = 1N$ and, just in simulation, an additive parameter variation of the rigidity coefficient of spring 2, given as $k_2 + \delta(t)$, where $\delta(t)$ is presented in Figure 2. It is important to note that the generalized approach to the functional observer allows it to have different matrices as degrees of freedom as well as some robustness to parametric uncertainties. These uncertainties may come as a variation in the parameter values due to different physical processes, and may or may not be time-variant. In this example, we choose a time-varying uncertainty with a sinusoidal form. The generalized observer is capable of estimating the function $z(t)$ with minor impact on performance.

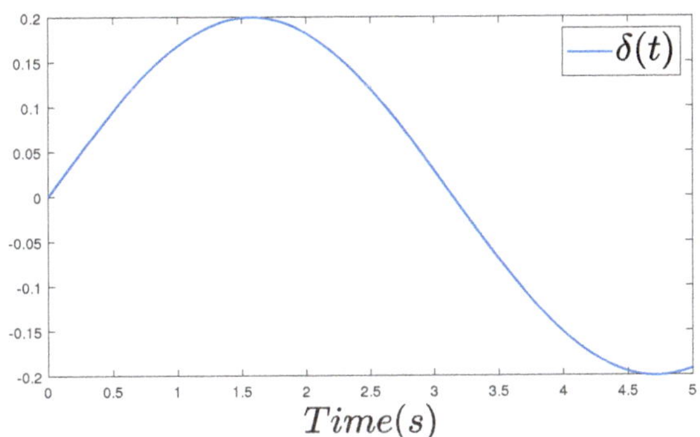

Figure 2. Parametric uncertainty.

This parameter variation allows us to observe the robustness characteristics of the GFO compared with a simplest observer structure as the PFO.

The simulation is realized taking the input and output of the nonlinear system (53) and (54) in face to parametric uncertainty of Figure 2, the T-S system of (71) and (72) is just used for the observer design. Considering the initial conditions for the nonlinear system as $x(0) = \begin{bmatrix} \frac{\pi}{12} & 0 & \frac{\pi}{6} & 0 & 0.09 \end{bmatrix}$ and the initial condition for both observers as $\zeta(0) = 0$ and $v(0) = 0$, the results of the simulation are shown in Figures 3–5.

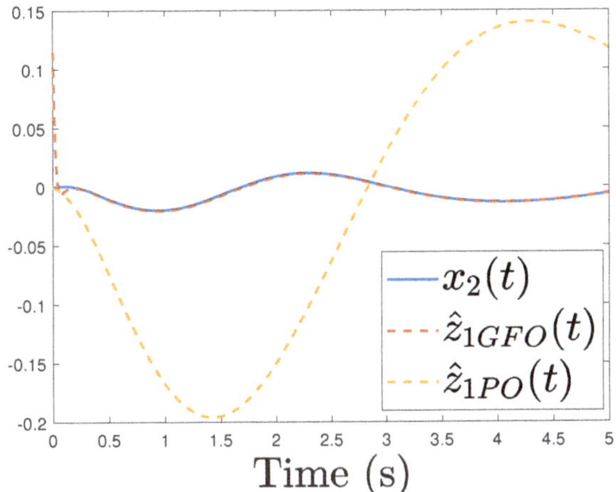

Figure 3. $x_2(t)$ and its estimates.

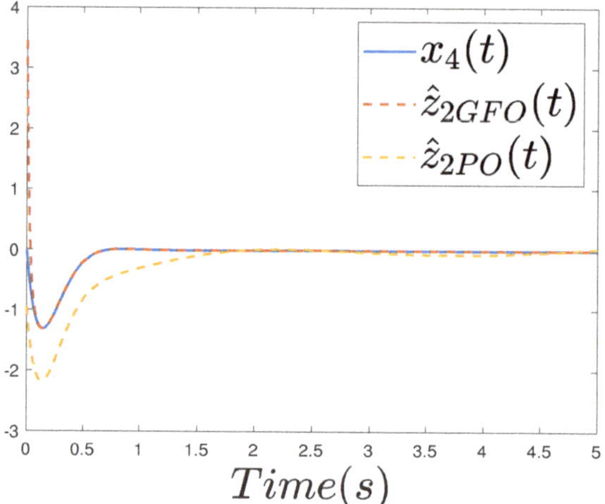

Figure 4. $x_4(t)$ and its estimates.

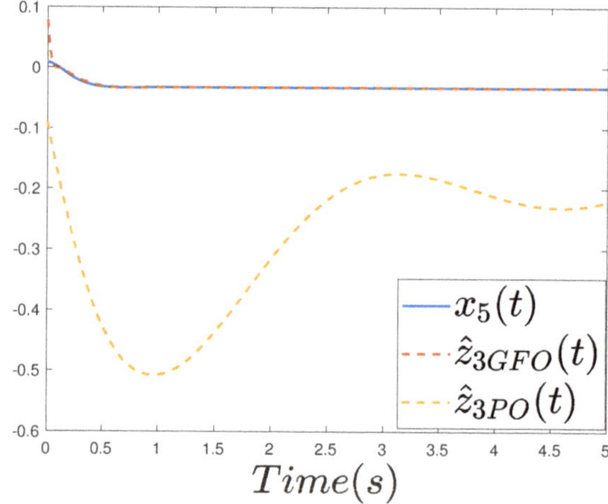

Figure 5. $x_5(t)$ and its estimates.

As can be seen in Figures 3–5, the generalized observer is capable of estimating the state of the system even in the presence of parametric uncertainties. This presents an advantage compared to the classical proportional observers. The comparison between the performance of the genrealized observer and the Proportional observer is conducted through the error indexes which provide important information about the estimation error. For the case of a parametric uncertainty, the generalized approach outperforms the classical proportional observer. The error indexes of the Integral of the Absolute Error (IAE) and the Integral of the Time weighted Absolute Error (ITAE) are shown in Table 2.

From Table 2, it can be seen that the proposed GFO exhibits better performance in comparison to the proportional observer. The generalized approach allows a better estimation of the function when parametric uncertainties are present in the system.

Table 2. Error indexes.

	GFO	PFO
IAE	$e_{z_1} = 0.0047$ $e_{z_2} = 0.0691$ $e_{z_3} = 0.0024$	$e_{z_1} = 0.8657$ $e_{z_2} = 0.8337$ $e_{z_3} = 1.2966$
ITAE	$e_{z_1} = 0.0048$ $e_{z_2} = 0.0031$ $e_{z_3} = 0.0009$	$e_{z_1} = 1.5015$ $e_{z_2} = 0.7656$ $e_{z_3} = 1.2966$

9. Conclusions

In this paper, a method for function estimation based on generalized observer for descriptor Takagi–Sugeno systems is presented. A function estimation can be used for two main objectives: The first, and used in the simulation example presented herein, is the estimation of the non-measurable states that could be considered as a reducer order observer, although an even smaller number of states could be estimated, given that the impulse functional observability presented in Assumption 1, is satisfied. The second objective for the function estimation is to estimate a control law based on the states without the need to estimate the states independently, but rather the function directly. The design and conditions of existence of the proposed observer were provided through a stability analysis based on Lyapunov. An example of a physical system was used to provide a comparison of the GFO with a PFO, both in the presence of parametric uncertainties. This paper can be extended in future works to include fault estimation, time delays, or to estimate a fault-tolerant control law. The descriptor approach of this paper can be used in future work since it is a powerful tool for representing a large number of mathematical systems.

Author Contributions: Conceptualization, C.R.-R., G.-L.O.-G. and M.D.; methodology, C.R.-R., G.-L.O.-G. and C.-M.A.-Z.; software, C.R.-R. and H.S.-A.; validation, C.R.-R., H.S.-A. and J.R.-R.; formal analysis, C.R.-R., G.-L.O.-G. and M.D.; investigation, C.R.-R., M.D. and C.-M.A.-Z.; writing—original draft preparation, C.R.-R., G.-L.O.-G. and H.S.-A.; writing—review and editing, C.R.-R., G.-L.O.-G., C.-M.A.-Z., M.D., H.S.-A. and J.R.-R.; supervision, G.-L.O.-G., C.-M.A.-Z. and M.D. All authors have read and agreed to the published version of the manuscript.

Funding: This research received no external funding.

Institutional Review Board Statement: Not applicable.

Informed Consent Statement: Not applicable.

Data Availability Statement: Not applicable.

Acknowledgments: The development of this project is the product of the support provided by the National Council of Science and Technology (CONACYT), the National Technological Institute of Mexico (TecNM)/CENIDET.

References

1. Fernando, T.L.; Trinh, H.M.; Jennings, L. Functional observability and the design of minimum order linear functional observers. *IEEE Trans. Autom. Control* **2010**, *55*, 1268–1273. [CrossRef]
2. Fernando, T.L.; Trinh, H.M. A system decomposition approach to the design of functional observers. *Int. J. Control* **2014**, *87*, 1846–1860. [CrossRef]
3. Darouach, M. Existence and design of functional observers for linear systems. *IEEE Trans. Autom. Control* **2000**, *45*, 940–943. [CrossRef]
4. Darouach, M. On the functional observers for linear descriptor systems. *Syst. Control Lett.* **2012**, *61*, 427–434. [CrossRef]
5. Darouach, M.; Fernando, T.L. On the existence and design of functional observers. *IEEE Trans. Autom. Control* **2019**, *65*, 2751–2759. [CrossRef]
6. Emami, K.; Fernando, T.L.; Nener, B.; Trinh, H.; Zhang, Y. A functional observer based fault detection technique for dynamical systems. *J. Frankl. Inst.* **2015**, *352*, 2113–2128. [CrossRef]

7. Hamdoun, M.; Abdallah, M.B.; Ayadi, M.; Rotella, F.; Zambettakis, I. Functional observer-based feedback controller for ball balancing table. *SN Appl. Sci.* **2021**, *3*, 1–9. [CrossRef]

8. Alhelou, H.H.; Golshan, M.E.H.; Hatziargyriou, N.D. A decentralized functional observer based optimal LFC considering unknown inputs, uncertainties, and cyber-attacks. *IEEE Trans. Power Syst.* **2019**, *34*, 4408–4417. [CrossRef]

9. Koenig, D.; Marx, B.; Varrier, S. Filtering and fault estimation of descriptor switched systems. *Automatica* **2016**, *63*, 116–121. [CrossRef]

10. Darouach, M.; Amato, F.; Alma, M. Functional observers design for descriptor systems via LMI: Continuous and discrete-time cases. *Automatica* **2017**, *86*, 216–219. [CrossRef]

11. Ríos-Ruiz, C.; Osorio-Gordillo, G.L.; Souley-Ali, H.; Darouach, M.; Astorga-Zaragoza, C.M. Finite time functional observers for descriptor systems. Application to fault tolerant control. In Proceedings of the 2019 27th Mediterranean Conference on Control and Automation (MED), Akko, Israel, 1–4 July 2019; pp. 165–170.

12. Zhang, J.; Wang, Z.; Chadli, M.; Wang, Y. On prescribed-time functional observers of linear descriptor systems with unknown input. *Int. J. Control* **2021**, 1–16. [CrossRef]

13. Huang, J.; Che, H.; Raissi, T.; Wang, Z. Functional Interval Observer for Discrete-time Switched Descriptor Systems. *IEEE Trans. Autom. Control* **2021**. [CrossRef]

14. Islam, S.I.; Lim, C.C.; Shi, P. Robust fault detection of TS fuzzy systems with time-delay using fuzzy functional observer. *Fuzzy Sets Syst.* **2020**, *392*, 1–23. [CrossRef]

15. Islam, S.I.; Lim, C.C.; Shi, P. Functional observer based controller for stabilizing Takagi-Sugeno fuzzy systems with time-delays. *J. Frankl. Inst.* **2018**, *355*, 3619–3640. [CrossRef]

16. Xie, W.B.; Wang, T.Z.; Lam, H.K.; Wang, X. Functional observer-controller method for unmeasured premise variables Takagi-Sugeno systems with external disturbance. *Int. J. Syst. Sci.* **2020**, *51*, 3436–3450. [CrossRef]

17. Eltag, K.; Aslam, M.S.; Chen, Z. Functional Observer-Based T-S Fuzzy Systems for Quadratic Stability of Power System Synchronous Generator. *Int. J. Fuzzy Syst.* **2020**, *22*, 172–180. [CrossRef]

18. Naifar, O.; Boukettaya, G.; Oualha, A.; Ouali, A. A comparative study between a high-gain interconnected observer and an adaptive observer applied to IM-based WECS. *Eur. Phys. J. Plus* **2015**, *130*, 1–13. [CrossRef]

19. Ayadi, M.; Naifar, O.; Derbel, N. High-order sliding mode control for variable speed PMSG-wind turbine-based disturbance observer. *Int. J. Model. Identif. Control* **2019**, *32*, 85–92. [CrossRef]

20. López-Estrada, F.R.; Theilliol, D.; Astorga-Zaragoza, C.M.; Ponsart, J.C.; Valencia-Palomo, G.; Camas-Anzueto, J. Fault diagnosis observer for descriptor Takagi-Sugeno systems. *Neurocomputing* **2019**, *331*, 10–17. [CrossRef]

21. Nguyen, A.T.; Campos, V.; Guerra, T.M.; Pan, J.; Xie, W. Takagi-Sugeno fuzzy observer design for nonlinear descriptor systems with unmeasured premise variables and unknown inputs. *Int. J. Robust Nonlinear Control* **2021**. [CrossRef]

22. Zhang, J.; Zhu, F.; Karimi, H.R.; Wang, F. Observer-based sliding mode control for T-S fuzzy descriptor systems with time delay. *IEEE Trans. Fuzzy Syst.* **2019**, *27*, 2009–2023. [CrossRef]

23. Jafari, E.; Binazadeh, T. Observer-based tracker design for discrete-time descriptor systems with constrained inputs. *J. Process Control* **2020**, *94*, 26–35. [CrossRef]

24. Li, J.; Zhai, D. A descriptor regular form-based approach to observer-based integral sliding mode controller design. *Int. J. Robust Nonlinear Control* **2021**, *31*, 5134–5148. [CrossRef]

25. Blandeau, M.; Guerra, T.M.; Pudlo, P. Application of a TS unknown input observer for studying sitting control for people living with spinal cord injury. In *Control Applications for Biomedical Engineering Systems*; Elsevier: Amsterdam, The Netherlands, 2020; pp. 169–195.

26. Reyero-Santiago, P.; Ocampo-Martinez, C.; Braatz, R.D. Nonlinear Dynamical Analysis for an Ethanol Steam Reformer: A Singular Distributed Parameter System. In Proceedings of the 2020 59th IEEE Conference on Decision and Control (CDC), Virtual, 14–18 December 2020; pp. 23–29.

27. Van Nguyen, T.; Ha, X.V. An Application of Observer Reconstruction to Estimate Actuator Fault for DC Motor Nonlinear System Under Effects of the Temperature and Disturbance. *Furth. Adv. Internet Things Biomed. Cyber Phys. Syst.* **2021**, 55.

28. Flores-Hernández, A.A.; Reyes-Reyes, J.; Astorga-Zaragoza, C.M.; Osorio-Gordillo, G.L.; García-Beltrán, C.D. Temperature control of an alcoholic fermentation process through the Takagi-Sugeno modeling. *Chem. Eng. Res. Des.* **2018**, *140*, 320–330. [CrossRef]

29. Astorga-Zaragoza, C.M.; Osorio-Gordillo, G.L.; Reyes-Martínez, J.; Madrigal-Espinosa, G.; Chadli, M. Takagi—Sugeno observers as an alternative to nonlinear observers for analytical redundancy. application to a steam generator of a thermal power plant. *Int. J. Fuzzy Syst.* **2018**, *20*, 1756–1766. [CrossRef]

30. Nguyen, A.T.; Dinh, T.Q.; Guerra, T.M.; Pan, J. Takagi-Sugeno fuzzy unknown input observers to estimate nonlinear dynamics of autonomous ground vehicles: Theory and real-time verification. *IEEE/ASME Trans. Mechatron.* **2021**, *26*, 1328–1338. [CrossRef]

31. Flores-Martínez, M.A.; Osorio-Gordillo, G.L.; Vargas-Méndez, R.A.; Reyes-Reyes, J. Fuzzy functional observer for the control of the glucose-insulin system. *J. Intell. Fuzzy Syst.* **2019**, *37*, 5085–5096. [CrossRef]

32. Chakrabarty, A.; Corless, M.J.; Buzzard, G.T.; Żak, S.H.; Rundell, A.E. State and unknown input observers for nonlinear systems with bounded exogenous inputs. *IEEE Trans. Autom. Control* **2017**, *62*, 5497–5510. [CrossRef]

33. Skelton, R.E.; Iwasaki, T.; Grigoriadis, K. *A Unified Algebraic Approach to Linear Control Design*; Taylor & Francis: Abingdon, UK, 1998.
34. Trinh, H.; Fernando, T. *Functional Observers for Dynamical Systems*; Springer Science & Business Media: Berlin, Germany, 2011; Volume 420.

MDPI

St. Alban-Anlage 66

4052 Basel

Switzerland

Tel. +41 61 683 77 34

Fax +41 61 302 89 18

www.mdpi.com

Processes Editorial Office

E-mail: processes@mdpi.com

www.mdpi.com/journal/processes